HEAT TRANSFER

A Modern Approach

HEAT TRANSFER

A Modern Approach

Martin Becker

Rensselaer Polytechnic Institute
Troy, New York

PLENUM PRESS • NEW YORK AND LONDON

Library of Congress Cataloging in Publication Data

Becker, Martin.
 Heat transfer.

 Includes bibliographical references and index.
 1. Heat—Transmission. I. Title.
 TJ260.B4   1986   621.402′2   86-18661
 ISBN 0-306-42316-2

© 1986 Plenum Press, New York
A Division of Plenum Publishing Corporation
233 Spring Street, New York, N.Y. 10013

Printed in the United States of America

PREFACE

There have been significant changes in the academic environment and in the workplace related to computing. Further changes are likely to take place. At Rensselaer Polytechnic Institute, the manner in which the subject of heat transfer is presented is evolving so as to accommodate to and, indeed, to participate in, the changes.

One obvious change has been the introduction of the electronic calculator. The typical engineering student can now evaluate logarithms, trigonometric functions, and hyperbolic functions accurately by pushing a button. Teaching techniques and text presentations designed to avoid evaluation of these functions or the need to look them up in tables with associated interpolation are no longer necessary.

Similarly, students are increasingly proficient in the use of computers. At RPI, every engineering student takes two semesters of computing as a freshman and is capable of applying the computer to problems he or she encounters. Every student is given personal time on the campus computer. In addition, students have access to personal computers. In some colleges, all engineering students are provided with personal computers, which can be applied to a variety of tasks.

In the workplace, personal computers, distributed terminals, and engineering work stations are bringing computing power to bear on a variety of elementary problems in addition to major "number-crunching" tasks. The engineering office is changing and the teaching of engineering subjects should reflect the way the engineer of the future will function.

In undertaking a review of the Heat Transfer course and of a variety of existing texts in the field, I found that typical presentations did not accommodate fully to the new conditions. For example, the student typically is led to use charts to analyze problems that often can be dealt with directly and more effectively without them.

In radiation heat transfer, students are asked to use charts to obtain shape factors for standard configurations and to use shape factor algebra to obtain the shape factors for nonstandard configurations. The accuracy with which the charts can be read and the amount of shape factor algebra involving small differences of large numbers, however, frequently leads to answers of dubious value. For many standard configurations, analytical expressions have been obtained for shapes factors. While some of these are complicated, their evaluation with a personal computer (or even with an electronic calculator) is not a formidable task. Analogous observations can be made in regard to charts used elsewhere in the course in dealing with fins, time-dependent heat conduction, and heat exchangers.

This text provide charts as illustrations of trends and as sources of information when the reader is at a location without access to computational aids. However, the emphasis in discussion and in conduct of sample problems is on working with the analytical procedures.

There is a distinction between providing graphs for the purpose of illustration and providing graphs as a source of information. Information graphs are oriented to cramming a great deal of information into a busy chart with a large number of parametric curves. They may not be, and often are not, useful for transmitting physical insight. Graphs in this text, for example, for radiation shape factors, are designed (in terms of axes and parameters) to convey physical feeling. The graphs are discussed in such terms, with observation, for example, of how the shape factor declines as parallel disks are separated and of how shape factors reach limiting values as the disks come closer together. Similarly, graphs for F factors of heat exchangers, when expressed explicitly in terms of heat exchanger effectiveness, facilitate discussion of heat exchangers in terms of two figures of merit, one relative to the theoretically best heat exchanger configuration, the other relative to the theoretical thermodynamic limit. Graphs for effectiveness are displayed so as to emphasize distinctions among heat exchanger types, with graphs for counterflow and parallel flow appearing both qualitatively and quantitatively different from one another.

By working with basic relationships instead of with chart lookup, the student also develops a better sense of the basis for the technique being followed and operates less in a handbook manner. The student is then in a good position after completing the course to improvise to deal with situations different from those treated extensively in the textbook, for example, in dealing with a situation where the space–time separability (or the restriction on the Fourier modulus) of Heisler charts does not apply.

While most individual computer routines for materials contained in charts are simple, the combination of them can place a burden on the student who has to program them all during a single semester. Accordingly, FORTRAN listings of selected subroutines used at Rensselaer for such purposes are included in Appendix C. Emphasis is placed on subroutines involving some subtlety and complexity in programming. The problems at the ends of chapters include cases that involve extension or revision of some of these "standard" programs.

A particular realistic concern of the engineer is design. The design environment calls for the engineer to consider a number of variations about a basic concept. Basic engineering science courses such as heat transfer frequently avoid exploratory problems because they can be very time consuming. However, a class of software commonly available for personal computers, the electronic spreadsheet, is well suited to exploratory "what if" problems. For a large number of sample problems worked out in the body of the text, duplicate solutions in electronic spreadsheet form are presented. Problems at the ends of chapters include exploratory problems that can be approached through variations on sample spreadsheets. For sample problems, the notation of the VisiCalc program, a common program available for many personal computers, has been used. The notations for other programs are similar.

Another important educational objective related to the "what if" type of problem is the development of a physical "feel" for numbers and relationships. What if oil were used instead of water? What if a gas were used instead of a liquid? How important are viscosity corrections for various fluids? What values of Nusselt numbers do we encounter in laminar flow and in turbulent flow? Problems at the ends of chapters include many cases of making tables or plots. Such numerical surveys can be prepared efficiently with spreadsheet software.

The spreadsheet has proved to be useful in standard problem solving too. It is demonstrated, by means of the sample problems, how the spreadsheet solution for a complicated problem can be performed by adapting the solution for a simpler problem. The result is a substantial reduction of effort on the part of the student and an ability to deal with more elaborate problems. This adaptability is very important in an educational environment. New material is introduced as simply as possible. Initial illustrative examples are made elementary to emphasize basic concepts. There is a natural progression from the simple to the complicated. The ability to adapt simple spreadsheets to more complicated situations blends effectively with the natural progression of the course.

Problems provided at the ends of the chapters take advantage of the ability to build on earlier results. Frequently, there is a set of problems which introduces variations and extensions relative to an original problem, which may be one of the sample problems in the text. With these sets of problems, the student is able to explore a variety of features relative to a basic situation with an economical use of time. Since these studies are subdivided into separate questions, the instructor has the capability to define the scope of the investigation of a single topic.

The spreadsheet has one clear advantage relative to a conventional computer program. The conventional computer program sometimes "hides" the theory and intermediate results within the coding so that the student enters input and receives output. It is not necessary for things to be hidden in standard coding, but it often works out that way. This is because specific steps are required to call for information to be displayed. It takes anticipation, conscious decisions, and additional effort to get many displays. Spreadsheet software displays the formulas being used and many of the intermediate results as a matter of course.

The spreadsheet has other advantages of importance to the course. The typical spreadsheet is designed to be very easy to learn, and my experience is that students, even without prior exposure to spreadsheets, become proficient very rapidly. The typical spreadsheet is designed to be insensitive to choices causing errors and delays with standard programming, for example, with respect to formatting of input numbers.

Since the spreadsheet sample problems duplicate conventional sample problems in the text, it is not necessary to use the spreadsheet approach. In my opinion, however, the spreadsheet is sufficiently useful and easy to learn and has such a variety of other benefits to those who use personal computers that it is highly desirable, from the viewpoint of the course as a whole and of the student as an individual, to make use of the spreadsheet.

Various levels of spreadsheet software are available, from basic to elaborate spreadsheets to sophisticated integrated packages including database management, graphics, and word processing. This text deals with common basic spreadsheets. Where more sophisticated options are available, the instructor and student can extend the computer applications introduced in the text.

For those who do not have spreadsheets available or who prefer to deal with standard programming, Appendix D shows how to convert the spreadsheet layouts to standard BASIC programs with explicit conversions provided for selected sample problems. Included in the selection of explicit conversion are cases involving use of particular spreadsheet functions (e.g., the LOOKUP function for a table) not provided automatically in BASIC or FORTRAN. Discussion is provided as to how FORTRAN programs would differ (e.g., via use of a DO loop instead of a FOR $\cdots$ NEXT loop).

With examples provided in the main text and in appendixes in spreadsheet, FORTRAN, and BASIC, it is deemed likely that all students will be able to use this book with no more than a modest adaptation from the computing mode with which they are most familiar. I have found use of the combination of languages to be convenient professionally, using spreadsheet and BASIC in primarily interactive modes and using FORTRAN for routines that might be incorporated into other and more sophisticated programs on a mainframe computer.

A difficulty encountered in the study of heat transfer is the existence of cases that are intrinsically iterative. Calculation of convection heat transfer coefficient requires knowledge of certain temperatures for evaluation of properties, but all required temperatures may not be known at the outset of the problem. Frequently, conditions are such that preliminary estimates of temperatures yield reasonable estimates of properties and, thereby, reasonable answers for the problems. However, there are situations, for example, when change in bulk temperature is large, where preliminary estimates are crude, and iteration is needed.

Because of the labor involved in carrying out calculations, there can be a tendency to assign cases where iteration is not necessary or to accept lack of iteration where it would be desirable. Access to a personal computer or comparable computational aid permits having routine iterative evaluations set up. Spreadsheet software can be particularly useful for this purpose. Thus, this text does not avoid classes of realistic situations that others may avoid or deemphasize.

Another feature of computer availability is that iteration can be more convenient than direct solution, for example, when a related problem computer solution has already been developed. For example, if we already have prepared a computer solution for the heat flow through a multilayered wall, that computer solution can be used iteratively to obtain the thickness of a particular layer that is needed to yield a specified heat flow. Such cases are pointed out. The student is encouraged to plan problem-solving strategy on both the logic of the situation and on the tools already available.

While the text uses samples involving elementary spreadsheets, some problems are included at ends of chapters which are made easier if database and graphing capabilities (either integrated with or separate from the spreadsheet) are available. Graphing results can help the student to visualize trends. Working with databases helps the student to get a feeling for properties of various materials, how properties vary with temperature, and so on.

To facilitate database work, I have organized Appendix A on data to place information in common formats. Thus, room temperature properties of all solids (e.g., metals, structural materials, insulators) are placed in one format. Properties of all fluids are put in a common format. (By contrast, many texts not concerned with database handling use different sets of properties for gases and liquids, put information for some materials in °C and some in °K with a lack of evaluation at common temperatures, or use different powers of 10 to multiply entries like viscosity for different fluids.) The organization of the data is thus well suited for direct entry into database programs and for subsequent database manipulation.

Part of recognizing the importance of modern computing is recognizing the need for appropriate approximation methods and numerical analysis. Heat transfer courses and texts traditionally have included discussion of approximation methods. However, some techniques still featured prominently in present-day texts are not techniques that would be used with computers. The Schmidt plot and other graphical approximation methods, valuable as they were when they were introduced, are in this category. In addition, numerical procedures are frequently presented and illustrated as if the iterative procedures would be done by hand.

I have found it useful to introduce finite differences approximations earlier than is typical in heat transfer texts, namely, in dealing with steady-state, one-dimensional heat conduction. The steady-state, one-dimensional finite difference problem, besides having intrinsic utility of its own, is a basic building block in solving other problems involving time dependence and involving multiple dimensions. Indeed, much numerical analysis development, including alternating-direction implicit methods for transients and line inversion methods for static problems, are oriented to capitalizing on the one-dimensional solution as a core component.

Since many heat transfer students will not have had much prior exposure to difference equations, the separation between the application to one-dimensional problems and to two-dimensional problems, which require use of iteration methods, provide times for the new differencing concepts to be digested. Analogous benefit comes from the separation between applications to static and dynamic problems. The early introduction of numerical approximation also makes it easier for the instructor to assign computer term projects that can build on the one-dimensional solution as a subroutine.

For efficiency and for ease of presentation, I have used the general conservation form of Laplacian representation to treat the conduction equation in the three principal geometries of interest—plane, cylindrical, and spherical. Using this approach makes analytical derivations simpler than those of many texts in curved geometries and is conducive to development of a single form of difference equation to apply to the three geometries. While the derivation of the difference approximation is more complicated than a simple statement (e.g., this is the central difference approximation in rectangular geometry), it provides a basis for the student in the future to deal with different types of geometry, such as triangular mesh cells for an irregular-shape surface.

While attention is given to suitable development of numerical approximations, due concern is given to limiting the amount of numerical analysis. The text is, first and foremost, a heat transfer text. Presentation of numerical approximation procedures has been designed to get the student off the ground in the right direction. Emphasis is typically on the simplest (not the most efficient) approach, with more elaborate approaches, as in iteration methods, provided in appendixes and contained in references.

This concern with limitation is important. From my contact with heat transfer courses, it appears that numerical approximations, while generally included in texts, frequently are omitted from course coverage. The numerical treatments in this text, while sophisticated and general, are brief and can be included in a course without detracting a great deal from the basic physical subject matter.

With more emphasis on explicit solutions as opposed to charts and with the importance of cylindrical geometry, explicit identification of Bessel equations and functions is made. Many heat transfer texts avoid any mention of Bessel functions. In my experience, the background of students taking heat transfer (junior and seniors from various engineering disciplines) is such that they are capable of dealing with this material. Supplementary information on Bessel functions is contained in Appendix B. This supplementary information includes tables for looking up values and also includes convenient algorithms for evaluating Bessel functions in computer routines.

Even where Bessel function solutions are obtained, convenient alternatives are presented where possible. Bessel function solutions are obtained for triangular and circumferential fin efficiencies. Simple analytical approximations in terms of hyperbolic tangents are presented and compared with these precise solutions. Problems are worked out in examples both ways. These analytical approximations can be put into spreadsheets by modest adaptation of the spreadsheet for the straight rectangular fin. The instructor can skip or skim over the Bessel function solutions for such situations.

Throughout this text, sample exercises are provided to illustrate the material. In addition, suggested computer projects are provided. The student who performs these computer projects on a personal computer should find that he or she has a useful diskette of heat transfer programs for use after completion of this course.

Some of the projects cited in the text can incorporated into a course computer program available for student use. A general-purpose utility code has been provided for student use at Rensselaer to deal with, for example, fins, time-dependent conduction, and shape factors. In addition, a set of sample problem spreadsheets on diskette (based on the sample problems in the text) is provided to the student as a base from which to work. However, even then the instructor may find it advisable to have the student undertake a project for which a course code is available to avoid making the computer a mysterious and inaccessible source of information.

One concern in preparing a text for a course is efficiency. The time available for the course is limited. As a result, particular attention has been given to opportunities for recurrent themes that can be applied in more than one situation. One such theme, already noted, is associated with the use of the conservation form of the Laplacian operator. Use of this form makes derivations in different geometries very similar, with integrals obtained easily by perfect differentials. This same type of integration over the general operator leads efficiently (and, by this time, in a familiar fashion) to construction of difference equations in various geometries. This pattern is helpful to students having no prior background in numerical analysis. The thematic concept is reinforced by informing the reader that a theme will be used again and by reminding the reader that a theme has been used before.

Another theme is the use of a general conservation equation for use in analyzing boundary layers for convection. The heat conduction equation, derived explicitly in Chapter 3, is generalized in Chapter 6 to a conservation equation for an unspecified property. Then, upon identifying properties of mass, momentum, and energy, the required individual equations are obtained readily.

One choice faced by authors of heat transfer texts is whether and/or how to treat mass transfer. Some texts include detailed coverage and other texts omit the subject altogether. I find that there is not time in a typical heat transfer course to give extensive coverage to mass transfer. However, it is deemed desirable for students to be made aware of the existence of analogies between the two subjects. Accordingly, a section is included in Chapter 6 to provide this awareness.

While the principal motivation leading to this text is the desire for compatibility with a modern computing environment, a number of other features have been incorporated based on my experience and on the suggestions of others.

To reinforce and integrate concepts, problems at ends of chapter include situations requiring application of knowledge gained in earlier chapters. For example, a problem in the natural convection chapter includes placement of a layer of insulation over a pipe. The problem can be approached by merging spreadsheets for natural convection and for conduction through a cylindrical layer. Iteration is needed to obtain the outside surface temperature. This type of problem, in addition to integrating concepts, gives practice in applying the computer to treat relatively simple problems that otherwise would be tedious and time consuming.

To facilitate an end-of-semester review and to provide an opportunity for integrating aspects of heat transfer, a chapter of sample problems is included at the end of the text. The problems include having the student determine what mode or combination of heat transfer modes applies. While there are sample problems in individual chapters, it tends to be clear that the problem is related to the mode just taught, although the proper procedure to use may depend on details (e.g., Reynolds number).

The multimode heat transfer problems in this chapter are more elaborate than typical of the earlier chapters and tend to involve nonlinearities and iteration. Computer usage is particularly useful in dealing with these problems.

Another aid to review is Appendix F which lists key formulas and equations from the entire text. The student thus has a convenient reference point for information.

Throughout the text, there is reiteration on the theme that it is important to identify the nature of the problem. It is the flow turbulent? Is flow fully developed? Is discussing high-speed flow, for example, it is noted that high-speed-flow effects (or effects of temperature increase associated with bringing a fluid to rest) can apply to flows even at relatively modest speeds, while most texts tend to leave the impression that high-speed-flow effects are to be considered only in connection with supersonic aircraft.

Indeed, a point emphasized throughout the text is the need to quantify qualitative statements. If the flow is to be taken as high speed, the student must ask, high compared to what? If the fin is to be taken as thin, the student must ask, thin compared to what? Criteria for accepting qualitative statements are obtained in the body of the text in some cases and left as assignment for the student to go through the process of assessment in others.

Another teaching consideration is to lead the student to question why things turn out the way they do. For example, the transition from laminar to turbulent flow occurs at a relatively low Reynolds number for flow in a tube and at a relatively high number for flow over a flat plate. Heat transfer texts tend to take this for granted, leaving students to conclude that it just turns out this way. In this text, it is shown why this relationship between transition Reynolds numbers is to be expected (and why an analogous relationship exists with the Grashof number of natural convection). This philosophy of explanation is carried through on topics throughout the text.

The above example illustrates the emphasis provided on understanding the physical processes involved. The emphasis provided on making use of modern computing environments does not detract from the emphasis on the role of the engineer in the design process or from the emphasis on developing insight into the physical processes involved and the nature of the results obtained.

Discussion of the nature of results is given particular attention in dealing with convection, where a conscious effort has been made to avoid providing just a catalogue of empirical formulas. For example, natural convection heat transfer rates from hot surfaces facing down in enclosed spaces (where heat transfer is by conduction) and in open spaces (where convection occurs) are compared and explained.

Where mathematical treatments are used, mathematical requirements are related to physical phenomena and engineering decisions. In Chapter 3, for example, interface conditions are discussed in terms of heat flow and of avoiding a contradiction to Fourier's Law. In addition, it is emphasized that the boundary conditions can represent requirements that the engineer may impose on a problem. Furthermore, where unusual results (e.g., infinite heat fluxes over infinitesimal times and infinite heat transfer coefficients over infinitesimal distances) occur, these are explained in terms of idealizations in modeling. For example, in discussing the infinite heat flux in the standard solution for a semi-infinite medium, it is explained that the boundary condition of a sudden change in surface temperature is a modelling idealization that implies an infinite rate of change.

Where treatments are mathematical, it is useful to recognize how much sophistication is needed for a particular application. Accordingly, the treatment of one-dimensional conduction is divided into two chapters. The first treats situations that can be addressed on the basis of Fourier's Law, without resort to the full heat conduction equation. That equation is not introduced until the next chatper, where it is needed to treat problems with sources.

In presenting general supporting information, emphasis has been placed on locating such information as close as possible to where it actually is needed. For example, in dealing with radiation heat transfer, discussion of wavelength dependence is concentrated in the section of transmission, the phenomenon most sensitive to wavelength.

The typical student enrolled in the RPI heat transfer course is a junior or senior who has 2 years of mathematics (including differential equations) and courses in both thermodynamics and fluid mechanics. The mathematics used in the text is part of the background of the typical RPI junior in engineering. The backgrounds in the other subjects are helpful, but not absolutely necessary. A student who understands boundary layer concepts from fluid mechanics will have an easier time mastering convection than one who sees the concepts in heat transfer for the first time, but prior knowledge is not assumed.

As to mathematics, students are expected to be familiar with Fourier series, with solution of second-order differential equations with constant coefficients, and with properties of hyperbolic functions. Readers are advised to review these items in their background at the outset of the course. Vector notation (gradient and divergence operators) is used occasionally to generalize some relationships obtained in one dimension and to provide compact statements of general relationships, but detailed knowledge of vector calculus is neither assumed nor required.

Heat transfer at Rensselaer is taught as a one-semester basic engineering science course of concern to all engineering disciplines. Thus, the presentation does not presuppose specific background associated with any particular field.

In summary, this text is designed to be used by students and by engineers whose workplaces (campus or office) provide access to personal computing, either through actual personal computers or through readily accessible terminals. This type of computing environment permits use of more sophisticated analytical formulas and procedures than are convenient when one is restricted to hand calculations, even when one does not resort to numerical procedures for solving equations. When numerical procedures are called for, the computing environment provides some natural selection among available approximation methods. The text should make it possible for students to approach the subject of heat transfer in a way that is both more sophisticated and more efficient (but that is appropriate to their level of background) than has been the case in the past.

The author wishes to acknowledge the encouragement and continued support of Dean Paul DeRusso and Dean Joseph H. Smith, Jr., for the introduction of computer-related concepts into the engineering curriculum. Part of the development effort was supported by a Harlan and Lois Anderson Grant for microcomputer courseware development. The program of Anderson grants stimulated a number of microcomputer-related innovations at RPI. The author appreciates the time spent and comments provided by colleagues at other institutions who reviewed early drafts.

Martin Becker

Troy, New York

CONTENTS

xviii

Contents

Chapter One

INTRODUCTION

Engineers in a variety of disciplines are concerned about heat transfer. There are many engineering objectives that are encountered in this variety of disciplines, which can be classified into a simplified set of engineering objectives as follows:

1. Move heat to where we want it.
2. Move heat away from where we do not want it.
3. Keep heat from leaving where we wish to retain it.
4. Keep heat from entering where we wish to avoid it.

Let us consider some examples.

Most power plants produce electricity by converting heat contained within a "working fluid" to work. This working fluid may not be the fluid in which heat was generated, however. Coal may be burned to produce hot combustion products. Heat may then be transferred from these hot gases to water which serves as the working fluid as illustrated in Fig. 1-1. Thus, the engineer must provide for transferring heat to where it is wanted.

Heat may be produced as a by-product of a process. If that heat is not removed, damage to the equipment may result. Heat must be removed from an automobile engine, for example, as illustrated in Fig. 1-2. Typically, a liquid coolant (water, antifreeze) will cool the engine in the first step of heat removal.

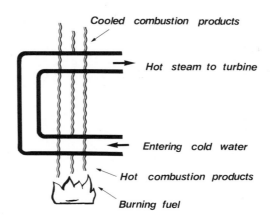

FIGURE 1-1. Example of moving heat to where we want it: power plant.

Then, air will cool the liquid coolant. The engineer provides for removal of the heat from where it is not wanted.

Heat must be provided to residential and commercial buildings to maintain a comfortable environment. For economy, it is desirable to retard the rate at which heat is lost from buildings as illustrated in Fig. 1-3. The engineer is concerned with preventing heat from leaving the area where the heat should be kept.

Superconducting magnets work properly at very low temperatures (near absolute zero). In some cases, these magnets may be near locations of very high temperature, as in proposed designs for nuclear fusion plants. The engineer must prevent heat from entering the superconducting magnet at too rapid a rate; that is, heat must be kept away from where it is not wanted as illustrated in Fig. 1-4.

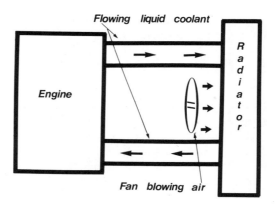

FIGURE 1-2. Example of moving heat away from where we do not want it: automobile engine.

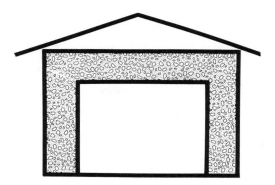

FIGURE 1-3. Example of keeping heat from leaving where we wish to retain it: insulated building.

All fields of engineering deal with problems of the type cited. Aerospace engineers are concerned with heat transfer in high-speed flow. Chemical engineers provide for input or removal of heat, depending on whether processes are endothermic or exothermic. Civil engineers and architects must be concerned with heat transfer in the design and construction of buildings. Electrical engineers are concerned with proper operating temperatures of equipment such as computers. Mechanical engineers are concerned with many heat transfer situations, such as the automobile engine example discussed above. Nuclear engineers are concerned with dynamic heat removal problems in fission reactors in addition to the fusion reactor example cited above. Thus, heat transfer is a basic engineering science of general concern to the engineering profession.

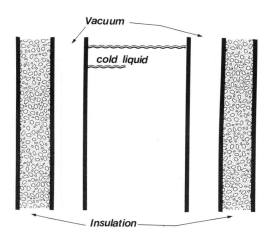

FIGURE 1-4. Example of keeping heat from where we wish to avoid it: cold liquid container.

Thus far, we have discussed why engineers should be concerned about heat transfer. Next, we introduce the subject itself.

From studying thermodynamics, we know that the internal energy of a mass of material is measured by its temperature and by knowledge of its phase (since internal energy can change without a temperature change while, for example, water boils). A basic premise of the subject of heat transfer is that heat tends to flow from a region of high temperature to a region of low temperature.

There are three basic mechanisms according to which heat can move from a high-temperature to a low-temperature region.

1. Heat can move through a static body by interaction with the internal structure of the body.
2. Heat can be carried from one place to another by movement of a fluid.
3. Heat can be transported through space even in the absence of any intervening material.

The first process is called conduction. We are not concerned with microscopic interactions with internal structure. We simply characterize materials by a property called thermal conductivity. The basic principle of heat conduction, called Fourier's Law, is that heat is transferred in proportion to the gradient of temperature. In one-dimensional plane geometry, this statement can be expressed as

$$\frac{q}{A} = -k\frac{dT}{dx} \tag{1-1}$$

where q is heat flow (in watts or in Btu/hr), A is area, and T is temperature (see Fig. 1-5). The parameter k, the constant of proportionality, is called the thermal conductivity. Table 1-1 contains representative thermal conductivities

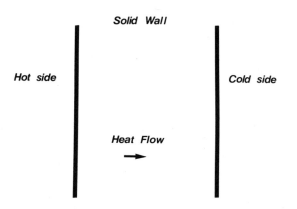

FIGURE 1-5. Heat is conducted through a solid body.

TABLE 1-1
Thermal Conductivity (W/m · ° C) of Selected
Materials at 300° K

Material	Conductivity
Solid metals	
Silver	427
Copper	386
Aluminum	204
Iron	73
Common solid materials	
Wood (maple)	0.17
Concrete block	1
Glass fiber insulation	0.04
Common brick	0.73
Granite	3
Liquids	
Water	0.61
Ethylene glycol (antifreeze)	0.25
Glycerin	0.29
Engine oil	0.15
Gases	
Water vapor	0.020
Helium	0.15
Hydrogen	0.18
Air	0.026

for various types of material. More complete data are provided in Appendix A.

The second process is called convection, since heat is carried or conveyed by a fluid. Convection may occur naturally or may be stimulated. Natural convection occurs because fluids generally expand on being heated and become lighter (Fig. 1-6). Gravity effects then cause warmer, lighter fluid to move up and cooler, heavier fluid to move down. Convection may be stimulated by having an external agent, such as a pump, provide motion to the fluid. The resulting heat transfer is called forced convection. The mechanisms involved in convection are complicated. As a result, equations for convection tend to characterize overall effects, as opposed to local detail, and are expressed in the form

$$\frac{q}{A} = h(T_S - T_F) \tag{1-2}$$

where h is the convective heat transfer coefficient and T_S and T_F are the surface temperature and a temperature characterizing the fluid, respectively. T_F may be the asymptotic temperature in the case of a flat plate exposed to a very large volume of fluid, or it may be the mixed mean (or bulk) temperature if the fluid is contained in a closed pipe.

Convective heat transfer coefficients can vary substantially, depending on the conditions. Coefficients tend to be higher for forced convection than for

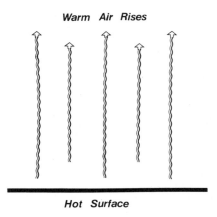

Hot Surface

FIGURE 1-6. Heat is carried away by natural convection.

natural convection, for turbulent flow than for laminar flow, for a two-phase fluid (as in boiling) than for a single-phase fluid. Some representative values are given in Table 1-2. Convection coefficients depend on flow characteristics (geometry, flow rate), so the values in Table 1-2 should be taken as indicative of general trends and not as specific values to be used.

The third process is called radiation. This is a familiar process in that this is the mechanism whereby the sun transmits heat to the earth (Fig. 1-7). Actually, all bodies radiate heat. The heat radiated is proportional to the fourth power of absolute temperature. Thus, we tend to become concerned increasingly with radiation heat transfer as temperatures increase. The equation for the radiant heat emitted by a body is

$$\frac{q}{A} = \varepsilon \sigma T^4 \qquad (1\text{-}3)$$

where σ is a constant of proportionability (the same for all bodies) and ε is the emissivity of the surface. For a black body, ε is 1. For other surfaces, ε is less than 1. From equilibrium considerations, it is demonstrated that the emissivity must be equal to the absorptivity, that is, to the fraction of incident radiation

TABLE 1-2
Selected Typical Convection Coefficients

Type	h (W/m^2 · ° C)
Natural convection, air	7
Natural convection, water	500
Forced convection, air	50
Forced convection, water	10,000
Boiling water	30,000

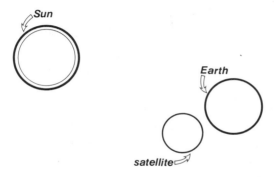

FIGURE 1-7. Heat is exchanged through empty space by radiation.

that is absorbed by the body. Thus, a black body is a perfect absorber of radiation. All radiant heat must go somewhere. Thus, heat that is not absorbed must be either reflected or transmitted. The radiant heat leaving a surface is not just the heat given by Eq. (1-3), but also the heat reflected from and transmitted through the body.

This text is considered with macroscopic aspects of heat transfer. Accordingly, we are not concerned with phenomena of the atomic, molecular, and crystalline structure of matter which determine thermal conductivity and radiation. Our attention is focused on the macroscopic effects associated with the three basic mechanisms of heat transfer.

While this text is concerned primarily with heat transfer, we may observe that many of the principles and procedures involved apply also to mass transfer. Indeed, many results of importance can be obtained by direct analogy. Key analogies of importance are developed and noted.

Now that we have discussed why we should be concerned with heat transfer and what the basic mechanisms are that have to be studied, let us consider briefly how this subject is to be studied.

This text is oriented, through presentation and through sample problems, to take advantage of the availability of a computing environment either through personal computers or through interactive terminals. When faced with problems of a general character (i.e., of a kind that the student is likely to encounter other similar problems), the student is guided toward entering the sequence of calculations into a computer (e.g., via a spreadsheet program, which is actually a powerful and flexible programmable calculator with many convenient features). The student is guided to building solutions for new problems onto solutions for earlier, simpler problems.

This orientation toward a computing environment is designed to:

1. Maximize the personal productivity of the student (the effort needed for the second problem is much less than the effort needed for the first).
2. Minimize the fraction of time spent on tedious arithmetic, thereby increasing the fraction of time devoted to concentrating on the basic

aspects of the subject and of the problems.
3. Use assignments devoted to design survey, that is, to problems of the type found in realistic engineering situations .
4. Accustom the student to operating in a mode increasingly characteristic of modern engineering offices.

In this introduction, we have discussed why the subject of heat transfer should be studied, what the subject contains, and how the subject is approached. Let us now proceed to the next chapter to begin the first major topic area, conduction.

REFERENCES

Below is a list of other introductory texts and references that the reader may wish to consult.

1. J. P. Holman, *Heat Transfer*, 5th ed., McGraw-Hill, New York, 1981.
2. F. P. Incoprera, and D. P. DeWitt, *Introduction to Heat Transfer*, Wiley, New York, 1985.
3. M. Jakob, *Heat Transfer*, Vol. 1, Wiley, New York, 1949.
4. F. Kreith, and W. Z. Black, *Basic Heat Transfer*, Harper & Row, New York, 1980.
5. S. S. Kutateladze, *Fundamentals of Heat Transfer*, Academic Press, New York, 1963.
6. S. S. Kutateladze, and V. M. Borishanskii, *A Concise Encyclopedia of Heat Transfer*, Pergamon, New York, 1966.
7. J. H. Lienhard, *A Heat Transfer Textbook*, Prentice-Hall, Englewood Cliffs, NJ, 1981.
8. M. N. Ozisik, *Heat Transfer: A Basic Approach*, McGraw-Hill, New York, 1985.
9. W. M. Rohsenow, and J. P. Hartnett, *Handbook of Heat Transfer*, McGraw-Hill, New York, 1973.
10. J. Sucec, *Heat Transfer*, W. C. Brown, Dubuque, IA, 1985.
11. F. M. White, *Heat Transfer*, Addison-Wesley, Reading, MA, 1984.
12. H. Wolf, *Heat Transfer*, Harper & Row, New York, 1983.

Chapter Two

ONE-DIMENSIONAL HEAT CONDUCTION IN SOURCE-FREE MEDIA

2-1. INTRODUCTION

Fourier's Law of heat conduction states that heat flows in proportion to the temperature gradient. A number of problems of interest can be analyzed just on the basis of Fourier's Law. A more complete description of heat conduction involves, in addition, application of the principle of conservation of energy, leading to a second-order differential equation for temperature. Such situations are dealt with in Chapter 3.

This chapter introduces Fourier's Law and applies it to an elementary problem in plane geometry, heat transfer across a wall. We then proceed to more complicated situations in which we add layers to the wall. We specify convection conditions at surfaces instead of known temperatures; that is we stipulate that the wall is adjacent to a fluid of known temperature. We then apply the approaches followed in plane geometry to cylindrical geometry, and note the significant differences associated with curved geometries. Next, we demonstrate how to set up examples on the computer. As we proceed, we show how to build on the simple examples to deal with more complex situations.

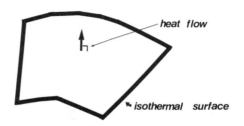

FIGURE 2-2-1.   Fourier's law of conduction: Heat flows normal to isothermal surface.

2-2.   FOURIER'S LAW

A general mathematical statement of Fourier's Law is

$$\mathbf{q} = -kA \nabla T \qquad (2\text{-}1)$$

where ∇ is the gradient operator and $\mathbf{q}$ is a vector indicating the magnitude and direction of heat flow. In one-dimensional geometries—plane, cylindrical, and spherical—Eq. (2-1) becomes

$$q = -kA \frac{dT}{dx} \qquad (2\text{-}2)$$

where x denotes the appropriate length coordinate in each geometry. The area A is different in each geometry. The negative sign means that heat flows from high to low temperature. Note that a positive q means heat is flowing in the direction of increasing x and a negative q means heat is flowing in the direction of decreasing x. (See Fig. 2-2-1.)

2-3.   PLANE GEOMETRY

Suppose a wall of thickness H has temperatures T_L and T_R on its left and right faces and has a thermal conductivity k. Let us find the heat flowing through this wall (see Fig. 2-3-1). Equation (2-2) may be rewritten

$$\frac{q}{kA} dx = -dT \qquad (2\text{-}3)$$

Let us integrate Eq. (2-2) over the wall. We assume that k is independent of position and temperature. The heat flow q must be constant, because we are dealing with a source-free medium, and, in steady state, heat cannot accumulate anywhere in the wall. We obtain, upon integrating,

$$\frac{qH}{kA} = -(T_R - T_L) = T_L - T_R \qquad (2\text{-}4)$$

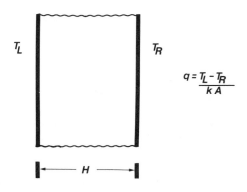

$$q = \frac{T_L - T_R}{kA}$$

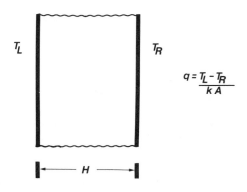

$$q = \frac{T_L - T_R}{kA}$$

high-conductivity material is used. Whether one wishes high or low thermal resistance depends on the particular situation.

2-4. MULTILAYERED WALLS

Heat transfer is not the only consideration in design. Sometimes all considerations can be accommodated by a simple wall of a single material. It also happens that multiple layers may be needed to deal with multiple objectives. These objectives may include, in addition to heat transfer, such considerations as

1. structural strength,
2. resistance to corrosion,
3. radiation shielding, and
4. decorative character.

The wall may be required to have enough strength not only to stand by itself but also to support something above. One side of the wall may contain a chemical reaction involving corrosive materials. One side of the wall may contain a source of ionizing radiation for which the wall must provide shielding. One side of the wall may have office space which must be made attractive as a working environment and as a location to receive visitors. The types and thicknesses of the materials selected for these various objectives were not selected for heat transfer, but they do have heat transfer properties.

The analysis of Section 2-3 may be applied to each layer of wall individually, yielding, as per Eq. (2-3),

$$T_{Li} - T_{Ri} = q\left(\frac{H}{kA}\right)_i \tag{2-7}$$

for layer i. If there are I layers, we may evaluate the sum

$$\sum_{i=1}^{I} (T_{Li} - T_{Ri}) = q\sum_{i}\left(\frac{H}{kA}\right) di \tag{2-8}$$

We note that the right-side temperature of one layer is the same as the left-side temperature of the next layer (we defer the question of contact resistance until later), so that

$$\sum_{i=1}^{I} (T_{Li} - T_{Ri}) = T_{L1} - T_{RI} \tag{2-9}$$

Thus, we may relate the overall temperature difference across I layers of wall to heat flow by

$$q = \frac{T_{L1} - T_{RI}}{\sum_{i=1}^{I}(H/kA)_i} \tag{2-10}$$

Note that the denominator contains a sum of thermal resistances, in keeping with the analogy to a series electric circuit.

We may observe that the temperature change in one layer of the wall is related to the overall temperature change by the ratio of thermal resistances. By the ratio of results of Eqs. (2-9) and (2-6), we see that

$$\frac{T_{Li} - T_{Ri}}{T_{L1} - T_{RI}} = \frac{(H/kA)_i}{\sum_{i=1}^{I}(H/kA)_i} \tag{2-11}$$

We also may observe that the results obtained indicate that the relationship between overall temperature change and heat flow is independent of the sequence in which the layers are placed. This is because in plane geometry, area does not vary with location. In curvilinear geometry, we find the situation to be somewhat different.

EXAMPLE 2-1.   A three-layered wall of a structure consists of (from exterior to interior) 10 cm of brick ($k_1 = 0.7$ W/m $\cdot°$ C), 10 cm of insulating material ($k_2 = 0.05$ W/m $\cdot°$ C), and 1 cm of wood panel ($k_3 = 0.1$ W/m $\cdot°$ C) (see Fig. 2-4-1). The temperature on the inside surface of the wood panel is 20°C and the temperature on the outside surface of the brick is -15°C. What is the heat loss per unit area from the wall?

SOLUTION.

*where
H = wall thickness*

$$\frac{q}{A} = \frac{\Delta T_{\text{overall}}}{\Sigma H/k}$$

$$= \frac{20°\text{C} - (-15°\text{C})}{0.1 \text{ m}/(0.7 \text{ W/m} \cdot° \text{C}) + 0.1 \text{ m}/(0.05 \text{ W/m} \cdot° \text{C}) + 0.01 \text{ m}/(0.1 \text{ W/m} \cdot° \text{C})}$$

$$\frac{q}{A} = \frac{35}{0.14 + 2.0 + 0.10} = \frac{35}{2.24} = 15.6 \text{ W/m}^2$$

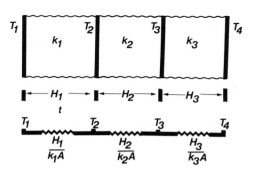

FIGURE 2-4-1.   A multilayered wall and its electrical analogue.

Note that the thermal resistance is dominated by the insulation resistance. Where thermal resistances are in series, a single large resistance determines the overall resistance. It should be noted that heat loss from a building can involve parallel paths with conduction and a different resistance (e.g., windows) and with air exchange with the outside (e.g., from opening and closing doors).

In performing Computer Project 2-1 below and other projects provided in this text, those having personal computers and "spreadsheet" software may wish to use the spreadsheet software instead of preparing a new program from scratch. The electronic spreadsheet is well suited for setting up frequently performed calculations of a relatively simple nature.

Indeed, the student with a personal computer would be well advised to set up homework assignments on a spreadsheet. Subsequent problems may involve a very similar sequence of steps. Thus, time can be saved in subsequent problems. In addition, the likelihood of error is reduced. Below, we illustrate how Example 2-1 nay be done on a spreadsheet using VisiCalc notation (analogous notation is used by other available spreadsheets).

If the student does not have a spreadsheet program available, it is a simple matter to make an explicit computer code corresponding to the spreadsheet. Illustrative computer code preparation based on the spreadsheets given for Examples 2-1 and 2-2 is provided in Appendix D. While the programming of arithmetic operations is very similar in the two approaches, the spreadsheet has a number of advantages associated with display and editing. On the other hand, there are situations where explicit programming has advantages, as noted in Appendix D.

EXAMPLE 2-1 (Spreadsheet). The spreadsheet is an array of rows and columns in which text, values and arithmetic operations may be entered. The columns are denoted by letters and the rows by numbers. In this example, we have the following spreadsheet:

	A	B
1	HIGHTEMP	20
2	LOW TEMP	−15
3	DELTA − T	+B1 − B2
4	DELTA − X − 1	.1
5	K1	.7
6	R1	+B4 / B5
7	DELTA − X − 2	.1
8	K2	.05
9	R2	+B7 / B8
10	DELTA − X − 3	.01
11	K3	.1
12	R3	+B10 / B11
13	R − TOTAL	+B6 + B9 + B12
14	Q / A	+B3 / B13

The spreadsheet program typically displays the numbers associated with the indicated operations in Column B. Note that we have used Column A to explain the variables and operations, using K and R to denote conductivity and resistance. Now, if this sheet is saved on disk, whenever another layered wall problem arises, the student can simply enter different numbers where appropriate. If there are fewer layers in the new problem, for example two instead of three, the student can enter zero for one or more layer thicknesses. If there are more than three layers, the student can insert rows after Row 12 (three rows for each additional layer) and modify the calculation of R-TOTAL. An example involving modification of this spreadsheet by insertion of rows is given later in this chapter.

The spreadsheet approach to Example 2-1 is useful for considerations related to design. The designer frequently needs to ask "what if" questions such as:

1. Suppose we were to use a different exterior material, for example, vinyl siding instead of brick (different conductivity and thickness of Material 1)?
2. Suppose we were to use a different thickness of insulation?
3. Suppose we were to use wallboard instead of paneling?

These questions, separately or together, can be addressed easily from the spreadsheet given. Indeed, a large number of questions can be addressed with relatively little additional effort. Some of the problems at the end of the chapter are of a design survey type.

Those who have not used spreadsheets previously may observe that the spreadsheet above is essentially self-explanatory. For example, $+B3/B13$ means the entry in Column B, Row 3 divided by the entry in Column B, Row 13. For modest scale calculations that are tedious to do by hand, it can be easier to use spreadsheet software than to prepare a dedicated program.

Another advantage we may observe from this spreadsheet example is that all the steps are laid out and a variety of intermediate information is presented. This layout makes it relatively easy to check for logical errors (the computer should not make computational errors) if the calculated final answer does not appear reasonable.

When working with computer routines, we might approach problems differently than when working with pencil and paper. Suppose, for example, that you were asked to find the insulation thickness that would reduce the heat loss in Example 2-1 to 14 W/m^2. With pencil and paper and a calculator, you might start from Eq. (2-10), derive an explicit expression for the thickness of the layer of insulation, and evaluate that expression. However, with a spreadsheet or interactive program already set up for Example 2-1, you might prefer to vary insulation thickness by trial and error until you obtained the desired heat loss. There is more arithmetic involved in trial and error, but the computer does this arithmetic. Your total effort, your time spent, and your likelihood of error may all be smaller using a "less elegant" approach based on an existing routine that was tailored originally to a different objective.

COMPUTER PROJECT 2-1. Make a program to analyze a plane wall of up to 10 layers. The materials and thicknesses of the layers are to be specified as input. You should have the option of specifying overall leftmost and rightmost temperatures and solving for heat flow, or specifying heat flow and one temperature and solving for other temperatures. You should solve for the temperatures at all interfaces in either option. Check your program by applying it to Example 2-1.

2-5. CONVECTION AT SURFACES

Equation (2-9) provides a relationship between overall temperature difference across a multilayered wall and heat flow through the wall. In an actual practical situation, we may not know the temperature at the wall, but we may know the temperature far from the wall in surrounding fluid (see Fig. 2-5-1). For example, the weather report may tell us the outside air temperature, but we may not know the outside wall temperature of the building we are in. We may relate these temperatures by the convection equation

$$q = h_R A (T_{RI} - T_{\infty R}) \qquad (2\text{-}12)$$

which may be put in the form

$$T_{RI} - T_{\infty R} = \frac{q}{h_R A} \qquad (2\text{-}13)$$

Equation (2-12) may be added to Eq. (2-7) to yield

$$T_{L1} - T_{\infty R} = q \left[\sum_i \left(\frac{H}{kA} \right)_i + \frac{1}{h_R A} \right] \qquad (2\text{-}14)$$

Thus, convection provides a thermal resistance

$$R_{th} = \frac{1}{h_R A} \qquad (2\text{-}15)$$

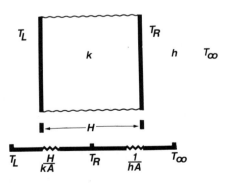

FIGURE 2-5-1. Convection at surface with electrical analogue.

If we have convection conditions prevailing on both sides of the wall, we obtain

$$q = \frac{T_{\infty L} - T_{\infty R}}{1/h_L A + \sum_{i=1}^{I}(H/kA)_i + 1/h_R A} \qquad (2\text{-}16)$$

EXAMPLE 2-2.   Consider the wall of Example 2-1. Suppose that the outside and inside air temperatures (instead of wall surface temperatures) were specified to be $-15°C$ and $20°C$ and that the convection heat transfer coefficients inside and outside are 10 W/m^2 $\cdot°$ C. Find the heat loss per unit area form the wall.

SOLUTION

$$\frac{q}{A} = \frac{\Delta T_{\text{overall}}}{1/h_L + \sum_{i=1}^{I}(H/k) + 1/h_R} = \frac{35}{1/10 + 2.24 + 1/10} = \frac{35}{2.44}$$

$$\frac{q}{A} = 14.3 \text{ W/m}^2$$

EXAMPLE 2-2 (Spreadsheet).   This problem may be approached by modifying the spreadsheet already developed for Example 2-1. The following lines are inserted after Line 12:

	A	B
13	LEFT CONVECTION	10
14	R - LEFT	1 / B13
15	RIGHT CONVECTION	10
16	R - RIGHT	1 / B15

The original line 13 becomes line 17 after the insertion. The entry for B17 is now modified so that line 17 is

17	R - TOTAL	+B6 + B9 + B12 + B14 + B16

The original line 14 automatically becomes the correct line 18

18	Q / A	+B3 / B17

This example illustrates the value of saving even simple cases performed with spreadsheets, since subsequent, more elaborate problems can be treated by extending and modifying the simple calculations.

It should be recognized that the spreadsheet of Example 2-2 includes other situations as special cases. If there are fewer than three regions, then the thicknesses of one or more regions may be entered as zero. If there is a specified temperature at one surface, then the convection coefficient for that surface can be entered as a very large number (so that the associated thermal resistance obtained by taking the reciprocal is, for all practical purposes, zero).

COMPUTER PROJECT 2-2.   Extend your computer program of Project 2-1 to allow for an option to apply convective conditions at either surface or at both surfaces.

2-6. CYLINDRICAL GEOMETRY

In cylindrical geometry, area is not a constant. For cylinder length L, Fourier's Law becomes

$$q = -kA\frac{dT}{dr} = -k(2\pi rL)\frac{dT}{dr} \tag{2-17}$$

Equation (2-16) may be rearranged in the form

$$\frac{q}{2\pi kL}\frac{dr}{r} = -dT \tag{2-18}$$

Integrating from inner radius r_i to outer radius r_o yields

$$\frac{q}{2\pi kL}\ln\frac{r_o}{r_i} = T_i - T_o \tag{2-19}$$

which may be cast in the thermal resistance form

$$q = \frac{T_i - T_o}{R_{th}} \tag{2-20}$$

$$R_{th} = \frac{\ln(r_o/r_i)}{2\pi kL} \tag{2-21}$$

As in plane geometry, this result may be applied separately to each of several layers. The result is

$$q = \frac{T_{i1} - T_{oN}}{\sum_{n=1}^{N}\left[\ln(r_o/r_i)/2\pi kl\right]_n} \tag{2-22}$$

Similarly, if convection conditions are applied, thermal resistance is again given by Eq. (2-14), although the area must be evaluated for cylindrical geometry. We obtain

$$q = \frac{T_{FL} - T_{\infty R}}{1/h_L A_L + \sum_{n=1}^{N}\left[\ln(r_o/r_i)/2\pi kL\right]_n + 1/h_R A_R} \tag{2-23}$$

We have used a fluid temperature at the left since the inside of the cylinder cannot contain an infinite amount of fluid.

We note that in Eq. (2-23), unlike in the situation for plane geometry, the sequence in which the layers appear matters. Interchanging layers will affect results. This is because heat transfer depends on area which varies with radius.

Sometimes thicknesses of material will be small compared to radius. Noting that

$$\ln \frac{r_o}{r_i} = \ln\left(1 + \frac{\Delta r}{r_i}\right) \approx \frac{\Delta r}{r_i} \tag{2-24}$$

Eq. (2-2) becomes

$$R_{th} \approx \frac{\Delta r}{2\pi r_i L k} = \frac{\Delta r}{k A_i} \tag{2-25}$$

This is of the same form as the resistance for plane geometry, except that the area depends on location.

Suppose we have layers of cylindrical shells, and we wish one to be an insulator to retard heat flow. This material is likely to dominate overall thermal resistance. Thus, the overall thermal resistance, for given thicknesses of all layers, is enhanced of the insulating material is placed at the inside. (Other design objectives, of course, may prevent this.)

Similar considerations apply to spherical geometry. The derivation of appropriate equations for spherical geometry is called for in a problem at the end of the chapter.

EXAMPLE 2-3.   A steel pipe 0.3 m in outside diameter contains a hot fluid. The inside temperature of the pipe is 280°C. The pipe is 5 cm thick. The outside of the pipe is in contact with air at 20°C. The convection coefficient to air is 7 W/m² · ° C. The conductivity of the steel is 43 W/m · ° C. How much heat is lost per unit length of pipe?

SOLUTION

$$\frac{q}{L} = \frac{\Delta T}{\ln(r_o/r_i)/2\pi k + 1/2\pi r_o h_R} = \frac{260°}{\ln[15/(15-5)]/2\pi(43) + 1/2\pi(15)(7)}$$

$$\frac{q}{L} = \frac{260}{0.0015 + 0.1516} = 1698 \text{ W/m}$$

EXAMPLE 2-3 (Spreadsheet).   This can be approached in a similar manner as Examples 2-1 and 2-2. It involves use of built-in functions to evaluate logarithms. In

VisiCalc the @ sign is used to deal with functions. The following spreadsheet treats this problem.

	A	B
	A	B
1	INSIDE TEMP	280
2	OUTSIDE TEMP	20
3	OUTSIDE RADIUS	.15
4	THICKNESS	.05
5	INNER RADIUS	+B3 − B4
6	RADIUS RATIO	+B3 / B5
7	LOGARITHM	+@LN(B5)
8	CONDUCTIVITY	43
9	PIPE RESISTANCE	+B7 / (2*3.1416*B8)
10	H − RIGHT	7
11	RESISTANCE	1./(B10*2*3.1416)
12	R − TOTAL	+B9 + B11
13	Q / L	+(B1 − B2)/B12

Note that additional lines may be inserted for dealing with problems that involve multiple cylindrical layers (e.g., pipe wrapping insulation) and convection conditions at the inner surface of the pipe. Note also that this spreadsheet could have been accomplished with fewer rows. For example, we could have avoided a line with explicit evaluation of ratio of radii by placing this evaluation within the argument of the logarithm. It is a matter of choice as to how much intermediate information the analyst wishes to have displayed explicitly. The bias of this author is for a larger number of simpler steps, both to reduce likelihood of error and to gain insight when the problem is complicated and the result surprising.

The spreadsheet made use of a "built-in" function, the natural logarithm. Spreadsheets typically have a variety of built-in functions, as do electronic calculators and as are provided in programming languages such as BASIC and FORTRAN. Commonly used functions are given in Appendix E, along with common spreadsheet commands. Built-in functions in spreadsheets typically include not only common functions such as logarithms and cosines, but also other types of function, including logical functions. Built-in functions permit application of true–false tests, conditional choices, and so on. As we proceed in the text and deal with more complicated examples, we shall make use of the various types of function.

COMPUTER PROJECT 2-3.   Extend your computer program of Projects 2-1 and 2-2 to handle cylindrical and spherical geometries.

2-7. OVERALL HEAT TRANSFER COEFFICIENT

It is common to express heat transfer in terms of an overall heat transfer coefficient U in the form

$$q = UA \, \Delta T \tag{2-26}$$

where ΔT denotes overall temperature difference. For plane geometry, Eq. (2-15) may be adapted so as to yield

$$U = \left[\frac{1}{h_L} + \sum_{i=1}^{I} \left(\frac{H}{kA} \right)_i + \frac{1}{h_R} \right]^{-1}$$ (2-27)

Equation (2-26) may be defined uniquely because all areas in Eq. (2-15) are the same.

In curvilinear geometry, the definition of an overall heat transfer coefficient is not unique, because there is not a unique area to use with Eq. (2-26). Thus, definition of an overall heat transfer coefficient must be done consistently with specifications of a reference area. If the inside area is selected, then Eq. (2-22) would yield

$$U = \left\{ \frac{1}{h_L} + A_L \sum_{n=1}^{N} \left[\frac{\ln(r_o/r_i)}{2\pi kL} \right]_n + \frac{1}{h_R} \frac{A_L}{A_R} \right\}^{-1}$$ (2-28)

2-8. CRITICAL RADIUS OF INSULATION

Suppose we wish to retard heat flow from a cylindrical surface by adding insulation to it. Referring to Eq. (2-28), we can add layer $N + 1$ of material. An additional conduction thermal resistance is added which tends to reduce the overall heat transfer coefficient as desired. However, there is a second effect. The addition of another layer causes the outside area to increase from A_R. Inspection of Eq. (2-28) shows that the thermal resistance associated with convection will be reduced.

The effect of insulation on convection is associated with curved geometry. In plane geometry, all areas are the same and there is no effect. In cylindrical and spherical geometries, area is a function of radius. Increasing outer radius tends to enhance convection.

It is possible that the effect on convection could override the effect of insulation. Consider a simple example with a cylinder of specified surface temperature T_s. Adding insulation, we may express heat flow by

$$q = \frac{T_s - T_\infty}{\ln(r_o/r_i)/2\pi kL + 1/2\pi r_o Lh}$$ (2-29)

Effects will balance if

$$\frac{dq}{dr_o} = 0$$ (2-30)

Applying Eq. (2-29) to Eq. (2-28) yields

$$r_o = \frac{k}{h}$$ (2-31)

If the outer radius exceeds r_o, adding insulation will reduce heat transfer.

The most desirable (least–limiting) situation occurs when the insulating material is good (i.e., when k is very small) and when convection is very effective (i.e., when h is very large). The greatest potential difficulty for the engineer occurs with insulation of small diameter tubing, for which it may be difficult to get a critical radius small enough for practical purposes.

2-9. CONTACT RESISTANCE

In discussing multi-layered walls, we have assumed that there is no space between layers. This is an idealization that may not apply in realistic situations. Even if we try to make two surfaces fit flush against one another, the roughness of each surface may lead to the presence of gaps as illustrated in Fig. 2-9-1. In addition, we may deliberately permit spaces to exist to allow for such factors as expansion upon heating, change in dimension during operation, and provision for insertion or removal of sections. It is not always straightforward to evaluate contact resistance.

An idealization of the problem is illustrated in Fig. 2-9-2. Here the gap region is specified as a zone of finite thickness. Part of the region is filled with a gap fluid (usually air or an inert gas). The fraction of area associated with gap fluid is $A_g/(A_g + A_c)$. In the contact portion, with H the thickness of the gap

$$\frac{q_c(H/2)}{k_{\mathrm{I}} A_c} = T_{\mathrm{I}} - T_c \qquad (2\text{-}32)$$

$$\frac{q_c(H/2)}{k_{\mathrm{II}} A_c} = T_c - T_{\mathrm{II}} \qquad (2\text{-}33)$$

Adding these two equations yields

$$q_c = \left(\frac{1}{k_{\mathrm{I}}} + \frac{1}{k_{\mathrm{II}}}\right)^{-1} \frac{2A_c}{H}(T_I - T) = \frac{2k_{\mathrm{I}}k_{\mathrm{II}}}{k_{\mathrm{I}} + k_{\mathrm{II}}}\frac{A_c(T_{\mathrm{I}} - T_{\mathrm{II}})}{H} \qquad (2\text{-}34)$$

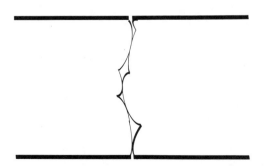

FIGURE 2-9-1. Roughness at interface between two regions.

In the gap portion

$$q_g = \frac{k_g A_g (T_{\mathrm{I}} - T_{\mathrm{II}})}{H} \qquad (2\text{-}35)$$

The total heat transfer is

$$\frac{q}{A} = \frac{q_c + q_g}{A} = \left(\frac{2 k_{\mathrm{I}} k_{\mathrm{II}}}{k_{\mathrm{I}} + k_{\mathrm{II}}} \frac{A_c}{A} + k_g \frac{A_g}{A} \right) \frac{(T_{\mathrm{I}} - T_{\mathrm{II}})}{H} \qquad (2\text{-}36)$$

Since the conductivity for materials I and II is usually greater than the conductivity of the gap material, the heat flow across the gap is determined primarily by degree of contact.

While the above analysis is useful for visualizing the nature of the gap conductance situation, it is clearly an idealization. It is also difficult to estimate actual contact area for general situations.

The gap conductance can be represented in the form

$$q = h_g A (T_{\mathrm{I}} - T_{\mathrm{II}}) \qquad (2\text{-}37)$$

In the idealized model above

$$h_g = \frac{2 k_{\mathrm{I}} k_{\mathrm{II}}}{k_{\mathrm{I}} + k_{\mathrm{II}}} \frac{A_c}{A} + k_g \frac{A_g}{A} \qquad (2\text{-}38)$$

The temperature change across a gap will be modest if pressure is applied to minimize the gap. On the other hand, if a gap is deliberately maintained for clearance for insertion or removal, the temperature change may be larger. In such circumstances, an attempt may be made to increase conductance by using a gas, like helium, that has a relatively high conductivity.

While there can be considerable variation in contact resistance depending on the specific circumstances, some representative values are listed in Table 2-9-1 for types of interfaces frequently encountered. The value of contact

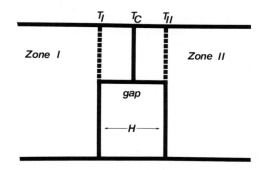

FIGURE 2-9-2.   Idealization of contact at interface between zones.

TABLE 2-9-1
Typical Gap Conductance (W/m^2 · ° C)

Material	Contact Pressure	
	100 kN/m^2	10,000 kN/m^2
Stainless steel	600	5,000
Aluminum	3,000	30,000
Copper	2,000	30,000
Magnesium	4,000	30,000

resistance depends on the roughness of the material and on the interface pressure. For further information on contact resistance, the reader should consult the references at the end of the chapter.

PROBLEMS

2-1. Suppose that the thermal conductivity is not constant with temperature, but varies according to

$$k = k_0 + k_1(T - T_0)$$

where T_0 is some reference temperature. Derive an equation to use instead of Eq. (2-5).

2-2. Based on the data in Appendix A, select suitable values for the parameters of Problem 2-1 (k_0, k_1, T_0) to use in the range 300–400°C for iron. With these parameters and the equation derived in Problem 2-1, evaluate the heat conduction through a cast iron slab 20 cm thick with a 1m^2 area with temperatures maintained at 400°C and 300°C at the ends. What single average thermal conductivity used with Eq. (2-5) would yield the same result?

2-3. You wish to determine the cost implications of energy loss. Starting form the heat flow of Eq. (2-10) and given the total exterior surface area A and the cost of energy (e.g., in dollars per kilowatt hour), derive an expression for the cost of energy loss over a period of 1 month (30 days), assuming that the temperatures stay constant over this time period.

2-4. Assuming that the conditions of Example 2-1 apply for a period of 1 month (30 days) for a structure having a surface area of 250 m^2 and that the cost of energy is $0.03/kW · hr, evaluate the energy cost for the month.

2-5. Extend the spreadsheet provided for Example 2-1 to include energy cost, as per Problem 2-4. Evaluate the monthly energy cost for
a. prices in the range of $0.03–.09/kW · hr in steps of $0.02;
b. insulation thicknesses of 5 and 15 cm at the reference price;
c. structures with areas of 150 and 350 m^2; and
d. using insulation with conductivity of 0.04 W/m ·° C.

2-6. As will be studied in more detail in connection with convection, heat transfer coefficients tend to increase with speed of air. Repeat Example 2-2 with exterior heat transfer coefficents of 20, 50, and 100 W/m^2 ·° K (using the Example 2-2 spreadsheet should be helpful).

2-7. A toaster oven in which food is baking at a temperature of 200°C has a glass window 2 cm thick, 10 cm high, and 15 cm wide through which the food can be viewed. If the inside glass temperature is 200°C, the room temperature is 20°C, and h is 5 W/m^2 ·° K, at what rate is heat lost through the window?

2-8. Extend the spreadsheet of Example 2-3 to include multiple layers of cylindrical shells.

2-9. Pipe-wrapping insulation ($k = 0.04$ W/m $\cdot°$ K) is available to you with a thickness of 1 cm. Evaluate the heat loss per unit length of pipe of Example 2-3 if one, two, or three layers of insulation are wrapped around the pipe.

2-10. Derive an expression for the heat transfer through a spherical shell of conductivity k with inside and outside surface temperatures T and T_o.

2-11. Extend the result of Problem 2-10 to treat cases with convection conditions at the inside and outside.

2-12. Modify the spreadsheet of Example 2-2 to treat the spherical geometry case by substituting formulas for thermal resistance in spherical geometry for original plane geometry formulas.

2-13. Derive an expression for the critical radius of insulation for a sphere.

2-14. A 1 cm diameter tube is in an environment where the heat transfer coefficient is 5 W/m^2 $\cdot°$ K. It is proposed that the tube be insulated with a 1 cm layer of material with $k = 0.1$ W/m $\cdot°$ K.
a. Will the heat loss from the tube be reduced?
b. What is the minimum thickness of insulation needed to get a heat loss lower than that for a bare tube?

2-15. Modify the spreadsheets developed for multilayered walls, cylinders, and spheres to obtain the overall heat transfer coefficients (note the relationship with overall thermal resistance).

2-16. Consider again the situation of Example 2-1, except that the thickness of insulation is not known, but the heat loss is 12 W/m^2.
a. What is the thickness of insulation?
b. What are the temperatures at the two ends of the insulation?

2-17. A device generates a heat flux of 20 W/cm^2. What is the temperature change in a 2 cm thick plate as a function of the conductivity of the plate? Analyze a range of conductivities from insulators (0.01 W/m $\cdot°$ K) to high-conduction metal (400 W/m $\cdot°$ K). In your table or plot, note the location of copper, aluminum, silicon, and fiberglass.

2-18. How will the heat loss in Example 2-1 be affected if the paneling is replaced by an equally thick (1 cm) strip of wallboard ($k = 0.17$)?

2-19. Some buildings have foundations serving as parts of walls. What is the heat loss per unit area through 15 cm of concrete ($k = 1$–2 W/m $\cdot°$ K) when the temperature is 20°C on the inside surface and -15°C on the outside surface.

2-20. A window is placed in the wall of Example 2-2. The window is glass ($k = 0.8$) 7 mm thick. What is the heat loss per unit area through the window?

2-21. A storm window of the same glass and thickness is placed outside the window of Problem 2-20, leaving an air space of 10 cm between the two windows. Assuming the air between the windows to be stagnant (a significant assumption of windy days because of openings in the frame for draining condensation), what is the conduction heat loss through the composite of the two windows? How does this compare with the loss through the wall?

2-22. A thermopane window is a sealed unit of two panes of glass separated by an air space. The panes are 7 mm thick and separated by a 7 mm space. What is the heat loss for the temperature conditions of Example 2-2?

2-23. Normal body temperature is 98.6°F, or 38.1°C. Assume that the interior of the body is insulated from the outside by 3 mm of skin and tissue ($k = 0.2$ W/m $\cdot°$ K). The outside air temperature is -15°C. Make a table or plot of the heat loss per unit area of exposed body surface as a function of heat transfer coefficient, with h varying from 5 to 100 W/m^2 $\cdot°$ K.

2-24. Take the case of $h = 25$ W/m^2 $\cdot°$ K as representative of a calm day. For $25 < h < 100$ in your table for Problem 2-23, find the outside air temperature that will yield the same heat transfer if h were 25 (i.e., how much colder does it feel on a windy day?).

2-25. A conic section is of nonuniform circular cross-section. The ends have radii R_1 and R_2. The length (along the axis) is H. The radius obeys the formula $r = R_1 + (R_2 - R_1)x/H$. The ends are maintained at temperatures T_1 and T_2 and the side is insulated (no heat flow out the sides). Derive an expression to use in place of Eq. (2-5).

2-26. Use the formula obtained in Problem 2-25 to find the heat flow in an aluminum section with one end at $60°C$ with a radius of 5 cm and the other end at $20°C$ with a radius of 10 cm. The length of the section is 5 cm.

2-27. An 0.5 cm diameter electrical cable carrying 500 A has an electrical resistance of 0.06 ohms/cm of length.
  a. What is the heat generated in the cable per meter of length? (Recall that heat goes as I^2R, where I and R are current and resistance, respectively.)
  b. The cable is to be in a heat transfer environment with $h = 20$ W/m$^2 \cdot °$ K and temperature of $20°C$.

2-28. Gap conductance between nuclear fuel and its cladding (the can containing the fuel) varies between 5,500 and 11,000 W/m$^2 \cdot °$ C over the time period energy is produced within a nuclear reactor. If the cladding is 0.01 cm thick and has a conductivity of 12 W/m $\cdot °$ K, the outside of the cladding sees a heat transfer coefficient of 20,000 W/m$^2 \cdot °$ K and a fluid temperature of $300°C$. What is the range of temperatures expected at the surface of the fuel if the heat flux is 2.7×10^5 W/m^2 and the fuel diameter is 1 cm.

2-29. Consider again the case of Example 2-2. A bookcase is to be built into the wall facing the outside. This floor-to-ceiling bookcase is filled with books. Model this situation by assuming that the filled bookcase is equivalent to a 15 cm thick layer of paper ($k = 0.18$ W/m $\cdot °$ C). What is the heat loss per unit area.

2-30. You are considering building a house with the walls of Example 2-2 and designing for the temperature conditions of that example. You wish to have 10% of the wall area of the house used for windows. Find the overall heat loss and the percentage due to loss through windows for the window options identified in Problems 2-20, 2-21, and 2-22.

2-31. A steam pipe has an inner diameter of 10 cm and a thickness of 1 cm. Its thermal conductivity is 40 W/m $\cdot °$ C. The heat transfer coefficient from the steam to the inside of the pipe is 200 W/m$^2 \cdot °$ C. The heat transfer coefficient from the outside of the pipe to the surroundings is 25 W/m$^2 \cdot °$ C.
  a, What is the heat loss per meter of pipe?
  b. What is the heat loss per meter of pipe if 2.5 cm of insulation ($k = 0.04$ W/m $\cdot °$ C) is put around the pipe?

2-32. How sensitive is the heat loss of Problem 2-31b to
  a. the heat transfer coefficients;
  b. the pipe thickness;
  c. the conductivity of the pipe material;
  d. the insulation thickness; and
  e. the conductivity of the insulation material.
  Express sensitivity as the percentage change in heat loss caused by a 1% change in the parameter of interest.

2-33. Consider the database of thermal conductivities of metals and alloys in Appendix A. From this database:
  a. Sort the listing of materials in rank order from highest to lowest thermal conductivity.
  b. Note specifically the substances with (1) the five highest conductivities and (2) the five lowest conductivities.
  Note: This problem can be handled most easily if the student has a database program or a spreadsheet with integrated database features.

2-34. Consider the nonmetallic solids listed in Appendix A. Search the listing to see which, if any, of these have thermal conductivities greater than that of
  a. aluminum and
  b. 304 stainless steel.

2-35. Rank order building structural materials—brick, stone, wood, concrete, steels—in terms of thermal conductivity, that is, which materials would be least likely to lead to heat loss.

REFERENCES

1. M. G. Cooper, B. B. Mikic, and M. M. Yovanovich, "Thermal Contact Conductances," *Int. J. Heat Mass Transfer*, **12**, 279 (1969).
2. C. J. Moore, Jr., H. A. Blum, and H. Atkins, "Studies Classification Bibliography for Thermal Contact Resistance Studies," ASME Paper 68-WA/HT-18, 1968.
3. P. J. Schneider, *Conduction Heat Transfer*, Addison-Wesley, Reading, MA, 1955.

Chapter Three

ONE-DIMENSIONAL HEAT CONDUCTION EQUATION

3-1. INTRODUCTION

In Chapter 2, we considered situations that could be treated only by use of Fourier's Law of heat conduction. In this chapter, we combine Fourier's Law with the principle of conservation of energy to obtain the heat conduction equation. We then apply the equation to situations involving sources and sinks of energy.

In reality, we did consider conservation of energy in Chapter 2, but application of the principle was trivial and did not complicate equations beyond Fourier's Law. In the absence of sources in one dimension, we could conclude, on the basis of conservation of energy, that heat flow is independent of position (i.e., a constant). When there is a heat source, the heat flow is no longer constant and a more elaborate representation of the principle of conservation of energy becomes appropriate.

The chapter begins with the application of conservation to obtain the heat conduction equation. The equation is then applied to the case of a plane wall with an internal uniform source. The set of steps in the solution process are abstracted to facilitate subsequent solution in other geometries. Types of boundary and interface conditions are considered.

A particular class of heat conduction problems that can be represented as one dimensional with a source is the set of cases corresponding to fins. Fins are thin extensions that may be added to surfaces to promote heat transfer.

Heat is conducted along the fin (the one-dimensional heat conduction) and lost through the sides by convection. The convective loss may be modeled as a negative source.

If the form of the heat source is such that the heat conduction equation cannot be solved analytically, numerical analysis becomes appropriate, especially when a computer is available. Formation of a difference equation from a differential equation is discussed. In essence, this is a conversion from a differential equation to a set of algebraic equations. In addition, consideration is given to approximating boundary conditions.

3-2.   CONSERVATION OF ENERGY

We must account for all energy. If we consider a region in space, we can introduce energy to the region by flow in across boundaries and by sources within the boundaries. We can remove energy by flow out across boundaries and by sinks within the region. The difference between what is introduced and what is removed is the rate of change of energy content.

Consider a plane geometry region bounded by location x at the left and by location $x + dx$ at the right as illustrated in Fig. 3-2-1. The net heat flow in through location x is given by Fourier's Law as

$$q(x) = -kA\frac{dT}{dx}(x)$$
(3-1)

The net heat flow out through location $x + dx$ is

$$q(x + dx) = -kA\frac{dT}{dx}(x + dx)$$
(3-2)

The net flow into the region is the difference between these two quantities, that is,

$$q(x) - q(x + dx) = -kA\frac{dT}{dx} + kA\frac{dT}{dx}(x + dx)$$
(3-3)

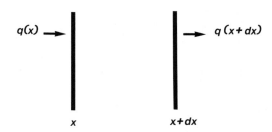

FIGURE 3-2-1.   Heat entering and leaving a small interval.

If the region is thin, that is, if dx is very small, then

$$q(x) - q(x + dx) = -kA \frac{dT}{dx}(x) + kA \frac{dT}{dx}(x) + \frac{d}{dx}\left(kA \frac{dT}{dx}\right) dx \quad (3\text{-}4)$$

If there is a heat source per unit volume q''' within the region, then the total energy source is

$$S = q'''A \, dx \qquad (3\text{-}5)$$

If there is a heat sink per unit volume, it may be considered as a source having a negative value. The rate of change of energy in the region, if there is no change in phase, is manifested as a temperature change via

$$\rho c \frac{\partial T}{\partial t} A \, dx \qquad (3\text{-}6)$$

where ρ and c are density and specific heat, respectively. The heat conduction equation in plane geometry becomes

$$\rho c \frac{\partial T}{\partial t} A \, dx = \frac{\partial}{\partial x}\left(kA \frac{\partial T}{\partial x}\right) dx + q'''A \, dx \qquad (3\text{-}7)$$

If the area is constant, we obtain

$$\rho c \frac{\partial T}{\partial t} = \frac{\partial}{\partial x}\left(k \frac{dT}{\partial x}\right) + q''' \qquad (3\text{-}8)$$

$q''' = \frac{per}{unit \, volume}$

A convention used commonly in heat transfer is to represent heat sources by a number of primes appropriate to the nature of the source. A heat source per unit area would be represented as q'' and one per unit length would be represented as q'.

Generally, we deal with situations involving heat sources and not with heat sinks. However, certain situations can be represented as involving heat sinks as we shall see later when we discuss heat transfer from fins.

3-3. HEAT CONDUCTION EQUATION IN GENERAL GEOMETRIES

In vector notation, as discussed in Chapter 2, Fourier's Law may be written

$$q = -kA \nabla T \qquad (3\text{-}9)$$

In general geometry, we may write Eq. (3-8) as

$$\rho c \frac{\partial T}{\partial t} = \nabla \cdot (k \nabla T) + q''' \qquad (3\text{-}10)$$

In one-dimensional geometries, we may write

$$\nabla \cdot k \nabla T = \frac{1}{r^n} \frac{d}{d^r} \left(kr^n \frac{dT}{dr} \right) \tag{3-11}$$

where $n = 0, 1, 2$ in plane, cylindrical, and spherical geometries, respectively. As we shall see, this form of equation is convenient for solving equations. It is also a convenient basis for setting up difference approximations for numerical solution. The explicit derivation of Eq. (3-11) for cylindrical and spherical geometries is left for problems at the end of the chapter. In the remainder of this chapter, we consider only steady-state problems.

3-4. HEAT CONDUCTION IN A PLANE WALL WITH A SOURCE

Consider a wall with center at $x = 0$ with a uniform source q'''. At the outer surfaces of the wall $(x = \pm H)$, the temperature is known to be T_w. The steady-state heat conduction equation is

$$\frac{d}{dx} \left(k \frac{dT}{dx} \right) + q''' = 0 \tag{3-12}$$

Let Eq. (3-12) be integrated from 0 to x. Then

$$k \frac{dT}{dx}(x) - k \frac{dT}{dx}(0) + q'''x = 0 \tag{3-13}$$

Because the problem is symmetric about $x = 0$, it must be true that

$$\frac{dT}{dx}(0) = 0 \tag{3-14}$$

Equation (3-13) can now be integrated from x to H, yielding

$$k[T(H) - T(x)] + \tfrac{1}{2} q'''(H^2 - x^2) = 0 \tag{3-15}$$

Noting that

$$T(H) = T_w \tag{3-16}$$

the temperature distribution becomes

$$T(x) = T_w + \frac{1}{2k} q'''(H^2 - x^2) \tag{3-17}$$

Let us summarize the steps in the derivation because they can be applied directly to analyzing other geometries:

1. Multiply by the volume element (a trivial step in plane geometry).
2. Integrate from 0 to x.
3. Apply a symmetry condition at $x = 0$.
4. Integrate from x to H.
5. Apply the condition $T(H) = T_w$.

Step 1 has more substance for curved geometries because of the r^n terms in Eq. (3-11). In curved geometries, division by r^n may be needed prior to Step 4 to facilitate integration. Keep these steps in mind when reading Section 3-6.

EXAMPLE 3-1. Heat is generated in large plate ($k = 0.6$ W/m · °C) at the rate of 4000 W/m³. The plate is 20 cm thick. What is the temperature rise in the plate?

SOLUTION

$$T(0) - T_w = \frac{q'''}{2k} H^2 = \frac{4000}{(2)(0.6)} (0.10)^2 = 33°C$$

3-5. BOUNDARY CONDITIONS

In solving the heat conduction equation in Section 3-4, we applied two conditions, Eqs. (3-14) and (3-16). This is to be expected, since the heat conduction equation is a second-order equation. If the problem has a multi-layered wall, then two conditions will be required for each layer. This is because each layer may be viewed initially as an independent problem. The boundary conditions supply coupling for these independent problems. Boundary conditions that couple regions are called interface conditions. These conditions are determined by the following consideration. When heat leaves one region, it must enter the next region. At an interface x_I, we may write

$$\lim_{\varepsilon \to 0} k(x_I - \varepsilon) \frac{dT}{dx}(x_I - \varepsilon) = \lim_{\delta \to 0} k(x_I + \delta) \frac{dT}{dx}(x_I + \delta) \qquad (3\text{-}18)$$

Thus, in a multilayered problem, we do not, in general, expect the slope of the temperature to be continuous, since we expect different regions to have different thermal conductivities. The change in slope is illustrated in Fig. 3-5-1.

While the derivative of the temperature need not be continuous, the temperature should be. Energy will flow in response to a difference in temperature. If there is finite temperature difference associated with an infinitesimal distance, an infinite heat flow would be implied by Fourier's Law. This

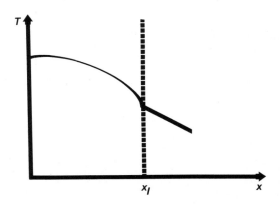

FIGURE 3-5-1. Temperature distribution in a two-layer wall.

is a contradiction unless there is a singular source (an artifice sometimes used in modeling) present, so temperature should be continuous; that is,

$$\lim_{\varepsilon \to 0} T(x_I - \varepsilon) = \lim_{\delta \to 0} T(x_I + \delta) \tag{3-19}$$

Boundary conditions express what the engineer knows about a situation, or what the engineer has imposed on the situation. For example, if a boundary condition like Eq. (3-16) is used, it may mean that the engineer knows what the temperature is at a surface. It also may mean that the engineer has decided that this is the temperature appropriate for that surface and has taken steps to have the temperature monitored, with heating or cooling activated as needed to maintain that surface temperature.

Boundary conditions other than Eqs. (3-14) and (3-16) may be used. For example, one may specify the heat flux at a surface; that is,

$$-k\frac{dT}{dx}(x_s) = q'' \tag{3-20}$$

The heat source per unit are q'' at the surface may be determined by radiation heat transfer from the sun.

Convection provides another boundary condition

$$-k\frac{dT}{dx}(x_s) = h\left[T(x_s) - T_\infty\right] \tag{3-21}$$

Sometimes, when this boundary condition is used, we find it convenient to change variables, letting

$$\Theta = T - T_\infty \tag{3-22}$$

$$-k\frac{d\Theta}{dx}(x_s) = h\Theta(x_s) \qquad (3\text{-}23)$$

and the heat conduction equation becomes

$$\frac{d}{dx}\left(k\frac{d\Theta}{dx}\right) + q''' = 0. \qquad (3\text{-}24)$$

This transformation is possible because the derivative of the constant T_∞ is zero. We will find several occasions to use this transformation.

The form of the condition of Eq. (3-14) may be used in another context. Sometimes we assume that insulation at a surface is perfect; that is, no heat is transferred across that surface. Other boundary conditions may be appropriate for other situations, for example, mixed radiation and convection conditions at a surface. Some other situations will be dealt with in problems at the end of the chapter.

EXAMPLE 3-2.   The outside surfaces of the plate of Example 3-1 are exposed to ambient air at 20°C with a convective heat transfer coefficient of 20 W/m² · °C. What is the temperature at the center of the plate?

SOLUTION.   The temperature change between the center and the surface calculated previously still applies. We now have to apply the boundary condition of Eq. (3-23):

$$\Theta(x) = T(x) - T_\infty \equiv T(x) - T_w + T_w - T_\infty = \frac{1}{2k}q'''\left(x_s^2 - x^2\right) + \Theta(x_s)$$

$$\frac{d\Theta(x)}{dx} = -\frac{q'''x}{k}$$

Applying Eq. (3-23),

$$\Theta(x_s) = \frac{-k}{h}\frac{d\Theta}{dx}(x_s) = \frac{q'''x_s}{h} = \frac{(4000)(0.1)}{(20)} = 20°C = T_w - T_\infty$$

It was given that

$$T_\infty = 20°C$$

We therefore find that

$$T_w = T_\infty + 20°C = 40°C$$

$$T(0) = T_w + 33°C = 73°C$$

EXAMPLE 3-2 (Spreadsheet).   Examples 3-1 and 3-2 can illustrate how use of a personal computer with spreadsheet software can be used to increase personal efficiency.

Example 3-1 could be solved with the following spreadsheet:

	A	B
1	HEAT SOURCE	4000
2	THICKNESS	.2
3	HALF THICKNESS H	+B2 / 2
4	H SQUARE	+B3 ∧ 2
5	CONDUCTIVITY	.6
6	DELTA T	+B1*B4 / (2*B5)

If this spreadsheet were saved, then for Example 3-2 it could be recalled and added to as follows:

	C	D
1	CONVECTION COEFFICIENT	20
2	DELTA T SURFACE	+B1*B3 / D1
3	AMBIENT TEMPERATURE	20
4	CENTER TEMPERATURE	+D3 + D2 ⓐ B6

Thus, the solution of Example 3-1, when saved, could be used as a building block for Example 3-2. Naturally, the saved spreadsheet also could be used to deal with similar problems with different numbers.

3-6. CYLINDRICAL GEOMETRY

The heat conduction equation in a cylinder of radius R is

$$\frac{1}{r} \frac{d}{dr} \left(rk \frac{dT}{dr} \right) + q''' = 0 \tag{3-25}$$

The temperature at the surface is specified as T_w. We shall follow the sequence of steps outlined in Section 3-4. First, we must multiply by the volume element. Note that an element of volume dV is given by

$$dV = 2\pi r L \, dr \tag{3-26}$$

Let Eq. (3-25) be multiplied by dV and, as per the second step, integrated from 0 to r. This yields

$$2\pi r L k \frac{dT}{dr}(r) - 2\pi r L k \frac{dT}{dr}(0) + 2\pi L k \left(\tfrac{1}{2} q''' r^2 \right) = 0 \tag{3-27}$$

We now see one aspect of the convenience of the form of Eq. (3-25). When the equation is multiplied by the volume element, the result is a perfect differential that integrates directly to Eq. (3-27).

There can be no net heat transfer at the center of the cylinder, so application of the symmetry condition (the third step) yields

$$2\pi r L k \frac{dT}{dr}(0) = 0 \tag{3-28}$$

Dividing the resulting equation by $2\pi rL$ and rearranging yields

$$k\,dT + \tfrac{1}{2}q'''r\,dr = 0 \tag{3-29}$$

Integrating from r to R (the fourth step) yields, upon setting the upper limit of the temperature integral to T_w (the fifth step),

$$\int_{T(r)}^{T_w} k\,dT + \tfrac{1}{4}q'''(R^2 - r^2) = 0 \tag{3-30}$$

Note that we have not assumed that conductivity is independent of temperature. Consider the condition at the center of the cylinder:

$$\int_{T_w}^{T(0)} k\,dT = \tfrac{1}{4}q'''R^2 = \frac{q'''}{4\pi}\pi r^2 = \frac{q}{4\pi} \tag{3-31}$$

Equation (3-31) is a significant result. The left side, the conductivity integral between surface and center temperatures, depends on the total heat per unit length q' but is independent of the radius of the cylinder. The conductivity integral is useful for characterizing fuel element materials for nuclear reactors. The higher the conductivity integral from surface to melting temperature, the more heat can be generated (if other factors are equal) per unit length of fuel rod. Fuel rod radius is important for heat transfer from the surface of the cylinder, because convection depends on area. Thus, uranium oxide fuels in different nuclear reactor types may have similar linear heat ratings but different rod diameters.

If the thermal conductivity is independent of temperature, Eq. (3-30) may be expressed as

$$T(r) = T_w + \frac{q'''}{4k}(R^2 - r^2) \tag{3-32}$$

Thus, in cylindrical geometry also, we find that temperature variation is quadratic.

The temperature distribution in a sphere with an internal source may be obtained by following a similar procedure to that of this section. The derivation is left as a problem at the end of this chapter.

EXAMPLE 3-3. A borosilicate glass ($k = 0.63$ W/m°C) cylinder 0.3 m in diameter contains a heat source of 4000 W/m³. What is the temperature rise from the surface to the center of the cylinder?

SOLUTION

$$T(0) - T_w = \frac{q'''}{4k} R^2 = \frac{4000}{(4)(0.63)} \left(\frac{0.3}{2} \right)^2 = 35.7°C$$

Note that the difference between the solutions for cylindrical and plane geometries is the presence of a 4 or a 2 in the denominator of the quantity to be evaluated. Thus, the spreadsheet solution for Example 3-1 could be used if diameter were used for thickness and if the last row were changed to

	A	B
6	DELTA T	+B1*B4 / (4*B5)

3-7. HEAT TRANSFER FROM FINS—NEGATIVE SOURCES

When convection from a surface is relatively weak, as in natural convection by air, an attempt may be made to enhance heat transfer by extending the surface by means of a fin, as illustrated in Fig. 3-7-1. Heat is conducted along the fin, and convection takes place transversely as shown.

It is common practice to represent the problem as one-dimensional with a loss term. In essence, it is assumed that the fin is so thin that the temperature variation inside the fin transversely is very small. (In engineering, when a postulate is made that something is thin or small, it also is necessary to provide a standard of comparison. This point will be explored further in a problem at the end of the chapter.) The governing equation for this problem is the steady-state version of Eq. (3-7):

$$\frac{d}{dx} \left(kA \frac{dT}{dx} \right) dx + q'''A \, dx = 0 \tag{3-33}$$

Note that A refers to cross-sectional area. The total heat source is negative

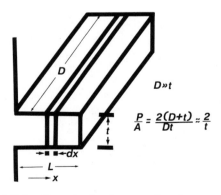

FIGURE 3-7-1. Rectangular fin extending from a plane wall.

and determined by convection out of the surface area element dS,

$$q'''A\,dx = -h\,dS(T - T_\infty) = -hP\,dx(T - T_\infty) \qquad (3\text{-}34)$$

where P is the perimeter of the fin.

Note that in characterizing the convective heat loss from the fins as a negative source, we have introduced a mathematical artifice to put the equation in the general form we derived earlier. There are situations in which the concept of a negative source has direct meaning, for example in endothermic chemical reactions.

In this problem, cross-sectional area is independent of x. However, for other fin shapes, for example, the triangular fin of Fig. 3-8-1, this is not the case. We shall turn to nonuniform cases later, but now we consider a fin of constant cross-section.

The heat conduction equation becomes

$$\frac{d^2T}{dx^2} - m^2(T - T_\infty) = 0 \qquad (3\text{-}35)$$

$$m^2 = \frac{hP}{kA} \qquad (3\text{-}36)$$

We introduce the change of variables

$$\Theta = T - T_\infty \qquad (3\text{-}37)$$

so the heat conduction equation becomes

$$\frac{d^2\Theta}{dx^2} - m^2\Theta = 0 \qquad (3\text{-}38)$$

At the base (or root) of the fin, we specify the temperature to be T_0, so

$$\Theta(0) = \Theta_0 = T_0 - T_\infty \qquad (3\text{-}39)$$

As a second boundary condition, we assume that the end of the fin is insulated, so

$$\frac{\partial T}{\partial x}(x = L) = \frac{\partial \Theta}{\partial x}(L) = 0 \qquad (3\text{-}40)$$

This is not really that case. However, it is a common practice to augment the length of the fin to a corrected length

$$L_c = L + \tfrac{1}{2}t \qquad (3\text{-}41)$$

to simulate the heat loss from the end while maintaining the one-dimensional character of the problem. When there is concern about this assumption, two-dimensional treatment is appropriate.

The solution to the heat conduction equation can be expressed in terms of exponentials or hyperbolic functions. A convenient representation is

$$\Theta = C_1 \cosh m(L_c - x) + C_2 \sinh m(L_c - x) \tag{3-42}$$

The insulated end condition yields

$$\frac{d\Theta}{dx}(L_c) = 0 = -C_1 m \sinh(0) - C_2 m \cosh(0) \tag{3-43}$$

Recall that $\sinh(0)$ is 0 and $\cosh(0)$ is 1. Therefore,

$$C_2 = 0 \tag{3-44}$$

The second condition is that

$$\Theta(0) = \Theta_0 = C_1 \cosh mL_c \tag{3-45}$$

The solution is therefore

$$\Theta(x) = T(x) - T_\infty = (T_0 - T_\infty) \frac{\cosh m(L_c - x)}{\cosh mL_c} \tag{3-46}$$

It should be clear why we chose the representation of Eq. (3-42) in terms of hyperbolic functions. When we have a zero boundary condition at a location (in this case L_c), it is convenient to use a function that satisfies this condition naturally, that is, $\sinh m(L_c - x)$. The algebra associated with evaluating undetermined constants turns out to be relatively simple. Had we used exponentials in Eq. (3-42), we would have obtained the same answer but would have done more complicated algebra.

The total heat transfer from the fin may be obtained by either of the following two conditions:

$$q = -kA \frac{d\Theta}{dx}(0) \tag{3-47}$$

$$q = \int_0^{L_c} dx \, hP \, dx \, \Theta(x) \tag{3-48}$$

This is because all the heat conducted into the fin is lost by convection along the fin. Thus,

$$q = \frac{hP\Theta_0}{m} \tanh mL_c \tag{3-49}$$

It is common practice to treat fins with the concept of fin efficiency, which is defined as the ratio of actual heat transfer to an ideal case where the entire fin is at the base temperature. This ideal heat transfer is

$$q_I = hPL_c\Theta_0 \tag{3-50}$$

The efficiency thus becomes

$$\eta = \frac{q}{q_I} = \frac{\tanh mL_c}{mL_c} \tag{3-51}$$

To determine how much is gained by adding a fin, we note that in the absence of the fin the reference heat removal by convection would be

$$q_{\text{ref}} = hA\Theta_0 \tag{3-52}$$

The ratio of heat transfer with and without the fin, called the effectiveness of the fin, is

$$\frac{q}{q_{\text{ref}}} = \frac{P}{Am} \tanh mL_c = \sqrt{\frac{k}{h} \frac{P}{A}} \tanh mL_c \tag{3-53}$$

For the case shown, the ratio P/A is (see Fig. 3-7-1)

$$\frac{P}{A} = \frac{2}{t} \tag{3-54}$$

For a long corrected length, the hyperbolic tangent tends to unity. Thus, we require for a gain from a fin that

$$\frac{2k}{ht} > 1 \tag{3-55}$$

For a fin of a high conductivity metal like copper ($k = 401$ W/m · °C) in still air ($h \approx 7$ W/m² · °C), this condition is easy to satisfy. However, with materials of low thermal conductivity or in situations (e.g., forced convection with water) with high values of h, fins could actually detract from heat transfer.

Note that there are diminishing returns from increasing fin length. When mL_c is 1.5, $\tanh mL_c$ is 0.9, while when mL_c is 2.0, $\tanh mL_c$ is 0.96, a 7% increase in heat flow for a 33% increase in effective fin length. Beyond an mL_c

of 2.0, very little increase in heat flow is possible. What constitutes the optimum fin length will depend on various design considerations, requirements for heat removal, cost of materials, significance of weight of materials, and so on.

EXAMPLE 3-4.   A metallic fin ($k = 385$ W/m · °C) 1 mm thick and 10 cm long extends from a plane wall whose temperature is 250°C. How much heat is transferred per meter of depth of the wall from the fin to the air of 20°C if $h = 7$ W/m² · °C.

SOLUTION.   The total heat lost by the fin per unit of depth D is

$$\frac{q}{D} = \frac{hP\Theta_0}{Dm} \tanh mL_c$$

The correct length is

$$L_c = L + \frac{t}{2} = 0.10 + \frac{0.001}{2} = 0.1005 \text{ m}$$

For a deep plane, D is much greater than t so $P \approx 2D$, as can be seen from Fig. 3-7-1. Therefore, the heat loss is

$$\frac{q}{d} = (7\text{W/m}^2 \cdot °\text{C})(2)(250°\text{C} - 20°)\frac{\tanh mL_c}{m}$$

$$m = \sqrt{\frac{hP}{kA}} = \sqrt{\frac{h}{k}\frac{2}{t}} = \sqrt{\frac{(7)(2)}{(385)(10^{-3})}} = 6.03$$

$$mL_c = 0.606$$

$$\tanh mL_c = 0.54$$

$$\frac{q}{D} = 288 \text{ W/m}$$

EXAMPLE 3-4 (Spreadsheet).   Let us prepare a spreadsheet to find the fin efficiency and heat transfer for this problem. This can be convenient since the spreadsheet can be adapted easily to treat approximate solutions for triangular and circumferential fins.

	A	B	C	D
1	FIN	EFFICIENCY		
2	GET	MLC	GET	ETA
3	H	7	EXP MLC	@ EXP (B9)
4	K	385	EXP (- MLC)	1 / D3
5	T	.001	TANH	(D3 - D4) / (D3 + D4)
6	M	((2*B3) / (B4*B5)) ∧ .5	ETA	+D5 / B9

	A	B	C	D
7	L	.1		
8	LC	+B7+(.5*B5)	GET	HEAT
9	MLC	+B6*B8	AREA	2*B8
10			TSURFACE	250
11			TINFINITY	20
12			HEAT	+B3*D6*D9*(D10-D11)

3-8. FINS OF NONUNIFORM AREA—TAPERED FINS

Fins of nonuniform cross-section also may be treated as one-dimensional with losses. The solution of the heat conduction equation for tapered fins can be somewhat complicated, involving use of Bessel functions. It will be shown, however, how a simple analytical approximation that avoids use of the Bessel functions can be prepared (Example 3-5).

Tapered fins provide a mechanism for reducing the amount of material in a fin. We have noted already that there are diminishing returns with increasing fin length. Tapering the fin increases the thermal resistance for heat to flow to the end of the fin. Since the end is less effective than the beginning area, the heat transfer penalty may be modest while the weight (and material cost) reduction may be substantial. Consider first the triangular fin of Fig. 3-8-1. We start from the equation

$$\frac{d}{dx}\left(kA\,\frac{d\Theta}{dx} \right)dx - hP\Theta\,dx = 0 \tag{3-56}$$

We have not considered the slight incline in evaluation of surface area $P\,dx$. This equation may be rearranged to

$$\frac{d^2\Theta}{dx^2} + \frac{1}{A}\frac{dA}{dx}\frac{d\Theta}{dx} - \frac{hP}{kA}\Theta = 0 \tag{3-57}$$

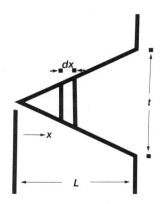

FIGURE 3-8-1.   Tapered fin extending from a plane wall.

Note that if A_b is the area of the fin at the base

$$A = A_b \frac{x}{L} \tag{3-58}$$

$$\frac{1}{A} \frac{dA}{dx} = \frac{1}{x} \tag{3-59}$$

and, for long transverse dimensions,

$$\frac{P}{A} = \frac{2}{t} \frac{L}{x} \tag{3-60}$$

which leads to

$$\frac{d^2\Theta}{dx^2} + \frac{1}{x} \frac{d\Theta}{dx} - \frac{2L}{xt} \frac{h}{k} \Theta = 0 \tag{3-61}$$

To put this equation into standard form, let

$$y = \sqrt{\frac{4L}{t} \frac{h}{k}} \sqrt{x} \tag{3-62}$$

This leads to the modified Bessel equation of order zero

$$\frac{d^2\Theta}{dy^2} + \frac{1}{y} \frac{d\Theta}{dy} - \Theta = 0 \tag{3-63}$$

This has the general solution

$$\Theta = C_1 I_0(y) + C_2 K_0(y) \tag{3-64}$$

where I_0 and K_0 denote zero-order modified Bessel functions of the first and second kind. Appendix B discusses properties of these functions. Transforming the solution back to x yields

$$\Theta(x) = C_1 I_0(2p\sqrt{x}) + C_2 K_0(2p\sqrt{x}) \tag{3-65}$$

$$p^2 = \frac{2L}{t} \frac{h}{k} \tag{3-66}$$

At the tip of the fin, $x = 0$, the K_0 function goes to infinity. This would yield a nonphysical solution, so one boundary condition permits us to set C_2 to zero. At the base, the wall temperature is specified so

$$\Theta(L) + C_1 I_0(2p\sqrt{L}) = \Theta_0 \tag{3-67}$$

The solution is therefore

$$\Theta(x) = \frac{\Theta_0 I_0(2p\sqrt{x})}{I_0(2p\sqrt{L})} \tag{3-68}$$

The total heat loss through the fin may be obtained from

$$q = kA \frac{d\Theta}{dx}(L) \tag{3-69}$$

The derivative of the Bessel function is

$$\frac{d}{dx} I_0(u) = I_1(u) \frac{du}{dx} \tag{3-70}$$

where $I_1(u)$ is the Bessel function of order one. For our problem

$$\frac{d}{dx} I_0(2p\sqrt{x}) = \frac{P}{\sqrt{x}} I_1(2p\sqrt{x})c \tag{3-71}$$

We thus obtain

$$q = \sqrt{\frac{k}{ht}} \, hA_b\Theta_0 \frac{I_1(2p\sqrt{L})}{I_0(2p\sqrt{L})} \tag{3-72}$$

To get fin efficiency, we divide by the ideal heat flow, $2h(A_b/t)L\Theta_0$ to get

$$\eta = \frac{I_1(2p\sqrt{L})}{I_0(2p\sqrt{L})} \frac{1}{mL} = \frac{1}{mL} \frac{I_1(2mL)}{I_0(2mL)} \tag{3-73}$$

where m has the same value $\sqrt{2h/kt}$ as for rectangular fins. Appendix B gives values of I_0, I_1, and the ratio I_1/I_0. We observe that the ratio $I_1(u)/I_0(u)$, like the hyperbolic tangent, goes to 1 for large values of u.

The Bessel functions and Bessel equations occur commonly in engineering analysis, especially in connection with cylindrical geometry, as will be seen in Section 3-9. However, Bessel functions are not as commonly provided for as convenience options (by comparison, for example, with trigonometric and hyperbolic functions) on electronic calculators and personal computers. Appendix B includes convenient, simple procedures for evaluating Bessel functions that can be used with a personal computer.

An alternative to the use of a Bessel function module is to construct a simple approximate analytical formula for efficiency. How this might be done is discussed below.

EXAMPLE 3-5. Construct a simple formula for efficiency of triangular fins for $mL \leq 1$ based on adapting the straight fin formula of Eq. (3-51).

SOLUTION.   The straight fin formula is

$$\eta = \frac{\tanh mL_c}{mL_c}$$

While the triangular fin formula is

$$\eta = \frac{I_1(2mL)/I_0(2mL)}{mL}$$

Examination of Eq. (B-12) shows that as mL gets small

$$\frac{I_1(2mL)/I_0(2mL)}{mL} \rightarrow \frac{2mL/2}{mL} = 1 = \frac{\tanh mL}{mL}$$

so we have a natural joining point for the straight fin formula. Below we evaluate, for several values of the inverse hyperbolic tangent, $I_1(z)/I_0(z)$.

mL	$2mL$	$I_1(2mL)/I_0(2mL)$	$\tanh^{-1}(I_1/I_0)$	f
0.1	0.2	0.0995	0.09983	0.9983
0.2	0.4	0.1962	0.1988	0.9939
0.3	0.6	0.2873	0.2956	0.9854
0.4	0.8	0.3711	0.3897	0.9742
0.5	1.0	0.4463	0.4800	0.9601
0.6	1.2	0.5129	0.5667	0.9444
0.7	1.4	0.5703	0.6480	0.9257
0.8	1.6	0.6170	0.7201	0.9002
0.9	1.8	0.6619	0.7962	0.8847
1.0	2.0	0.6979	0.8632	0.8632

We see that for $z > 0$,

$$\tanh^{-1}\left[\frac{I_1(z)}{I_0(z)}\right] = f\frac{z}{2} = fmL$$

where f is less than 1. Let us then take

$$\eta = \frac{\tanh fmL}{mL}$$

and develop a formula for f. This may be done as follows. Let us try to find an interpolation function that is exact at mL values of 0 and 2.0. Inspection of the data indicates that a linear fit would not be appropriate. Let us try

$$f = e^{C(mL)^2}$$

We evaluate C so that at $mL = 1$

$$C = \ln(0.8632) = -0.1471$$

We then get

mL	f exact	f approximate	tanh fml
0.1	0.9983	0.9985	0.0995
0.2	0.9939	0.9941	0.1962
0.3	0.9854	0.9868	0.2877
0.4	0.9742	0.9767	0.3719
0.5	0.9601	0.9638	0.4478
0.6	0.9444	0.9484	0.5147
0.7	0.9257	0.9304	0.5725
0.8	0.9002	0.9101	0.6219
0.9	0.8847	0.8877	0.6634
1.0	0.8632	0.8632	0.6979

We may observe that this simple interpolation procedure yields good accuracy (compare tanh fmL with I_1/I_0) and provides simple algorithm for evaluation with a calculator or for coding into a spreadsheet.

To modify the spreadsheet of Example 3-4 to apply to triangular fins, lines to compute f by the formula of Example 3-5 should be added to permit calculation of fmL. The accuracy of the approximate formula is sufficient for most practical purposes. (As will be discussed in Chapters 6–9 on convection, uncertainty in the convective heat transfer coefficient is generally greater than the small error associated with the approximation.)

EXAMPLE 3-6. Repeat Example 3-4 for a triangular fin.

SOLUTION. The heat per unit depth is

$$\frac{q}{D} = \sqrt{\frac{2k}{ht}} \frac{hA_b}{D} \Theta_0 \frac{I_1(2p\sqrt{L})}{I_0(2p\sqrt{L})}$$

$$p\sqrt{L} = \sqrt{2}\sqrt{\frac{h}{tk}} L = \sqrt{2}\sqrt{\frac{7}{10^{-3}(385)}}(0.1) = 0.6$$

$$\frac{I_1}{I_0}(2p\sqrt{L}) = 0.51$$

which is also obtained if we use

$$\frac{I_1}{I_0} = \text{tanh } fmL = \text{tanh } fp\sqrt{L} = 0.51$$

Noting that $A_b/D = t$, we obtain

$$\frac{q}{D} = \sqrt{2kht}\,\Theta_0(0.51) = \sqrt{(2)(385)(7)(10^{-3})}\,(230)(0.51)$$

$$\frac{q}{D} = 273 \text{ W/m}$$

The tapered fin delivers slightly less heat, but with only half the weight compared to the rectangular fin of Example 3-4.

EXAMPLE 3-6 (Spreadsheet).   As discussed earlier, it is a straightforward matter to modify the spreadsheet of Example 3-4 to treat triangular fins with the approximate efficiency formula. First, change LC to equal L (since there is no fin thickness at the fin tip) by changing the entry in B8 to be +B7. Then add

	A	B
10	ML SQUARE	+B9 $\wedge$ 2
11	C	-.1471
12	PRODUCT	+B11*B10
13	F	@EXP(B12)
14	FML	+B14*B9

To evaluate tanh *fmL*, change D3 from @EXP (B9) to @EXP (B14). This spreadsheet now computes the solution for a triangular fin.

Although personal computer spreadsheet programs may not contain Bessel functions as built-in functions, it is a simple matter to prepare a spreadsheet evaluation of Bessel functions as shown in Appendix B. This spreadsheet can then be used as a building block for a subsequent spreadsheet. For Example 3-5, a column could be inserted in the A column (before evaluating the Bessel functions) to evaluate the quantity $2p\sqrt{L}$. The entries in the columns evaluating the I_1 and I_0 functions could then use that value of $2p\sqrt{L}$. A column could then be added after the Bessel function columns to evaluate q/D. The new spreadsheet could be saved separately, and the Bessel function building block is still available for use in other situations.

3-9. FINS OF NONUNIFORM AREA—CIRCUMFERENTIALS FINS

In cylindrical geometry, heat conduction is governed by

$$\frac{1}{r}\frac{d}{dr}\left(dr\frac{d\Theta}{dr}\right) + q''' = 0 \tag{3-74}$$

For the circumferential fin shown in Fig. 3-9-1, the effective source in volume element $2\pi rt\, dr$ is the convection heat loss through the top and bottom surface area elements, each of which is $2\pi r\, dr$. Thus,

$$q'''(2\pi rt\, dr) = -h\Theta(4\pi r\, dr) \tag{3-75}$$

and the heat conduction equation becomes

$$\frac{1}{r}\frac{d}{dr}\left(kr\frac{d\Theta}{dr}\right) - \frac{2h}{t}\Theta = 0 \tag{3-76}$$

This again is the modified Bessel equation of order zero. Here, as in the last section, a complicated solution is obtained. Again, it is possible to introduce an analytical approximation that is easy to evaluate. The solution can be written

$$\Theta = C_1 I_0(mr) + C_2 K_0(mr) \tag{3-77}$$

$$m^2 = \frac{2h}{kt} \tag{3-78}$$

As with the plane geometry case, we define a corrected length (and also a corrected outer radius)

$$L_c = L + \frac{t}{2} \tag{3-79}$$

$$r_{2c} = r_2 + \frac{t}{2} \tag{3-80}$$

and impose the conditions

$$\frac{d\Theta}{dr}(r_{2c}) = 0 \tag{3-81}$$

$$\Theta(r_1) = \Theta_0 \tag{3-82}$$

The first condition yields

$$mC_1 I_1(mr_{2c}) - mC_2 K_1(mr_{2c}) = 0 \tag{3-83}$$

so the solution becomes

$$\Theta = C_1 \left[I_0(mr) + \frac{I_1(mr_{2c})}{K_1(mr_{2c})} K_0(mr) \right] \tag{3-84}$$

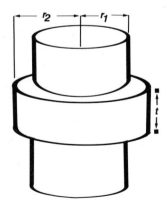

FIGURE 3-9-1. Circumferential fin extending from a cylinder.

We then find C_1 by

$$\Theta(r_1) = \Theta_0 = C_1\left[I_0(mr) + \frac{I_1(mr_{2c})}{K_1(mr_{2c})}K_0(mr_1)\right] \qquad (3\text{-}85)$$

The solution is thus

$$\theta(r) = \Theta_0\frac{I_0(mr) + [I_1(mr_{2c})/K_1(mr_{2c})]K_0(mr)}{I_0(mr_1) + [I_1(mr_{2c})/K_1(mr_{2c})]K_0(mr_1)} \qquad (3\text{-}86)$$

We note one difference between the circumferential fin case and the other cases considered so far. The circumferential fin case depends on inner and outer radius and not only on fin length (difference between these radii). This is a consequence of the curved nature of cylindrical geometry.

Total heat loss through the fin can be obtained from

$$q = -k(2\pi r_1 t)\frac{dT}{dr}(r_1) \qquad (3\text{-}87)$$

which can be evaluated as

$$q = 2\pi r_1 tkm\Theta_0\frac{K_1(mr_1) - I_1(mr_1)K_1(mr_{2c})/I_1(mr_{2c})}{K_0(mr_1) + I_0(mr_1)K_1(mr_{2c})/I_1(mr_{2c})} \qquad (3\text{-}88)$$

Putting in the expression for m,

$$q = 2\pi r_1\sqrt{\frac{2k}{ht}}\,th\Theta\frac{K_1(mr_1) - I_1(mr_1)K_1(mr_{2c})/I_1(mr_{2c})}{K_0(mr_1) + I_0(mr_1)K_1(mr_{2c})/I_1(rm_{2c})} \qquad (3\text{-}89)$$

To get efficiency, we divide by the ideal heat transfer

$$q_I = 2\pi(r_{2c}^2 - r_1^2)h\Theta_0 \qquad (3\text{-}90)$$

The efficiency is then

$$\eta = \left[\frac{r_1\sqrt{2k/ht}\,t}{r_{2c}^2 - r_1^2}\right]\left[\frac{K_1(mr_1) - I_1(mr_1)K_1(mr_{2c})/I_1(mr_{2c})}{K_0(mr_1) + I_0(mr_1)K_1(mr_{2c})/I_1(mr_{2c})}\right] \qquad (3\text{-}91)$$

which can be written

$$\eta = \left[\frac{2\sqrt{kt/2h}}{L_c(1 + r_{2c}/r_1)}\right]\left[\frac{K_1(mr_1) - I_1(mr_1)K_1(mr_{2c})/I_1(mr_{2c})}{K_0(mr_1) + I_0(mr_1)K_1(mr_{2c})/I_1(mr_{2c})}\right] \qquad (3\text{-}92)$$

It may be noted, upon inspection of the asymptotic formulas for Bessel functions in Appendix B, that for large values of mr_1 and mr_2c, the combination of Bessel functions tend to tanh mL_c which tends to unity. This type of condition may well be satisfied. Note that we are not discussing the size of $m(r_{2c} - r_1)$, that is mL_c.

It also may be noted that when the Bessel function expression in Eq. (3-92) and the hyperbolic tangent in Eq. (3-50) are each 1, then the fin efficiencies will be the plane values if r_{2c}/r_1 is unity. Thus, for finite fin lengths and thickness ($r_{2c}/r_1 > 1$) the circumferential fin will tend to provide lower efficiency than a rectangular fin.

An approximation that yields good results and avoids the need to evaluate Bessel functions is to evaluate efficiency with the straight fin formula—Eq. (3-51)—with m replaced by m^1 given by $m\sqrt{(1 - r_{2c}/r_1)/2}$. In effect, the combination of Bessel functions is replaced by $\sqrt{(1 + r_{2c}/r_1)/2}\tanh[\sqrt{(1 + r_{2c}/r_1)/2}\,mL_c]$. Table 3-9-1 shows comparisons of the approximation with the exact formula for selected values. This approximation is convenient when using electronic calculators. Scientific models of calculators typically have hyperbolic functions but not Bessel functions. Also, this approximation is easy to incorporate into a spreadsheet. Indeed, a spreadsheet for this case can be generated with minor modification of the spreadsheet of Example 3-4.

EXAMPLE 3-7.   Find the heat lost by a circumferential metallic fin 10 cm long and 1 mm thick about a cylinder of radius 20 cm whose surface is at 250°C. Air is at 20°C and $h = 7$ W/m² · °C. Metal conductivity is 385 W/m · °C.

SOLUTION.   Heat loss is

$$q = 2\pi r_1 \sqrt{\frac{2k}{ht}}\, th\Theta_0 \frac{K_1(mr_1) - I_1(mr_1)K_1(mr_{2c})/I_1(mr_{2c})}{K_0(mr_1) + I_0(mr_1)K_1(mr_{2c})/I_1(mr_{2c})}$$

As in Example 3-4, $m = 6.03$

$$r_1 = 0.2\,\text{m} \qquad mr_1 = 1.206$$
$$r_{2c} = r_1 + L_c = .2 + .1005 = .3005$$
$$mr_{2c} = 1.812$$

TABLE 3-9-1
Approximate and Exact Efficiencies of
Circumferential Fins

mr_1	mr_{2c}	η_{exact}	η_{approx}
0.5	0.75	0.974	0.975
0.5	1.50	0.644	0.628
1.0	1.50	0.908	0.907
1.0	3.00	0.344	0.351
2.0	3.00	0.722	0.722
2.0	4.00	0.393	0.402

The Bessel function combination is

$$\frac{K_1(mr_1) - I_1(mr_1)K_1(mr_{2c})/I_1(mr_{2c})}{K_0(mr_1) + I_0(mr_1)K_1(mr_{2c})/I_1(mr_{2c})} = \frac{0.4306 - 0.7195(0.1797/1.333)}{0.3159 + 1.395(0.1797/1.333)} = 0.66$$

$$q = 2\pi (0.2)\sqrt{2(385)(10^{-3})(7)} \, (230)(0.66) = 443 \text{ W}$$

If we evaluate the Bessel function combination by

$$\sqrt{\frac{(1 + r_{2c}/r_1)}{2}} \tanh m^1 L_c$$

$$= \sqrt{\frac{1}{2}\left(1 + \frac{0.3005}{0.2}\right)} \tanh\left[\sqrt{\frac{1}{2}\left(1 + \frac{0.3005}{0.2}\right)} \, 0.606\right] = 0.66$$

we see that we obtain the same result.

EXAMPLE 3-7 (Spreadsheet). To obtain a spreadsheet for the circumferential fin using the approximate efficiency formula, we again begin with the spreadsheet of Example 3-4. We add

	A	B
10	R1	.2
11	R2C	+B10 + B8
12	M'/ M	[(B10 + B11) / (2*B10)] ∧ .5
13	CYL MLC	+B12*B9

This gives us $M'L_c$. We then change D3 from @EXP (B9) to @EXP (B13). Thus, a circumferential fin spreadsheet has been obtained with very little effort from the straight fin spreadsheet.

COMPUTER PROJECT 3-1. Prepare a computer program to calculate the efficiency of and the heat removal from a fin, with options for rectangular, triangular, and circumferential fins.

3-10. DIFFERENCE APPROXIMATIONS FOR HEAT CONDUCTION

If the problem has complexity, via multiple layers, nonuniform sources, and nonuniform properties, numerical solution may be appropriate. The solution of a one-dimensional problem tends to be used as a building block in the solution of more elaborate problems.

Below, we shall go through a procedure for constructing a difference approximation based on integrating over a mesh interval. The approach followed is general and leads to equations that apply to plane, cylindrical, and spherical geometries. It generalizes directly to multidimensional cases, as will be noted in Chapter 4, and can be followed in dealing with different coordinate systems. No prior knowledge of difference approximations is assumed.

Because of the generality in including nonuniform properties and nonuniform mesh sizes, the derivation involves more algebra than if just the simplest case had been treated. However, the result will be an ability to cope with a variety of situations. Examples are given to illustrate how the resulting formulas can be applied.

Consider the heat conduction equation

$$\frac{1}{r^n} \frac{d}{dr} r^n k \frac{dT}{dr} + q''' = 0 \tag{3-93}$$

where $n = 0, 1, 2$ for plane, cylindrical, and spherical geometries, respectively. Multiply by r^n and integrate from $r_{i-1/2}$ to $r_{i+1/2}$ (see Fig. 3-10-1). Assuming that over a small interval properties and source can be taken as constant at the value at position r_i

$$r_{i+1/2}^n k_i \frac{dT}{dr}(r_{i+1/2}) - r_{i-1/2}^n k_i \frac{dT}{dr}(r_{i-1/2}) + \frac{q_i'''}{n+1}\left(r_{i+1/2}^{n+1} - r_{i-1/2}^{n+1}\right) = 0 \tag{3-94}$$

Let us approximate the derivatives

$$k_i \frac{dT}{dr}(r_{i+1/2}) = k_i \frac{T(r_{i+1/2}) - T(r_i)}{r_{i+1/2} - r_i} \tag{3-95}$$

Note that in the next interval we may approximate

$$k_{i+1} \frac{dT}{dr}(r_{i+1/2}) = k_{i+1} \frac{T(r_{i+1}) - T(r_{i+1/2})}{r_{i+1} - r_{i+1/2}} \tag{3-96}$$

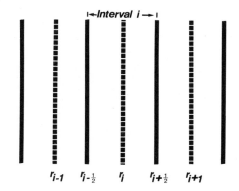

FIGURE 3-10-1. Intervals for difference approximation.

These two equations are equal because heat flow is continuous. We may solve for $T(r_{i+1/2})$

$$T(r_{i+1/2}) = \frac{2k_i/\Delta r_i}{2k_i/\Delta r_i + 2k_{i+1}/\Delta r_{i+1}} T(r_i) + \frac{2k_{i+1}/\Delta r_{i+1}}{2k_i/\Delta r_i + 2k_{i+1}/\Delta r_{i+1}} T(r_{i+1})$$

(3-97)

$$\frac{\Delta r_i}{2} = r_{i+1/2} - r_i$$

(3-98)

$$\frac{\Delta r_i}{2} = r_{i+1} - r_{i+1/2}$$

(3-99)

Similarly, we find that

$$T(r_{i-1/2}) = \frac{2k_i/\Delta r_i}{2k_i/\Delta r_i + 2k_{i-1}/\Delta r_{i-1}} T(r_i) + \frac{2k_{i-1}/\Delta r_{i-1}}{2k_i/\Delta r_i + 2k_{i-1}/\Delta r_{i-1}} T(r_{i-1})$$

(3-100)

We then obtain

$$\frac{2k_i}{\Delta r_i}\left[T(r_{i+1/2}) - T(r_i)\right] = \frac{(2k_i/\Delta r_i)(2k_{i+1}/\Delta r_i)}{2k_i/\Delta r_i + 2k_i/\Delta r_{i+1}}\left[T(r_{i+1}) - T(r_i)\right]$$ (3-101)

If the intervals Δr_i are all the same size, this simplifies to

$$\frac{2k_i}{\Delta r}\left[T(r_{i+1/2}) - T(r_i)\right] = \frac{2k_i k_{i+1}}{\Delta r(k_i + k_{i+1})}\left[T(r_{i+1}) - T(r_i)\right]$$ (3-102)

Thus, Eq. (3-95) may be written

$$C_{i,i+1}\left[T(r_{i+1}) - T(r_i)\right] - C_{i,i-1}\left[T(r_i) - T(r_{i-1})\right] + Q_i = 0$$ (3-103)

$$C_{i,i+1} = r^n_{i+1/2} \frac{(2k_i/\Delta r_i)(2k_{i+1}/\Delta r_{i+1})}{2k_i/\Delta r_i + \dfrac{2k_{i+1}}{\Delta r_{i+1}}}$$

(3-104)

with a similar expression for $C_{i,i-1}$, and

$$Q_i = \frac{q_i'''}{n+1}\left(r^{n+1}_{i+1/2} - r^{n+1}_{i-1/2}\right)$$

(3-105)

The differential equation has been replaced by a set of simultaneous algebraic equations given by Eq. (3-104). This set of equations is of a form called tridiagonal. In any individual equation, only three temperatures can be coupled. The simple form of the set of equations permits easy solution even for a very large number of unknowns. The solution technique, a method of forward substitution and backward elimination, is described in Appendix B.

The engineer concerned with writing a program on a personal computer for solving the one-dimensional equation should consult Appendix B. The engineer with access to a computer center generally will find standard library routines available for solving sets of simultaneous linear equations. The task in that case will be concerned primarily with constructing a suitable approximation.

The boundary conditions for the problem must be accounted for in the difference equations. Interface conditions of continuity of temperature and heat flow already have been incorporated into the equations. Our concern now is with conditions at outer boundaries.

If the heat flow is specified at one boundary and if the boundary is at the first interval, we would write, in Eq. (3-95)

$$k_i \frac{\partial T}{\partial r}(r_{-1/2}) = -q''$$ (3-106)

The heat flux (heat per unit area) q'' may be zero. Equation (3-104) becomes

$$C_{1,2}[T(r_2) - T(r_1)] + r_{-1/2}^n q'' + Q_1 = 0$$ (3-107)

If we wish to impose a convection boundary condition at mesh i instead of equating Eq. (3-97) with Eq. (3-96), we apply

$$-k_i \frac{T(r_{i+1/2}) - T(r_i)}{r_{i+1/2} - r_i} = h\left[T(r_{i+1/2}) - T_\infty\right]$$ (3-108)

We may solve for $T(r_{i+1/2})$

$$T(r_{i+1/2}) = \frac{hT_\infty + \left[k_i/(r_{i+1/2} - r_i)\right]T(r_i)}{k_i/(r_{i+1/2} - r_i) + h}$$ (3-109)

We then obtain

$$r_{i+1/2}^n \frac{2k_i}{\Delta r_i}\left[T(r_{i+1/2}) - T(r_i)\right] = r_{i+1/2}^n \frac{2k_i}{\Delta r_i} \frac{hT_\infty}{2k_i/\Delta r_i + h}$$

$$- r_{i+1/2}^n \frac{(2k_i/\Delta r_i)h}{2k_i/\Delta r_i + h}T(r_i)$$ (3-110)

The last equation in the set is modified to

$$C_{i,i-1}T(r_{i-1}) - C_{ii}T(r_i) + Q_i' = 0 \tag{3-111}$$

$$C_{ii} = C_{i,i-1} + r_{i+1/2}^n \frac{(2k_i/\Delta r_i)h}{2k_i/\Delta r_i + h} \tag{3-112}$$

$$Q_i' = Q_i + r_{i+1/2}^n \frac{2k_i}{\Delta r_i} \frac{hT_\infty}{2k_i/\Delta r_i + h} \tag{3-113}$$

When the boundary condition is imposed on temperature, that value is substituted directly. Thus, if we know wall temperature T_w, then in the last interval we set

$$T(r_{i+1/2}) = T_w \tag{3-114}$$

and use this information in Eq. (3-95) to yield

$$k_i \frac{dT}{dr}(r_{i+1/2}) = k_i \frac{T_w - T(r_i)}{r_{i+1/2} - r_i} \tag{3-115}$$

We again get the form of Eq. (3-112) with

$$C_{ii} = C_{i,i-1} + \frac{2k_i}{\Delta r_i} r_{i+1/2}^n \tag{3-116}$$

$$Q_i' = Q_i + \frac{2k_iT_w}{\Delta r_i} r_{i+1/2}^n \tag{3-117}$$

The same considerations apply at the leftmost interval in imposing boundary conditions. Similar expressions will be found for $T(r_{i-1/2})$ in that interval as were found above for $T(r_{i+1/2})$ in the rightmost interval.

Solution of a one-dimensional difference equation is used as a building block in more complex numerical solution problems. In time-dependent heat conduction, for example, each time step frequently involves solution of a one-dimensional static equation.

One decision that must be made in using a computer program based on finite differences is the specification of mesh intervals. In the way we have obtained difference equations, we have assumed that properties are uniform within a mesh interval. Therefore, we should choose interval sizes such that interfaces between zones of different materials are also interfaces between mesh spaces. This usually implies different mesh sizes in different regions. Beyond this requirement for consistency, interval size must be selected to obtain the proper degree of accuracy.

In obtaining the difference equation, we assumed that the derivative of the temperature could be replaced by a divided difference. The size of the mesh interval should be such that this is a reasonable assumption. The interval should be such that, in general, the percentage change in temperature over an interval is small. While this general guideline is helpful, one is faced with the difficulty that one cannot apply the criterion in advance, before one knows the solution to the problem.

One can, of course, make estimates of orders of magnitudes of temperature gradients and choose interval sizes accordingly. Since temperature gradient is likely to be greater in a zone of small conductivity, it is likely that more mesh spaces would be taken in zones of small conductivity. Success in estimating gradients is likely to improve with experience. It generally is a good practice to examine computer-generated results to see if changes occurring over intervals are large enough to cause results to be suspect. Should that be the case, another calculation with finer mesh spacing may be appropriate.

EXAMPLE 3-8. Solve Example 3-1 numerically using finite differences. Use three equal mesh intervals from the center to the surface.

SOLUTION. The temperatures and interval boundaries are illustrated in Fig. 3-10-2. Since the geometry is plane ($n = 0$), intervals are equal, and properties are uniform, Eq. (3-105) yields

$$C_{i,i+1} = \frac{k}{\Delta r} = \frac{0.6}{\frac{1}{3}(0.1)} = 18$$

Equation (3-106) yields

$$Q_i = \frac{q_i'''}{1}(\Delta r) = \frac{(4000)(0.1)}{3} = 133.33$$

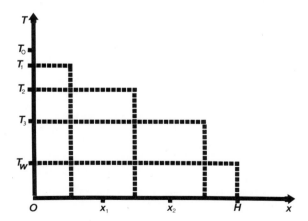

FIGURE 3-10-2. Temperatures and interval boundaries for Example 3-8.

For the first interval, where the zero current condition applies, Eq. (3-108) yields

$$18(T_2 - T_1) + 133.33 = 0$$

For the second interval, an interior point, Eq. (3-104) yields

$$18(T_3 - T_2) - 18(T_2 - T_1) + 133.33 = 0$$

For the third interval, at the end of which we have a specified temperature, we apply Eq. (3-112) with Eqs. (3-117) and (3-118) to obtain

$$18T_2 - \left[18 + \frac{2(0.6)}{\frac{1}{2}(0.1)}\right]T_3 + 133.33 + \frac{2(0.6)}{\frac{1}{3}(0.1)}T_w = 0$$

which also may be written

$$18(T_2 - T_3) - 36(T_3 - T_w) + 133.33 = 0$$

The problem calls for evaluating temperature rise, so let

$$\Theta_i = T_i - T_w$$

The resulting equations are

$$18(\Theta_2 - \Theta_1) - 133.33 = 0$$

$$-18(\Theta_2 - \Theta_1) + 18(\Theta_3 - \Theta_2) + 133.33 = 0$$

$$-36\Theta_3 - 18(\Theta_3 - \Theta_2) + 133.33 = 0$$

Adding the first two equations yields

$$18(\Theta_3 - \Theta_2) + 266.66 = 0$$

Substituting in the third equation

$$36\Theta_3 = 400$$

$$\Theta_3 = 11.1$$

Substituting back, we have

$$\Theta_2 = \frac{266.66}{18} + \Theta_3 = 25.9$$

and then

$$\Theta_1 = \frac{133.33}{18} + \Theta_2 = 33.32$$

The approximate and true solutions are tabulated below:

Location	Temperature—Approximate	Temperature—True
$x_{1/2} = 0.167$	33.3	32.4
$x_{3/2} = 0.5$	25.9	25.0
$x_{5/2} = 0.833$	11.1	10.2

The true solution at the center is 33.3°C from Example 3-1. The estimate we would make of the center temperature would be the closest value available, which would be for the first mesh interval. That value is also 33.3°C, so the difference approximation gives us a good estimate of temperature rise. In this example, we have taken only three mesh intervals but still have managed to achieve a fair degree of accuracy.

To solve the problem more accurately with more mesh points, the reader should consult the forward-substitution backward-elimination procedure given in Appendix B or use a standard computer routine for solving algebraic equations.

COMPUTER PROJECT 3-2.   Prepare a computer program to solve the one-dimensional heat conduction equation in plane, cylindrical, and spherical geometries. Check your program by applying it to Examples 3-1 and 3-2, for which you know the solutions. Vary the mesh size to see how the accuracy of the numerical solution changes with mesh size.

PROBLEMS

3-1. Starting with Fourier's Law in cylindrical geometry, follow the conservation of energy procedure of Section 3-2 to obtain the heat conduction equation in cylindrical geometry. Compare your result with Eq. (3-11) with $n = 1$.

3-2. Starting with Fourier's Law in spherical geometry, follow the conservation of energy procedure of Section 3-2 to obtain the heat conduction equation in spherical geometry. Compare your result with Eq. (3-11) with $n = 2$.

3-3. As an alternative to the procedure followed in Section 3-4, write the general solution to Eq. (3-12) with two undetermined coefficients and apply the boundary conditions to obtain the solution of Eq. (3-17).

3-4. Heat in a nuclear reactor fuel plate with conductivity k is generated nonuniformly, with the heat source lowest at the center. Assume that the heat source can be represented as

$$q''' = q_0''' + (q_1''' - q_0''')\frac{x^2}{H^2}$$

What is the temperature distribution in the fuel plate?

3-5. You wish to assess the importance of accounting for the uniformity in the heat source. Compare the center temperature (i.e., at $x = 0$) from Problems 3-4 with the center temperature obtained with a uniform heat source such that the total heat generated in the plate is the same in each case. Consider specifically the case of $q_1''' = 2q_0'''$.

3-6. The plate of Example 3-2 is encased in a container ($k = 40$ W/m · °K) 2.5 cm thick.
a. What is the temperature rise within the plate?
b. What is the temperature rise within the container?

3-7. Modify the spreadsheet of Example 3-2 to treat the case of the nonuniform heat source of Problem 3-4 with $q_0''' = 3000$ and $q_1''' = 6000$.

3-8. You wish to determine the sensitivity of the center temperature in Example 3-2 to nonuniformly of the heat source with parameters as defined in Problem 3-4. Keeping the total heat source fixed (as in Problem 3-5) vary q_1''' from 0.25 q_0''' to 4 q_0'''. Tabulate or plot the center temperature versus q_1''' / q_0'''.

3-9. The borosilicate glass cylinder of Example 3-3 is to be encased in a container ($k = 40$ W/m · °K) 2.5 cm thick. What is the temperature rise from the outside of the container to the center of the cylinder?

3-10. Borosilicate glass is a material considered for use as a medium to contain heat-generating radioactive material. Prepare a table or plot of the temperature rise from the outside of the container to the center of the cylinder with heat sources from 1,000 to 10,000 W/m³.

3-11. For the borosilicate glass of Problem 3-9, consider diameters from 0.15 to 0.6 m.

3-12. The conductivity integral for uranium dioxide from 630°C to the melting point is 63 W/cm. What is the maximum linear heat rating (in kilowatts/meter) that theoretically could be used in a cylindrical nuclear reactor fuel rod if center melting is to be avoided and if the surface temperature of the fuel is 630°C?

3-13. The average thermal conductivity of uranium carbide is 22.7 W/m · °K. For a fuel surface temperature of 630°C and a melting temperature of 2380°C, what is the maximum theoretical linear heat rating that would avoid melting?

3-14. Problem 2-28 called for finding a range of temperatures at the surface of nuclear fuel. For the conditions given in that problem, find the corresponding range of temperatures at the center of the fuel.

3-15. A sphere with a known surface temperature contains a uniform heat source. Derive an expression for the temperature rise between the surface and the center.

3-16. A cylinder contains a heat source of the form.

$$q''' = c_1 + c_2 r^2$$

Derive an expression for the temperature rise from the surface ($r = R$) to some radius r. Note that this problem will be encountered in finding the convective heat transfer coefficient in a tube in laminar flow in Chapter 7.

3-17. Electrical wiring with resistivity $\rho = 3.53 \times 10^{-6}$ Ω · m has a diameter of 3 mm and a thermal conductivity of 385 W/m · °K. It is covered with electrical insulation 1 mm thick having thermal conductivity $k = 0.5$ W/m · °K. The outside of the insulation sees an ambient temperature of 20°C and a convective heat transfer coefficient of 10 W/m² · °K. Find the temperatures at the surface and interior of the insulation and at the center of the wire for current flows between 0.5 and 15 A. Recall that resistance is given by

$$R = \rho \frac{L}{A}$$

3-18. What is the effectiveness of the fin of Example 3-4?

3-19. Add calculation of effectiveness to the spreadsheet of Example 3-4.

3-20. Starting with the information in Example 3-4, make a table or plot of the heat loss per unit depth as a function of length of fin from 5 to 35 cm.

3-21. Suppose that alternate materials had to be used for the fin. Repeat Problem 3-20 for aluminum ($k = 230$) and steel ($k = 43$).

3-22. The fins of Example 3-4 are placed on the wall with a pitch (center-to-center spacing) of 1 cm.
a. How much heat is transferred through the fins per square meter of wall?
b. How much heat is transferred from the wall between the fins per square meter of wall?

3-23. The environment is changed so as to increase the convective heat transfer coefficient in Problem 3-22. Evaluate for an h of 25 and 100 W/m² · °K,
a. the amount of heat transferred from the fins;
b. the amount of heat transferred between the fins;
c. the total heat transferred; and
d. the total heat transfer if fins were not present.

3-24. Construct a simple formula for efficiency of a triangular fin for the range $1 < mL < 2$.

3-25. Consider the data of Example 3-6. At what length is the same heat transfer achieved as with the rectangular fin of Example 3-4?

3-26. A pin fin consists of a small-diameter cylinder extending from the surface of a plate or wall.
 a. Define a corrected length of fin such that the new surface area of the side of the fin is equal to the sum of the areas of the actual side and of the end.
 b. Express the temperature distribution as a function of distance along the fin.
 c Develop expressions for the efficiency, effectiveness, and heat transfer of the pin fin.

3-27. Starting from the spreadsheet of Example 3-4, prepare a spreadsheet appropriate for a pin fin with diameter equal to the thickness of the fin in Example 3-4.

3-28. In developing formulas to analyze fins, it was assumed that the fin is so thin that the temperature variation inside the fin transversely is very small. A standard of comparison is needed, so let the criterion be

$$\frac{T_c - T_s}{T_s - T_\infty} << 1$$

where T_c and T_s denote center and surface temperatures, respectively. By approximate treatment of conduction within the fin, show that this criterion is satisfied if

$$\frac{h(t/2)}{k} << 1$$

The parameter on the left-hand side is called the Biot number, and it will be encountered again as a criterion for small dimension when we deal with time-dependent conduction.

3-29. For the conditions of Example 3-7, find the efficiency and heat flow for fins with lengths of 5 and 15 cm.

3-30. Circumferential fins as prescribed in Example 3-7 are placed on the cylinder on a pitch of 1 cm.
 a. How much heat is transferred from the fins per meter of length of cylinder?
 b. How much heat is transferred from the nonfinned area?
 c. What is the total heat transfer?

3-31. Straight fins as specified in Example 3-4 are placed on the cylinder of Example 3-7 instead of the circumferential fins. Eight such fins are used (placed uniformly 45° apart). What is the total heat transfer?

3-32. The straight rectangular fins of Example 3-4 are to be replaced by pin fins of the same material and length. The pins are to have a diameter of 3 mm and are to be placed on a square pitch of 1 cm.
 a. How much heat is transferred from one pin?
 b. How much heat is transferred per square meter of wall from all pins?
 c. How much heat is transferred from the area between pins?

3-33. The heat transfer coefficients used in Example 3-4 are based on a correlation having an uncertainty of $\pm 25\%$. What would the fin length have to be to assure that the calculated heat actually could be removed even allowing for uncertainty?

3-34. To allow for the $\pm 25\%$ uncertainty in heat transfer coefficient, it is suggested that the fin spacing relative to that specified in Problem 3-22 be adjusted to yield the same total heat transfer with the most disadvantageous heat transfer coefficient.

3-35. The heat source distribution in a nuclear reactor fuel plate has been found, by numerical calculation, to be

Interval (cm)	Heat Source $(W/m^3 \times 10^{-7})$
0 − 0.1	3.1
0.1 − 0.2	3.3
0.2 − 0.3	3.6
0.3 − 0.4	4.0
0.4 − 0.5	4.5

The heat source is symmetric. Surrounding the fuel plate is a coolant at 300°C with a heat transfer coefficient of 1000 $W/m^2 \cdot$ °C. Plate conductivity is 5 $W/m \cdot$ °K.

a. What is the temperature at the surface of the fuel?

b. What is the temperature distribution within the fuel?

REFERENCES

1. A. E. Bergles and R. L. Webb, *Augmentation of Heat and Mass Transfer*, Hemisphere, Washington, DC, 1983.
2. H. S. Carslaw and J. C. Jaeger, *Conduction of Heat in Solids*, 2nd. ed., Oxford University Press, London, 1959.
3. D. Q. Kern and A. D. Kraus, *Extended Surface Heat Transfer*, McGraw-Hill, New York, 1972.
4. F. Kreith, *Principles of Heat Transfer*, International Textbook, Scranton, PA, 1958.
5. W. M. Rohsenow and J. P. Hartnett, *Handbook of Heat Transfer*, McGraw-Hill, New York, 1973.
6. P. J. Schneider, *Conduction Heat Transfer*, Addison-Wesley, New York, 1955.

Chapter Four

STEADY-STATE CONDUCTION IN MORE THAN ONE DIMENSION

4-1. INTRODUCTION

In one-dimensional problems, it is possible to obtain analytical solutions for a variety of problems. In two or more dimensions, obtaining analytical solutions generally is not possible. Under some circumstances, transformation of coordinates or use of conformal mapping techniques of complex variable theory permits solution in terms of shape factors. Some simple cases can be treated by separation of variables and Fourier series. In general, when dealing with multiple dimensions, one should be prepared to apply numerical procedures.

In one-dimensional cases without sources, it is possible to treat the problems with the first-order equation of Fourier's Law. In more than one dimension, this simplification is no longer possible. The principle of conservation of energy is needed together with Fourier's Law to determine how much energy will flow in each direction. Thus, the second-order heat conduction equation is needed even without sources.

For those cases where analytical solutions are possible (even though the techniques involved are beyond the scope of this text), the forms of solutions can be cast into forms similar to the thermal resistance form of Chapter 2. By extracting the thermal conductivity from the resistance, we obtain a quantity depending only on geometric configuration. This quantity is called the shape factor. Situations are identified for which shape factors have been derived.

When shape factors are not available, it may be possible to solve the partial differential equation of heat conduction by the technique of separation of variables. This technique leads to a serious solution (a Fourier series in rectangular coordinates). The same technique will be used again in Chapter 5 for time-dependent problems.

For problems with sources, the separation of variables technique does not apply. However, it still may be possible to obtain solutions in terms of the types of series obtained when separating variables.

When none of the special procedures can be used, solution must be obtained numerically. It is shown how the procedures for developing difference equations to replace differential equations introduced in Chapter 3 can be generalized to multiple dimensions.

4-2. CONDUCTION SHAPE FACTORS

In Chapter 2, it was found that, for a one-dimensional case, we could represent heat flow by

$$q = \frac{\Delta T}{R_{th}} \tag{4-1}$$

When the thermal resistance is for conduction in a single region, the thermal resistance is inversely proportional to thermal conductivity, that is,

$$R_{th} = \frac{1}{kS} \tag{4-2}$$

where the quantity S, which we call the shape factor, depends on the geometry. Equation (4-1) then becomes

$$q = kS\,\Delta T \tag{4-3}$$

For a number of cases, it is possible to obtain relatively simple formulas for shape factor. These are provided with illustrations in Fig. 4-2-1. These cases are associated with constant temperatures on surfaces. The procedures used for these solutions are beyond the scope of this text. The references at the end of this chapter can direct the interested reader to more explicit solutions.

In applying shape factors to actual problems, one should give attention to the requirement of an isothermal surface. In principle, for many cases illustrated in Fig. 4-2-1, it is difficult for such a situation to prevail, although for practical purposes it may turn out reasonable to assume that the situation does indeed prevail.

Consider, for example, the case of a buried cylinder giving heat to an isothermal plane surface. One might expect the surface to be warmer near the cylinder than far from the cylinder. On the other hand, conditions at the surface may be such that the heating of the cylinder has only a small impact on surface temperature. It also may be that the evaluation of parameters (e.g.,

Configuration	Sketch	Formula	Limitations
1. Buried sphere	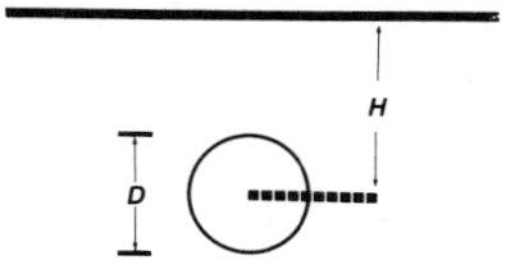	$\dfrac{2\pi D}{1 - D/4H}$	
2. Buried cylinder of length L	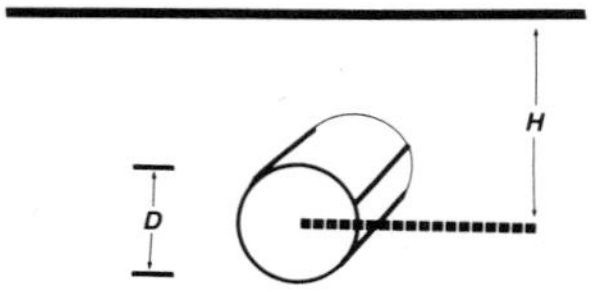	$\dfrac{2\pi L}{\cosh^{-1}(2H/D)}$	$L \gg D$
3. Two cylinders, infinite medium	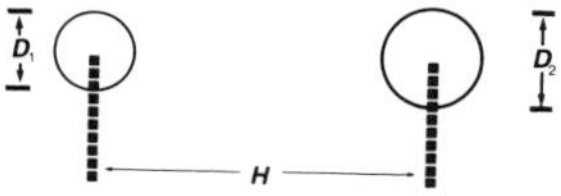	$\dfrac{2\pi L}{\cosh^{-1}\left[(4H^2 - D_1^2 - D_2^2)/2D_1 D_2\right]}$	$L_1 \gg D,$ D_2, H
4. Cylinder between plates	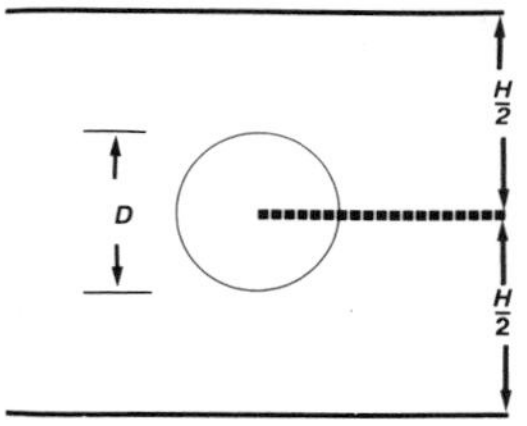	$\dfrac{2\pi L}{\ln(4H/\pi D)}$	
5. Cylinder in square	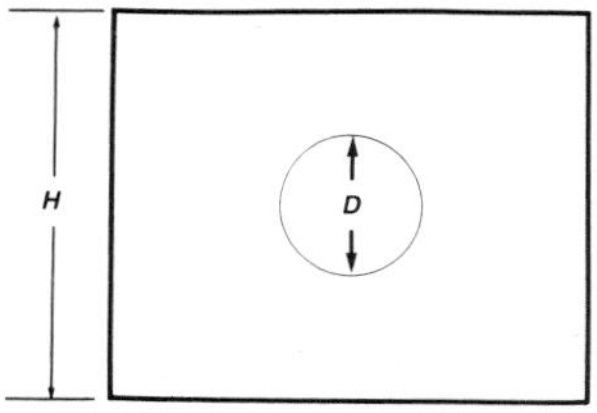	$\dfrac{2\pi L}{\ln(1.08\,H/D)}$	
6. Buried thin disk	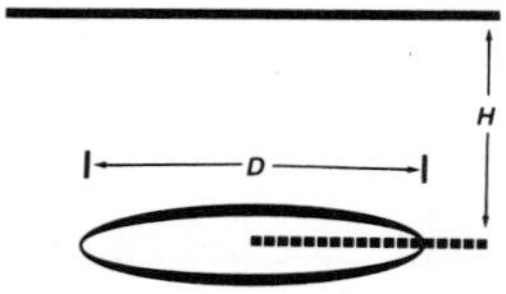	$4D$	$H \gg D$

FIGURE 4-2-1. Shape factor formulas.

7. Buried cavity

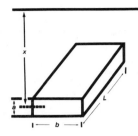

$$\frac{(5.7 + b/2a)L}{\ln(3.5x/b^{1/4}a^{3/4})}$$

$$L \gg a, b$$

8. Buried thin plate

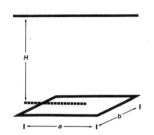

$$\frac{2\pi a}{\ln(4a/b)}$$

$$H \gg a$$

$$\frac{2\pi a}{\ln(2\pi H/b)}$$

$$a \gg b$$
$$H > 2b$$

9. Corner between walls, depth L

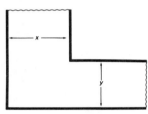

$$\frac{2L}{\pi}\left(\ln\frac{x^2 + y^2}{4xy} + \frac{x}{y}\right.$$

$$\left.\tan^{-1}\frac{y}{x} + \frac{y}{x}\tan^{-1}\frac{x}{y}\right)$$

$$0.559\ L$$

$$x = y$$

10. Parallelepiped shell:
inside area a_i,
outside area A_0

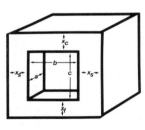

$$\frac{2c(a + b)}{x_s} + ab\left(\frac{1}{x_s} + \frac{1}{x_f}\right)$$

$$a, b, c > \frac{x_{max}}{5}$$

$$+ 2.16(a + b + c)$$

$$+ 0.2(4x_s + x_f + x_c)$$

$$\frac{A_i}{x} + 2.16(a + b + c)$$

$$x_s = x_f =$$

$$+ 1.2$$

$$x_c = x$$

$$a, b, c > \frac{x}{5}$$

$$\frac{A_i}{x} + 1.86(a + b)4.35x$$

$$a, b > \frac{x}{5} > c$$

$$\frac{6.41a}{\ln(A_0/A_i)}$$

$$a > \frac{x}{5} > b, c$$

$$\frac{0.79\sqrt{A_i A_0}}{x}$$

$$a, b, c < \frac{x}{5}$$

FIGURE 4-2-1. *Continued*

conductivity of earth, rock, etc.) may introduce as much uncertainty as the isothermal assumption. For the purposes of estimation, the shape factor assumption may be quite reasonable.

The thermal resistances associated with conduction shape factors can be integrated into problems involving several resistances. Consider, again, Fig. 4-2-1, Configuration 2, the buried cylinder giving heat to an isothermal plane surface. The cylinder may be a pipe of a given thickness whose inside temperature is specified. In addition, there may be a convection boundary condition at the surface with the ambient fluid temperature specified. The total thermal resistance is the sum of three resistances—two of which are obtained as in Chapter 2 and the third by use of the shape factor and Eq. (4-2). Indeed, solutions for problems such as these can be developed by adapting and building on spreadsheets or computer programs developed in dealing with problems in Chapter 2 and 3.

Note that several shape factor formulas include inverse hyperbolic cosines. These functions can be evaluated on many, but not all, electronic calculators. Typical spreadsheets do not have these functions built in. However, they can be expressed easily in terms of logarithms. Useful relationships involving hyperbolic functions are provided in Appendix B.

The parallelepiped shell is an important specific case with applicability to furnaces. Note that the analytical solution for an edge (Configuration 9 in Fig. 4-2-1) provides a coefficient of 0.559, while the commonly used empirical formulas for a full enclosure (Configuration 10) use 0.54 (2.16 for edges). Note that, in principle, the edge term for the unequal thickness situation in Configuration 10 should be more complicated than just 2.16 $(a + b + c)$, given the formula of Configuration 9. However, given the relative magnitudes of the various terms as discussed below, the consequences for the overall shape factor should be modest.

For common applications, where internal dimensions are substantially greater than wall thickness, the edge and corner effects are modest. Consider, for example, a cubical furnace of side a and thickness $x = \frac{1}{5}a$. The term A_i/x is $6a2/x = 30a$. The main corner or edge term 2.16 $(a + b + c)$ is $6.48a$. The outside corner term $1.20x$ is $0.24a$. Thus, the majority of heat transfer is predicted by the plane wall portion of the formula. A small though significant portion is associated with the edge increment to the shape factor, and a minor portion is associated with outer corners.

The parallelepiped shell is a good candidate for spreadsheet application. It is a type of problem encountered frequently, and it involves several steps for evaluation of shape factor, that is, for walls, edges, and corners. Preparation of such a spreadsheet is left as a homework problem at the end of the chapter.

EXAMPLE 4-1.   A long cavity 10 m high and 100 m across is placed 1000 m below the surface in a granite formation, having a thermal conductivity of 3 W/m · °C. Material containing a heat source is placed in this cavity so that its surfaces have a temperature of 250°C. The surface is at a temperature of 20°C. What is the heat transfer rate per meter of cavity length to the surface?

SOLUTION. The appropriate shape factor formula is

$$\frac{q}{L} = \frac{k\,\Delta T(b/2a + 5.7)}{\ln(3.5x/b^{1/4}a^{3/4})}$$

$$x = 1000, \qquad a = 10, \qquad b = 100$$

With these values

$$\frac{q}{L} = 2k\,\Delta T = 1380 W/M$$

COMPUTER PROJECT 4-1. Write a program to obtain the shape factor for a parallelepiped shell. The program should be such that you enter as input the appropriate dimensions and the program selects the proper formula to apply.

4-3. SEPARATION OF VARIABLES

In a uniform source-free medium, heat conduction may be treated by the technique of separation of variables. In one-dimensional cases, we could treat problems by dealing only with Fourier's Law. Here, however, we need the full heat conduction equation because we cannot perform a one-dimensional integral as was done in Chapter 2.

The technique of separation of variable is a common approach to treating partial differential equations. However, it is of restricted applicability. This technique is not useful when dealing with multiple layers of materials (i.e., nonuniform media) or with internal sources of heat. An analogous procedure for dealing with sources will be presented Section 4-4.

Despite the restricted applicability of this analytical approach, the results for the applicable cases are useful. In addition, it is important to be able to construct analytical solutions where possible, so that numerical (e.g., finite difference) techniques developed for more general cases can be tested in situations where the correct solution is known.

In general, the heat conduction equation is

$$\nabla \cdot k\nabla T + q''' = 0 \tag{4-4}$$

where the ∇ operator in rectangular geometry is

$$\nabla = \frac{\partial}{\partial x}\mathbf{i} + \frac{\partial}{\partial y}\mathbf{j} + \frac{\partial}{\partial z}\mathbf{k} \tag{4-5}$$

In two-dimensional x-y geometry in a source-free medium

$$\frac{\partial^2 T}{\partial x^2} + \frac{\partial^2 T}{\partial y^2} = 0 \tag{4-6}$$

Let us consider a problem where a source per unit area q'' is applied to the surface of a block (see Fig. 4-3-1). At other faces, $x = a$ and $y = +b$, temperature is specified to be T_w. Changing variables to

$$\Theta = T - T_w \qquad (4\text{-}7)$$

yields the equation and boundary conditions

$$\frac{\partial^2 \Theta}{\partial x^2} + \frac{\partial^2 \Theta}{\partial y^2} = 0 \qquad (4\text{-}8)$$

$$- k \frac{\partial \Theta}{\partial x}(0, y) = q'' \qquad (4\text{-}9)$$

$$\Theta(a, y) = \Theta(x, b) = \Theta(x, -b) = 0 \qquad (4\text{-}10)$$

By symmetry, we may replace the $\Theta(x, -b)$ condition with $(\partial \Theta / \partial y)(x, 0) = 0$. Let us assume a solution of the form

$$\Theta(x, y) = F(x)G(y) \qquad (4\text{-}11)$$

This is the separation of variable assumption. Then

$$G(y)\frac{d^2 F^{(x)}}{dx^2} + F(x)\frac{d^2 G}{dy^2} = 0 \qquad (4\text{-}12)$$

Dividing by FG yields

$$\frac{1}{F}\frac{d^2 F}{dx^2} = -\frac{1}{G}\frac{d^2 G}{dy^2} \qquad (4\text{-}13)$$

The left-hand side of the above equation is a function only of x. The right-hand side is a function only of y. This equation can be true only if each

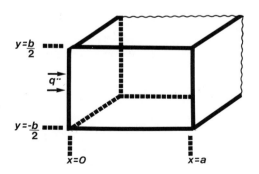

FIGURE 4-3-1. Heat flux on one face of a two-dimensional block.

function is a constant, for example, if

$$\frac{d^2F}{dx^2} = B^2F \tag{4-14}$$

We have assumed that B^2 is a positive number. This assumption is convenient, since the answer will turn out this way. However, had we assumed the number to be negative, our subsequent analysis would still give the correct solution and yield an imaginary value for B.

The equation for F can be solved in terms of exponential functions or hyperbolic functions. Thus, we let

$$F(x) = C_1 \sinh B(a - x) + C_2 \cosh B(a - x) \tag{4-15}$$

To apply the boundary condition that $\Theta(a, y)$ be zero, we require

$$F(a) = 0 \tag{4-16}$$

This forces C_2 to be zero. We defer application of the boundary condition at $x = 0$ and consider the equation for $G(y)$.

The equation for G is

$$\frac{d^2G}{dy^2} = -B^2G \tag{4-17}$$

for which the solution is

$$G_3(y) = C_3 \cos By + C_4 \sin By \tag{4-18}$$

The symmetry condition requires that C_4 be zero. We then require, for $\Theta(x, b) = 0$, that

$$G(b) = 0 \tag{4-19}$$

For the overall solution to be nonzero, we require that C_3 not be zero. Therefore, we obtain a condition on the constant B, namely,

$$\cos Bb = 0 \tag{4-20}$$

There is an infinite of values of B for which this condition is satisfied, that is,

$$B_n = (2n + 1)\frac{\pi}{2b} \qquad n = 0, \ldots, \infty \tag{4-21}$$

Our general solution then must be a superposition of all possible solutions, so

$$\Theta(x, y) = \sum_{n=0}^{\infty} (C_1C_3)_n \sinh B_n(a - x) \cos B_n y \tag{4-22}$$

We now apply the final condition

$$-k\frac{\partial}{\partial x}\Theta(x, y)\bigg|_{x=0} = q'' \qquad (4\text{-}23)$$

We thus obtain

$$q'' = \sum_{n=0}^{\infty} (C_1 C_3)_n k B_n \cosh B_n a \cos B_n y \qquad (4\text{-}24)$$

This has the form of a standard Fourier series

$$q'' = \sum_{n=0}^{\infty} \beta_n \cos B_n y \qquad (4\text{-}25)$$

$$\beta_n = (C_1 C_3)_n k B_n \cosh B_n a \qquad (4\text{-}26)$$

The coefficients β_n are given by

$$\beta_n = \frac{q'' \int_{-b}^{b} \cos B_n y \, dy}{\int_{-b}^{b} \cos^2 B_n y \, dy} = \frac{(-1)^2}{2n+1} \frac{4q''}{\pi} \qquad (4\text{-}27)$$

Then the unknown constants $(C_1 C_3)_n$ are obtained from

$$(C_1 C_3)_n = \frac{\beta_n}{k B_n \cosh B_n a} \qquad (4\text{-}28)$$

Applying the above procedure permits solution for certain specialized cases. Evaluation of the many terms in a series can be cumbersome by hand calculation, but the general form of solution can be programmed easily on a personal computer.

Some information may not require evaluation of many terms. For example, the modes attenuate at different rates with x, going as $\exp[-(2n + 1)(\pi/2)(x/b)]$. Thus, by the end of the block ($x = a$), high-order (high values of n) modes will have decayed much farther than low-order modes, and only a small number of terms may be needed to evaluate the heat leaving the end of the block. Similarly, few terms may be needed if a/b is large.

Let us summarize the steps taken in applying separation of variables so that these steps can be followed in other geometries.

Step 1. Assume a separable solution [as in Eq. (4-11)].
Step 2. Substitute into the equation [as in Eq. (4-12)].
Step 3. Divide by the total function [as in Eq. (4-13)].
Step 4. Infer the existence of separate equations [as Eqs. (4-14) and (4-17)].
Step 5. Infer boundary conditions for these equations [e.g., Eq. (4-16)].
Step 6. Solve the separate equations.
Step 7. Combine for the overall solution.

The same approach can be taken in other geometries. For the finite cylinder case, the heat conduction equation is

$$\frac{1}{r}\frac{\partial}{\partial r}kr\frac{\partial \Theta}{\partial r} + \frac{\partial}{\partial z}\left(k\frac{\partial \Theta}{\partial z}\right) = 0 \tag{4-29}$$

Let k be uniform and let

$$\Theta(r, z) = F(r)G(z) \tag{4-30}$$

which leads to

$$G(z)\frac{1}{r}\frac{d}{dr}\left(r\frac{dF}{dr}\right) + F(r)\frac{d^2G}{dz^2} = 0 \tag{4-31}$$

We divide by FG and find that we may write

$$\frac{d^2G}{dz^2} = B^2G \tag{4-32}$$

which has the solution

$$G(z) = C_1\sinh B(H - z) \tag{4-33}$$

Our other equation is

$$\frac{1}{r}\frac{d}{dr}\left(r\frac{dF}{dr}\right) = -B^2F(r) \tag{4-34}$$

which has solutions that are Bessel functions

$$F(r) = C_3 J_0(Br) + C_4 Y_0(Br) \tag{4-35}$$

The Y_0 function is infinite at $r = 0$, so the coefficient C_4 must be zero. Since the coefficient C_3 should not be zero, we require, for $\Theta = 0$ at the surface,

$$J_0(BR) = 0 \tag{4-36}$$

As discussed in Appendix B, the Bessel function J_0, a cylindrical analogue of the cosine, is an oscillating function with an infinite number of zeros. Thus, the solution for the finite cylinder is of the form

$$\Theta(r, z) = \sum_{n=0}^{\infty} (C_1 C_3)_n \sinh B_n(H - z) J_0(B_n r) \tag{4-37}$$

The heat flux at the end of the cylinder is

$$q'' = \sum_{n=0}^{\infty} (C_1 C_3)_n k B_n \cosh(B_n H) J_0(B_n r) \qquad (4\text{-}38)$$

which is of the form

$$q'' = \sum_{n=0}^{\infty} \beta_n J_0(B_n r) \qquad (4\text{-}39)$$

where the β_n may be evaluated using the integrals over Bessel functions given in Appendix B as

$$\beta_n = \frac{q'' \int_0^R r J_0(B_n r)\, dr}{\int_0^R r J_0^2(B_n r)\, dr} = \frac{2q''}{B_n R J_1(B_n R)} \qquad (4\text{-}40)$$

The above discussion has been provided for the case of a uniform heat flux on the surface. If the heat flux is not uniform, then we may write for the rectangular block

$$q'' = q''(y) \qquad (4\text{-}41)$$

and for the cylinder

$$q'' = q''(r) \qquad (4\text{-}42)$$

The procedures followed are the same, except that we evaluate

$$\beta_n = \frac{\int_{-b}^{b} q(y) \cos B_n y\, dy}{\int_{-b}^{b} \cos^2 B_n y\, dy} \qquad (4\text{-}43)$$

and

$$\beta_n = \frac{\int_0^R q(r) J_0(B_n r) r\, dr}{\int_0^R J_0^2(B_n r) r\, dr} \qquad (4\text{-}44)$$

when the coefficients are not uniform.

The procedures discussed so far for two dimensions also apply in three dimensions. For rectangular geometry, we would assume

$$\Theta(x, y, z) = F_1(x) F_2(y) F_3(z) \qquad (4\text{-}45)$$

The solution for this case is left as a problem at the end of the chapter.

COMPUTER PROJECT 4-2.   Write a program to calculate the temperature at any point in a block with sides of L meters and M meters (the third dimension is assumed infinite). The program also should provide for the calculation of components of heat flux at any point and the total heat loss through each face of the block.

EXAMPLE 4-2.   A block infinite in the z direction has sides in the x and y dimensions of 1 m. A uniform heat flux of 1 kW/m² enters the block at the left face ($x = 0$). Find the temperature at the center of the block if the temperatures on all faces other than the left are maintained at 20°C. The conductivity of the block is 40 W/m · °K. Repeat the problem for a y dimension of 10 m.

SOLUTION.   The temperature is given by

$$\Theta(x, y) = \sum_{n=0}^{\infty} \frac{(-1)^n}{2n+1} \frac{4q''}{\pi k B_n} \frac{\sinh B_n(a-x)\cos B_n y}{\cos h B_n a}$$

$$\Theta = T - 20°C$$

$$B_n = (2n+1)\frac{\pi}{2b} = (2n+1)\frac{\pi}{1}$$

$$\Theta(x, y) = \sum_{n=0}^{\infty} \frac{(-1)^n}{2n+1} \frac{(4)(1000)}{(\pi)(40)(2n+1)\pi} \frac{\sinh(2n+1)(\pi)(1-x)\cos(2n+1)\pi y}{\cosh(2n+1)\pi}$$

At the midpoint of the block, x is $\frac{1}{2}$ and y is 0, so

$$\Theta = \frac{4(1000)}{40\pi^2} \sum_{n=0}^{\infty} \frac{(-1)^n}{(2n+1)^2} \frac{\sinh[(2n+1)\pi/2]}{(2n+1)^2\cosh[(2n+1)\pi]}$$

Evaluating the hyperbolic functions yields a temperature of 22.011°C if only the first term is used in the expansion and 22.001°C if the first two terms are used. The series thus converges very rapidly.

For the second case, with the y dimension of 10 m, we have

$$B_n = (2n+1)\frac{\pi}{10}$$

and

$$\Theta = \frac{(4)(1000)}{40\pi^2} 10 \sum_{n=0}^{\infty} \frac{(-1)^n}{(2n+1)^2} \frac{\sinh[(2n+1)\pi/20]}{\cosh[(2n+1)\pi/10]}$$

Evaluating these terms demonstrates that, as expected, the answer changes, and also that the number of terms required in the expansion changes. Temperature estimates are 35.224, 31.550, 32.903, and 32.298°C for one, two, three, and four terms used. In general, we should expect slower convergence as the ratio of y dimension to x dimension increases, since the exponential and hyperbolic functions depend on this ratio.

EXAMPLE 4.2 (Spreadsheet). This example illustrates use of the spreadsheet to repetitively modify an earlier spreadsheet to the extent needed. The spreadsheet below will evaluate the lowest-order approximation (one term) for the temperature. To evaluate the second term, we replicate Column D in Column E. We then change N in Column E from 0 to 1. We also fix E17 so the increment of Column E is added to D17 instead of B8. We proceed by continuing to replicate columns until we conclude that we have a converged solution.

When evaluating a series by hand, taking additional terms implies a laborious process. When working at a spreadsheet on a personal computer, there is little additional effort involved beyond the first term.

	A	B	C	D
1	2D HEAT	CONDUCTION	CENTER	TEMPERATURE
2	EVALUATE	CONSTANT	EVALUATE	EXPANSION
3	HEAT FLUX	1000	N	0
4	CONDUCTIVITY	40	2N + 1	2 * D3 + 1
5	PI	@ PI	Y - SIDE	1
6	CONSTANT	+ B3 * B4 / B5	BN	+ D4 * @ PI / D5
7	WALL TEMP	20	X - SIDE	1
8			X	.5
9			Y	0
10			COS TERM	@ COS (D6 * D9)
11			A - X	+ D7 - D8
12			EXP BN (A - X)	@ EXP (D6 * D11)
13			SINH TERM	(D12 - (1 / D12)) / 2
14			EXP BNX	@ EXP (D6 * D8)
15			COSH TERM	(D14 + (1 / D14)) / 2
16			TERM	(- 1) ∧ D3 * D13 * D10 / (D6 * D15 * D4)
17			EST. TEMP	+ D16 * B6 + B7

The above spreadsheet can be modified easily to handle problems with other boundary conditions. For example, if a temperature T_L were to be specified at the left face instead of the heat flux q'', the solution can be shown by the separation of variables method (the proof is one of the problems at the end of the chapter) to be

$$\Theta = \Theta_L \sum_{n=0}^{\infty} \frac{(-1)^n}{2n+1} \frac{4}{\pi} \frac{\sinh B_n(a-x)\cos B_n y}{\sinh B_n a}$$

This series is similar to that obtained for the constant heat flux case. To obtain the proper coefficient, the entry in B6 is changed to 4/@PI. To obtain a hyperbolic sine in the denominator, the plus sign in D15 is changed to a minus sign. To get the actual temperature, we enter the specified left side temperature in B8. We then change D17 to + D16 * B6 * (B8 − B7) + B7. Extending more terms in the series proceeds in similar fashion with the constant heat flux case.

4-4. PROBLEMS WITH INTERNAL HEAT SOURCES

When we applied the method of separation of variables in Section 4-3, we obtained solutions that turned out to be in the form of Fourier series. In this section, we treat the case of an internal heat source by making Fourier series expansions in multiple dimensions. The orthogonality properties of sines and cosines (and of Bessel functions in cylindrical geometry) simplify the solution procedure considerably.

Consider the problem

$$\frac{\partial^2 T}{\partial x^2} + \frac{\partial^2 T}{\partial y^2} + \frac{q'''}{k} = 0 \tag{4-46}$$

with temperature specified as T_w on the surface. Again, let

$$\Theta = T - T_w \tag{4-47}$$

Let us assume that we can expand the solution

$$\Theta(x, y) = \sum_{n=1}^{\infty} \sum_{n=1}^{\infty} A_{nm} \cos B_n x \cos C_m y \tag{4-48}$$

$$B_n = (2n + 1)\frac{\pi}{2a} \tag{4-49}$$

$$C_m = (2m + 1)\frac{\pi}{2b} \tag{4-50}$$

Let us expand the source in a Fourier series

$$\frac{q'''}{k} = \sum_{n=1}^{\infty} \sum_{m=1}^{\infty} Q_{nm} \cos B_n x \cos C_m y \tag{4-51}$$

where the coefficients are given by

$$Q_{nm} = \frac{\int_0^a dx \int_0^b dy \cos B_n x \cos C_b y (q'''/k)}{\int_0^a dx \cos^2 B_n x \int_0^b dy \cos^2 C_m y} \qquad (4\text{-}52)$$

The differential equation for Θ then becomes

$$\sum_{n=1}^{\infty} \sum_{m=1}^{\infty} A_{nm}(-B_n^2 - B_m^2) \cos B_n x \cos C_m y \qquad (4\text{-}53)$$

$$+ \sum_{n=1}^{\infty} Q_{nm} \cos B_n x \cos C_m y = 0 \qquad (4\text{-}54)$$

Multiplying by $\cos B_p x \cos C_q y$ and integrating yields, upon noting orthogonality properties,

$$\int_0^a dx \cos B_n x \cos B_p x \int_0^b dy \cos C_m y \cos C_q y = 0 \qquad p \neq n, q \neq m \quad (4\text{-}55)$$

the values for A_{nm} are given by

$$A_{nm} = \frac{Q_{nm}}{B_n^2 + B_m^2} \qquad (4\text{-}56)$$

Analogous solutions may be obtained in other geometries. Note, however, that for a cylindrical geometry, we would have a series of Bessel functions, which have orthogonality properties noted in Appendix B, instead of one of the cosines.

Fourier series approach has usefulness for simple geometries, for which it is easy to define a set of orthogonal functions, and for simple sources. The solution is easy to code for a personal computer.

4-5. DIFFERENCE APPROXIMATION FOR MORE THAN ONE DIMENSION

The range of situations for which we have standard published solutions (shape factors) or for which we can obtain analytical solutions by separation of variables or Fourier series in more than one dimension is quite restricted. Accordingly, we frequently must resort to numerical procedures. In this section, we generalize the one-dimensional differencing discussed in Chapter 3.

Consider the equation

$$\frac{1}{r^n} \frac{\partial}{\partial r}\left(r^n k \frac{\partial T}{\partial r}\right) + \frac{\partial}{\partial z}\left(k \frac{\partial T}{\partial z}\right) + q''' = 0 \qquad (4\text{-}57)$$

This two-dimensional equation includes rectangular and cylindrical geometries as special cases, depending on whether n is 0 or 1. The equation is to be integrated over a two-dimensional mesh interval. Thus, we multiply by r^n and integrate from $r_{i-1/2}$ to $r_{i+1/2}$ and from $z_{j-1/2}$ to $z_{j+1/2}$.

Integration leads to

$$
(z_{j+1/2} - z_{j-1/2}) \left[r_{i+1/2}^n k_{ij} \frac{\partial T}{\partial r}(r_{i+1/2}, z_j) - r_{i-1/2}^n k_{ij} \frac{\partial T}{\partial r}(r_{i-1/2}, z_j) \right]
$$

$$
+ \frac{r_{i+1/2}^{n+1} - r_{i-1/2}^{n+1}}{n+1} \left[k_{ij} \frac{\partial T}{\partial z}(r_i, z_{j+1/2}) - k_{ij} \frac{\partial T}{\partial z}(r_i, z_{j-1/2}) \right]
$$

$$
+ \frac{q_{ij}'''}{n+1} \left(r_{i+1/2}^{n+1} - r_{i-1/2}^{n+1} \right)(z_{j+1/2} - z_{j-1/2}) = 0 \tag{4-58}
$$

All the procedures followed previously in Chapter 3 for one-dimensional differencing can be applied separately to the terms in brackets containing r derivatives and containing z derivatives. This leads to

$$
C_{i,i+1}^j \left[T(r_{i+1}, z_j) - T(r_i, z_j) \right] - C_{i,i-1}^j \left[T(r_i, z_j) - T(r_{i-1}, z_j) \right]
$$

$$
+ C_{j,j+1}^i \left[T(r_i, z_{j+1}) - T(r_i, z_j) \right] - C_{j,j-1}^i \left[T(r_i, z_j) - T(r_i, z_{j-1}) \right]
$$

$$
+ Q_{ij} = 0 \tag{4-59}
$$

where the coefficients are given by

$$
C_{i,i+1}^j = (z_{j+1/2} - z_{j-1/2}) r_{i+1/2}^n \frac{2k_{ij} 2k_{i+1,j}/\Delta r_i \, \Delta r_{i+1}}{2k_{ij}/\Delta r_i + 2k_{i+1,j}/\Delta r_{i+1}} \tag{4-60}
$$

with analogous definitions for the other quantities. Similarly, boundary conditions extend directly.

For three dimensions and for other geometries, we may proceed in similar fashion by integrating over the appropriate area or volume element. In polar coordinates, for example, the heat conduction equation is

$$
\frac{1}{r} \frac{\partial}{\partial r} \left(kr \frac{\partial T}{\partial r} \right) + \frac{1}{r^2} \frac{\partial}{\partial \phi} \left(k \frac{\partial T}{\partial \phi} \right) + q''' = 0 \tag{4-61}
$$

The element of area to be multiplied by is $r \, dr \, d\phi$, and integration would be from $r_{i-1/2}$ to $r_{i+1/2}$ and from $\phi_{j-1/2}$ to $\phi_{j+1/2}$.

While the two-dimensional difference equation obtained involves only five temperatures (compared with three in the one-dimensional equation), enough complexity has been added that direct solution by the forward elimination and backward substitution of Appendix B is not an option. We could use direct solution by standard linear equation solution procedures (e.g., matrix inversion, Gauss reduction) if the number of unknowns is not too large. However, when dealing with multiple dimensions, the number of unknowns for even simple problems can be large. If 10 invervals were to be used for a one-dimensional problem, then a two-dimensional problem having the same overall size in each dimensions would have 10^2 or 100 unknowns. A three-dimensional problem would be 1000 unknowns. When working with large numbers of unknowns, it generally is attractive to use iterative methods.

Solution by iteration means making a guess of the solution and successively improving on it. The simplest form of iteration is the method of simultaneous displacements. If we denote the initial guess by superscript 0 and the first improvement by superscript 1, we rewrite the difference equation is

$$C_{i,I+1}^{j}\left[T^0(r_{i+1},z_j) - T^1(r_i, z_j)\right] - C_{i,I-1}^{j}\left[T^1(r_i, z_j) - T^0(r_{i-1}, z_j)\right]$$

$$+ C_{j,j+1}^{i}\left[T^0(r_i, z_{j+1}) - T^1(r_i, z_j)\right] - C_{j,j-1}^{i}\left[T^1(r_i, z_j) - T^0(r_i, z_{j-1})\right]$$

$$+ Q_{ij} = 0 \tag{4-62}$$

The only unknown in this equation is $T^1(r_i, z_j)$, so it may be solved for directly. Thus, the next estimate of the temperature distribution may be obtained easily. However, it is not likely that the next estimate will be acceptable. Therefore, the new estimate is considered as the guess and another estimate is obtained.

One question of concern in iterative methods is how long to keep upgrading estimates. Criteria for stopping usually are expressed in terms of how rapidly the solution is changing. For example, one may specify that iteration will end when

$$\frac{\max_{ij}\left|T^m(r_i, z_j) - T^{m-1}(r_i, z_j)\right|}{(1/IJ)\sum_{i=1}^{I}\sum_{j=1}^{J}T^m(r_i, z_j)} < \varepsilon \tag{4-63}$$

This criterion states that the maximum change in temperature between iterations m and $m - 1$ is less than some fraction of the average temperature.

A general concern in dealing with numerical methods is the question as to whether the iteration will converge to the true solution (or will converge to anything at all). The reader should consult references given at the end of Appendix B for discussions of this question in the literature.

The iterative method given here was used because of its simplicity. Some improvements over this simple scheme and the possibility of accelerating convergence are discussed in Appendix B.

The degree to which the engineer should be concerned with speed of solution depends on the situation. If faced with an isolated problem which requires solution, and for which a standard code is not available, the engineer is likely to try to minimize the time spent in writing and checking the program and is not likely to be overly concerned with cost of computer time. However, if faced with a situation of developing a computer code that is to be used many times on a routine basis, the engineer would be more inclined to incorporate time-saving procedures in the computer code.

COMPUTER PROJECT 4-3.   For the computer program you developed to solve the one-dimensional heat conduction equation by forward substitution–backward elimination, add on option to solve the equation by iteration using the method of simultaneous displacements.

EXAMPLE 4-3.   Set up difference equations for the numerical solution of Example 4-2. Use mesh intervals of 0.1 in both x and y directions.

SOLUTION.   The difference equations simplify because of the uniform mesh spacing andd uniform properties. Because of the rectangular geometry, we use Eqs. (4-59) and (4-60) with $n = 0$. Equation (4-60) becomes boundaries

$$C_{i,i+1}^{j} \frac{\Delta y \, k}{\Delta x} = \frac{(0.1)(40)}{0.1} = 40$$

The same value is obtained for all other C coefficients at interior points. Since there are no interior sources, the Q_{ij} terms are zero. We thus obtain for interior points i, j

$$40\big[T(x_{i+1}, y_j) - T(x_i, y_j)\big] - 40\big[T(x_i, y_j) - T(x_{i-1}, y_j)\big]$$

$$+ 40\big[T(x_i, y_{j+1}) - T(x_i, y_j)\big] - 40\big[T(x_i, y_j) - T(x_i, y_{j-1})\big] = 0$$

For boundary conditions, we note that surface temperature are given on three faces. We proceed as in Chapter 3, Eq. (3-116). For the rightmost x point, we evaluate from Eq. (4-58) the term

$$\Delta y \, k \frac{\partial T}{\partial x}(x_{I+1/2}, y_j) = \Delta y \, k \frac{T(y_{I+1/2}, y_j) - T(x_I, y_j)}{\Delta x/2}$$

which leads to the equation

$$80\big[20 - T(x_I, y_j)\big] - 40\big[T(x_I, y_j) - T(x_{I-1}, y_j)\big] + 40\big[T(x_I, y_{j+1}) - T(x_I, y_j)\big]$$

$$- 40\big[T(x_I, y_j) - T(x_I, y_{j-1})\big] = 0$$

The term $(80)(20) = 1600$ is an equivalent source analogous to Eq. (3-118). The same type of boundary condition leads to similar equations at $j = 0$ and $j = 1$, that is, the top and bottom y locations.

For the left boundary condition, when the heat flux is specified, we evaluate from Eq. (4-58) the term

$$-\Delta y \, k \frac{\partial T}{\partial x}(x_0, z_j) = \Delta y \, q'' = (0.1)(10^3) = 100$$

The equations at the left face become

$$40\left[T(x_2, y_j) - T(x, y_j)\right] + 100 + 40\left[T(x_1, y_{j+1}) - T(x_1, y_j)\right]$$

$$-40\left[T(x_1, y_j) - T(x_i, y_{j-1})\right] = 0$$

Because we have two dimensions, we have corners that are outers boundaries in both x and y. For example, at the left face when the heat flux condition is applied, at the top point we have

$$40\left[T(x_2, y_J) - T(x_1, y_J)\right] + 100 + 80\left[20 - T(x_1, y_J)\right]$$

$$-40\left[T(x_1, y_J) - T(x_1, y_{J-1})\right] = 0$$

while at the bottom point we have

$$40\left[T(x_2, y_1) - T(x_1, y_1)\right] + 100 + 40\left[T(x_1, y_2) - T(x_1, y_1)\right]$$

$$-80\left[T(x_1, y_1) - 20\right] = 0$$

Similarly, at the right face, we obtain at the top

$$80\left[20 - T(x_I, y_J)\right] - 40\left[T(x_I, y_J) - T(x_{J-1}, y_j)\right] + 80\left[20 - T(x_I, y_J)\right]$$

$$-40\left[T(x_I, y_J) - T(x_I, y_{J-1})\right]$$

while at the bottom, we obtain

$$80\left[20 - T(x_I, y_1)\right] - 40\left[T(x_I, y_1) - T(x_{I-1}, y_1)\right] + 40\left[T(x_I, y_2) - T(x_I, y_1)\right]$$

$$-80\left[T(x_I, y_1) - 20\right] = 0$$

We now have obtained the full set of difference equations.

EXAMPLE 4-4. Set up an iterative sequence for solving the equations obtained in Example 4-3.

SOLUTION. We use the method of simultaneous displacements. For our 1 by 1 block with mesh intervals that are 0.1 by 0.1, we have 10 mesh intervals in each dimension. This means that there are $10 \times 10 = 100$ equations. We rewrite the equations for $i = j = 1$ in the form

$$(80 + 40 + 40)T^{n+1}(x_1, y_1) = 100 + 1600 + 40T^n(x_2, y_2) + 40T^n(x_1, y_2)$$

where n is the iteration index. The value $n = 0$ corresponds to the initial guess that starts the iteration process.

We have choices as to how we may "sweep" through the grid to calculate values. Let us choose to evaluate temperatures at all x points for the first y point and then proceed to the next y point. We then would have

$$120T^{n+1}(x_i, y_1) = 40[T^n(x_{i-1}, y_1) + T^n(x_{i+1}, y_1)] + 40T^n(x_i, y_2) + 1600$$

$$i = 2, \ldots, 9$$

$$240T^{n+1}(x_{10}, y_1) = 40T^n(x_9, y_1) + 40T^n(x_{10}, y_2) + 3200$$

We then increase the y index by 1. Since the set of equations obtained for $j = 2$ to $j = 9$ will be of the same form, let us keep the j index. Then, for the first x interval,

$$120T^{n+1}(x_1, y_j) = 40[T^n(x_2, y_j) + T^n(x_1, y_{j+1}) + T^n(x_1, y_{j-1})] + 100$$

For the interior intervals, we have

$$160T^{n+1}(x_i, y_j) = 40[T^n(x_{i+1}, y_j) + T^n(x_{i-1}, y_j) + T^n(x_i, y_{j+1}) + T^n(x_i, y_{j-1})]$$

then, for $i = 10$, we have

$$200T^{n+1}(x_{10}, y_j) = 40[T^n(x_9, y_j) + T^n(x_{10}, Y_{j-1}) + T^n(x_{10}, y_{j+1})] + 1600$$

For $j = 10$, the last y interval, we have, at the first x interval,

$$160T^{n+1}(x_1, y_{10}) = 40[T^n(x_2, y_{10}) + T^n(x_1, y_9)] + 1700$$

For $i = 2$ to $i = 9$ we have

$$200T^{n+1}(x_i, y_{10}) = 40[T^n(x_i, y_9) + T^n(x_{i-1}, y_{10}) + T^n(x_{i+1}, y_{10})] + 1600$$

Finally, for $i = j = 10$, we have

$$240T^{n+1}(x_{10}, y_{10}) = 40[T^n(x_9, y_{10}) + T^n(x_{10}, y_9)] + 3200$$

We may start the iterative process by assuming an initial guess $T^0(x, y)$ for the temperature distribution. One crude way of doing this would be to take $T^0(x, y) = 20$, that is, equal to the surface temperature.

In this section and in the above examples in particular, we have referred to and generalized the procedure developed in one dimension in Chapter 3. This was done deliberately, since it is not practical to provide explicit formulas for all coordinate systems in any number of dimensions. By going through the above examples, the reader should be able to generalize to constructing difference approximations under a variety of circumstances.

PROBLEMS

4-1. Copper tubing ($k = 385$ W/m · °C) 3.0 cm O.D. and 1.5 mm thick is used to carry hot and cold water. The hot water is at 60°C and the cold water is at 10°C. The two tubes are parallel and pass through a large region filled with fiberglass insulation ($k = 0.035$

W/m · °C). The tube centers are 5 cm apart. At what rate is heat transferred from the hot water tube to the cold water tube per unit length.

4-2. Repeat Problem 4-1 for separations of 4 cm and 10 cm. Tabulate or plot your results.

4-3. A copper tube ($k = 385$ W/m · °C) 3.0 cm O.D. and 1.5 mm thick carrying hot water at 60°C passes through the center of a space between two walls. The space is 8 cm thick and is filled with fiberglass insulation ($k = 0.035$ W/m · °C). The walls are at 20°C. How much heat is lost per unit length.

4-4. A spherical container has its surface temperature regulated to stay at 80°C. It is placed in granite. The surface temperature is 20°C. The diameter of the sphere is 1 m. Find the heat transfer to the surface as a function of depth below the surface. Tabulate or plot your results. What is the implication of your result for a decision as to how deeply below the surface to bury the sphere to reduce heat loss.

4-5. The copper tube of Problem 4-3 passes through the center of a square block of concrete 10 cm on a side whose outside surface temperature is 10°C. What is the heat loss per unit length?

4-6. The concrete block in Problem 4-5 is in a region surrounded by air at 10°C (instead of the surface temperature being known). If the heat transfer coefficient is 10 W/m² · °K, what is the heat loss per unit length of pipe.

4-7. Derive an expression for the temperature distribution in a rectangular block with sides a, $2b$, and $2c$ in the x, y, and z directions, respectively. A uniform source q'' is incident on the left face ($x = 0$). At other faces, $x = a$, $y = +b$, $z = +c$, the temperature is specified to be T_w.

4-8. Repeat Example 4-2 except that the block is finite in the z direction with a side of 2 m. Use the solution derived in Problem 4-7.

4-9. Repeat Example 4-2 for the same total heat source on the left face, with the source having a distribution $q'' = q_0'' y^2$.

4-10. For Example 4-2 find the maximum temperature at the left face of the block.

4-11. For the block of Example 4-2, suppose that the maximum temperature at the left face were specified to be 50°C and the heat flux were unknown. Find the heat flux distribution at the left face.

4-12. A cylindrical block is made of the source material used in Example 4-2. The block is 1 m in diameter and 1 m long. A heat flux of 1 kW/m² is applied to one end. Find the temperature at the center of the block.

4-13. A long borosilicate glass block of square cross-section 0.3 m on a side contains a heat source of 4000 W/m³. The surface of the glass is maintained at a uniform temperature. What is the temperature rise from the surface to the center?

4-14. Derive an expression for the temperature distribution in a three-dimensional rectangular block with a uniform heat source.

4-15. Suppose the block of Problem 4-13 were 1 m long. Use the results of Problem 4-14 to find the temperature rise from the surface to the center.

4-16. For Example 4-2, find the average temperature on the left face of the block.

4-17. Consider again Problem 4-3. How much heat would be lost if the space occupied by fiberglass insulation were instead filled with air?

4-18. In Problems 4-3 and 4-17, consider the walls to be wallboard ($k = 0.17$ W/m²) 2 cm thick with the temperature at the outside of the wall maintained at 20°C. Find the heat losses.

4-19. In Problem 4-18, suppose that at the outside walls there are convection boundary conditions ($h = 10$ W/m² · °C) with the ambient air temperature being 20°C. Find the heat losses.

4-20. A solid cylinder with diameter D and thermal conductivity k_c has a uniform heat source q'''. The cylinder center is a distance H below the surface of a medium with thermal conductivity k_m. The surface loses heat by convection (coefficient h) to a fluid with asymptotic temperature T_∞. Derive an expression for the temperature at the center of the cylinder. State any assumptions made.

4-21. Find the temperature distribution in the block of Example 4-2 using a finite difference computer code.

4-22. Find the temperature distribution in the cylinder of Problem 4-12 with a finite difference code.

4-23. Find the temperature distribution in the block of Problem 4-13 with a finite difference code.

4-24. Find the temperature distribution in the three-dimensional block of Problem 4-15 with a finite difference code.

REFERENCES

1. V. S. Arpaci, *Conduction Heat Transfer*, Addison-Wesley, Reading, MA, 1966.
2. H. S. Carslaw and J. C. Jaeger, *Conduction of Heat in Solids*, 2nd. ed., Oxford University Press, London, 1959.
3. E. Hahne and U. Grigulll, "Formfactor und Forweiderstand der stationaren mehrdi-mensionsionalen Warmeleitung," *Int. J. Heat Mass Transfer*, **18**, 751 (1955).
4. S. S. Kutateladze, *Fundamentals of Heat Transfer*, Academic Press, New York, 1963.
5. S. S. Kutateladze, and V. M. Borishanskii, *A Concise Encyclopedia of Heat Transfer*, Pergamon, New York, 1966.
6. G. E. Meyers, *Analytical Methods in Conduction Heat Transfer*, McGraw-Hill, New York, 1971.
7. M. N. Ozisik, *Boundary Value Problems of Heat Conduction*, International Textbook, Scranton, PA, 1968.
8. M. N. Ozisik, *Heat Conduction*, Wiley, New York, 1980.
9. W. M. Rohsenow and J. P. Hartnett, *Handbook of Heat Transfer*, McGraw-Hill, New York, 1973.
10. P. J. Schneider, *Conduction Heat Transfer*, Addison-Wesley, Reading MA, 1955.
11. J. E. Sunderland and K. R. Johnson, "Shape Factors for Heat Conduction Through Bodies with Isothermal or Convective Boundary Conditions," *Trans. ASHRAE*, **10**, 237 (1964).
12. H. Y. Wong, *Heat Transfer for Engineers*, Longman, New York, 1973.
13. M. M. Yovanovich, *Advanced Heat Conduction*, Hemisphere, Washington, D.C., 1981.

Chapter Five

TRANSIENT HEAT CONDUCTION

5-1. INTRODUCTION

So far, we have considered only steady-state problems. In this chapter, we consider cases in which the temperature can vary with time. We have seen in Chapter 4 that when problems have more than one dimension, it can become difficult to solve the heat conduction equation. Time is a dimension, so introducing time as a variable introduces difficulties analogous to those introduced in Chapter 4. Accordingly, as in Chapter 4, we seek specific situations which can be treated relatively easily and introduce numerical procedures to handle other cases.

The simplest case we can envision is the one-dimensional problem. If we can have a one-dimensional problem with time as the dimension instead of space, then solutions may be expected to be straightforward. This is the approach followed in lumped parameter modeling. The conditions under which such modeling may be used are determined by a parameter called the Biot number.

When the lumped parameter model cannot be used, it may be possible to apply separation of variables, as was done in Chapter 4. As in Chapter 4, series solutions are obtained. Convergence rates improve with time, and after a waiting period, very few terms, possibly only one, may be needed. There also are special cases involving infinite and semi-infinite media where analytical solutions are possible. Situations of this type are identified.

For other cases, it is necessary to construct difference equations. It is found that the solution to a time-dependent problem involves solution of static problems at each step; that is the time-dependent problem involves repeated solution of the types of numerical procedures developed in Chapters 3 and 4.

5-2. LUMPED PARAMETERS

Consider a block of material initially at a uniform temperature T_0 placed in an environment of temperature T_∞, where it will cool with a convection heat transfer coefficient h. The heat conduction equation in the block is

$$\rho c \frac{\partial T}{\partial t} = \nabla \cdot k \nabla T \tag{5-1}$$

Let us integrate over the volume of the block

$$\rho c \frac{\partial}{\partial t} \int dV \, T = \int dV \, \nabla \cdot k \nabla T \tag{5-2}$$

We may define an average temperature in the block according to

$$T_{av} = \frac{1}{V} \int dV \, T \tag{5-3}$$

We may observe that the volume integral on the right-hand side may be converted to a surface integral

$$\int dV \, \nabla \cdot k \nabla T = \int d\mathbf{A} \cdot k \nabla T \tag{5-4}$$

The convection at the surface is

$$\int h \, dA (T_s - T_\infty) = - \int d\mathbf{A} \cdot \nabla T \tag{5-5}$$

We may derive an average surface temperature (although this would not be necessary in a simple geometry where we would know that the surface temperature is uniform)

$$T_{s,\,av} = \frac{1}{A} \int dA \, T_s \tag{5-6}$$

We thus obtain an equation

$$\rho C V \frac{dT_{av}}{dt} = -hA(T_{s,\,av} - T_\infty) \tag{5-7}$$

By integrating and averaging, we have removed spatial derivatives from the problem. However, we have an unknown relationship between the average temperature inside the block and the average temperature on the surface of the block.

The simplest form of lumped parameter model occurs when one assumes that the temperature variation in the block is negligible by comparison with the temperature difference between the block and the environment. Then, we assume

$$T_{s,\text{av}} = T_{\text{av}} \tag{5-8}$$

We let

$$\Theta_{\text{av}} = T_{\text{av}} - T_{\infty} \tag{5-9}$$

and obtain the equation

$$\frac{d\Theta_{\text{av}}}{dt} = -\frac{hA}{\rho c V}\Theta_{\text{av}} \tag{5-10}$$

which has the solution

$$\Theta_{\text{av}} = \Theta_{\text{av}}(0)e^{-(hA/\rho c V)t} \tag{5-11}$$

A second level of lumped parameter model might assume something about the relationship between average and surface temperatures. For example, for plane geometry, we will find later that, as time passes, the solution tends to the form

$$\Theta(x,\, t) = A \cos Bx\left(e^{-\beta t}\right) \tag{5-12}$$

We might then assume that

$$\frac{\Theta_s}{\Theta_{\text{av}}} = f \tag{5-13}$$

$$f = \frac{\cos Bx_s}{(1/x_s)\displaystyle\int_0^{x_s}\cos Bx\, dx} = Bx_s \cot Bx_s \tag{5-14}$$

We then would obtain

$$\Theta_{\text{av}}(t) = \Theta_{\text{av}}(0)e^{-f(hA/\rho c V)t} \tag{5-15}$$

with f less than 1. Thus, we expect that the simplest lumped parameter model will overestimate the rate at which heat is removed from the block.

We said above that the lumped parameter model tends to be applied when the temperature change in the block is small compared with the temperature change form the block to the environment. A rough estimate of this ratio may be obtained form the boundary condition at the surface, evaluated approximately

$$hA(T_{s,av} - T_\infty) \approx -kA\frac{T_{s,av} - T_c}{L_{eff}} \tag{5-16}$$

where L_{eff} is some effective length characterizing the block. We then can rearrange the equation to

$$\frac{T_{s,av} - T_c}{T_{s,av} - T_\infty} = \frac{hL_{eff}}{k} << 1 \tag{5-17}$$

Imposing a postulate about the ratio of temperatures imposes a requirement on the combination of parameters, which we call the Biot number Bi,

$$\mathrm{Bi} = \frac{hL_{eff}}{k} \tag{5-18}$$

It is common to use for the characteristic length

$$L_{eff} = \frac{V}{A} \tag{5-19}$$

A guideline frequently found in the literature is that the lumped parameter approach may be used when the Biot number is less than 0.1. In this matter, as will be the case regarding other criteria (e.g., for the Fourier number later in this chapter), the precise numerical criterion used should relate to the accuracy requirements of the situation. Thus, it is important to retain an appreciation of the physical implication of the guideline used, as per the ratio of temperature differences in Eq. (5-17).

If we examine our lumped parameter solution and regroup parameters, we obtain

$$\Theta_{av}(t) = \Theta_{av}(0)\exp\left[-\frac{hV}{kA}\frac{k}{\rho c}\left(\frac{A^2}{V}\right)t\right] \tag{5-20}$$

which may be written

$$\Theta_{av}(t) = \Theta_{av}(0)e^{-\mathrm{Bi\,Fo}} \tag{5-21}$$

where Fo is the Fourier modulus

$$\mathrm{Fo} = \frac{\alpha t}{(V/A)^2} \tag{5-22}$$

and where

$$\alpha = \frac{k}{\rho c} \qquad (5\text{-}23)$$

is called the thermal diffusivity. These combinations of terms, Bi and Fo, are found to occur in various transient situations.

The thermal diffusivity α expresses the ratio of the tendency of heat to be conducted away to the tendency of heat to be accommodated internally. The Fourier modulus, which involves multiplying α by a characteristic time and dividing by the square of a characteristic length, puts this ratio into a dimensionless from. The Fourier modulus is thus a measure of the tendency of the temperature to change. If the conductivity is high and the specific heat is low, the Fourier modulus will tend to be high and a substantial change in temperature may occur in a relatively short time. If the conductivity is low and the specific heat is high, heat will tend not to be conducted away and will be accommodated internally without much temperature change.

EXAMPLE 5-1. A copper cylinder 5 cm in diameter and 10 cm long is taken form a hot environment (initial temperature 100°C) and placed in room temperature air (20°C). The convection heat transfer coefficient is 10 W/m² · ° C. Is it appropriate to use a lumped parameter analysis to study the cooling of the cylinder?

SOLUTION. Evaluate the Biot number

$$Bi = \frac{hL_{eff}}{k}$$

$$L_{eff} = \frac{V}{A} = \frac{\frac{1}{4}\pi D^2 H}{\pi DH + 2\left(\frac{1}{4}\pi D^2\right)} = \frac{D/4}{1 + \frac{1}{2}D/H}$$

$$Bi = \frac{(10)(0.01)}{386} = 2.6 \times 10^{-4} << 1$$

Lumped parameter analysis is appropriate.

5-3. SEPARATION OF VARIABLES

When lumped parameter analysis is not appropriate, then explicit representation of space–time effects is needed. One way of obtaining this explicit representation is through separation of variables. This technique is basically the same as that used for two-dimensional spatial analysis in Chapter 4.

Consider the time-dependent heat conduction equation in a plane wall of thickness $2L$

$$\rho c \frac{\partial T}{\partial t} = k \frac{\partial^2 T}{\partial x^2} \tag{5-24}$$

with boundary conditions

$$-k \frac{dT}{dx}(0) = 0 \tag{5-25}$$

$$-k \frac{dT}{dx}(L) = h[T(L) - T_\infty] \tag{5-26}$$

We change variables to

$$\Theta = T - T_\infty \tag{5-27}$$

Our equation becomes

$$\frac{1}{\alpha} \frac{\partial \Theta}{\partial t} = \frac{\partial^2 \Theta}{\partial x^2} \tag{5-28}$$

We assume separability in the form

$$\Theta(x, t) = F(t)G(x) \tag{5-29}$$

This leads to

$$\frac{1}{\alpha} \frac{1}{F} \frac{dF}{dt} = \frac{1}{G} \frac{d^2G}{dx^2} \tag{5-30}$$

Each side of this equation, being a function of a different variable, must be a constant. We assume the form

$$\frac{d^2G}{dx^2} = -B^2 G \tag{5-31}$$

which, upon application of the condition

$$\frac{dG}{dx} = 0 \tag{5-32}$$

leads to the solution

$$G(x) = C_1 \cos Bx \tag{5-33}$$

The convection boundary condition

$$\frac{dG}{dx}(L) = -\frac{h}{k} G(L) \tag{5-34}$$

$$-B \sin BL = -\frac{h}{k} \cos BL \qquad (5\text{-}35)$$

Regrouping parameters leads to

$$BL \tan BL = \frac{hL}{k} = \text{Bi} \qquad (5\text{-}36)$$

The separation constant B is determined from this transcendental equation. The quantity $z \tan z$ for z between 0 and $\pi/2$ is tabulated in Table 5-3-1. One proceeds to find the value of $z \tan z$ that corresponds to the Biot number. One then reads off the corresponding value of BL. Since the tangent is a periodic solution, there is an infinite number of values B_n that are admissible.

The function $BL \tan BL$ is sketched in Fig. 5-3-1, along with a horizontal line for the Biot number. Note that there are intersections in alternate intervals of thickness $\pi/2$, that is, in intervals where the tangent is positive. Note also that $BL \tan BL$ has a zero derivative at the origin, so that intersections always exist.

It is a simple matter, especially with access to a computer, to generate a more detailed version of Table 5-3-1 and to construct similar tables for intervals $n\pi$ to $n\pi + \pi/2$. Alternatively, if one is preparing a general computer code, one may use a systematic search. A sample subroutine that does this is provided in Appendix C. The search is illustrated in Fig. 5-3-2. Initial and final values (e.g., 0, $\pi/2$) x_{I1} and x_{F1} bound the desired value. An initial guess is made that the value is at the midpoint x_{M1}. For the case chosen in the illustration, the midpoint x_{M1} is to the left of the desired value (since $BL \tan BL$ is less than Bi). We therefore set a new initial value equal to this value x_{M1} leaving the final value at x_{F1}. Another midpoint value x_{M2} is chosen, which is found to be greater than the desired value. Thus, a new final

TABLE 5-3-1
Function for Plane Geometry
Separation Constant

BL	$BL \tan BL$
0.157	0.0249
0.314	0.102
0.471	0.240
0.628	0.457
0.785	0.785
0.942	1.297
1.100	2.158
1.257	3.868
1.414	8.926
1.571	∞

value x_{F2} is chosen equal to x_{M2}. As this process continues, the interval in which the solution exists shrinks. The process is continued until the value $BL \tan BL$ is within some prespecified limit ε of the Biot number.

The time-dependent portion of the equation is

$$\frac{dF_n}{dt} = -\alpha B_n^2 F_n \qquad (5\text{-}37)$$

which leads to the overall solution

$$\Theta(x, t) = \sum_{n=0}^{\infty} A_n \cos B_n x \left(e^{-\alpha B_n^2 t} \right) \qquad (5\text{-}38)$$

The coefficients A_n are evaluated as Fourier coefficients of the series expansion

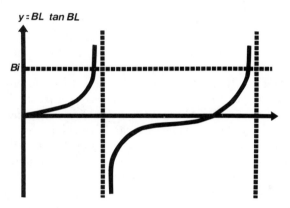

FIGURE 5-3-1. Sketch of $BL \tan BL$ with Biot number intersections.

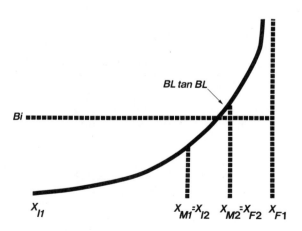

FIGURE 5-3-2. Sketch illustrating systematic search.

of the initial condition

$$\Theta(x, 0) = \Theta_0(x) = \sum_{n=0}^{\infty} A_n \cos B_n x \qquad (5\text{-}39)$$

If the initial temperature distribution is constant, then

$$A_n = \Theta_0 \frac{\int_0^L dx \cos B_n x}{\int_0^L dx \cos^2 B_n x} = \Theta_0 \frac{2 \sin B_n L}{B_n L + \sin B_n L \cos B_n L} \qquad (5\text{-}40)$$

Note that for a thin plate and low-order terms, $B_n L$ will be small and A_n will be close to Θ_0, since $\sin B_n L$ and $B_n L$ will be close in value and $\cos B_n L$ will be close to 1.

A common practice in heat transfer is to analyze transients on the basis of the first term in the expansion, that is, to apply

$$\Theta(x, t) = A_0 \cos B_0 x \left(e^{-\alpha B_0^2 t} \right) \qquad (5\text{-}41)$$

One approach in the literature has been to express this solution in two parts. The first represents the variation of temperature at the center

$$\frac{\Theta(0, t)}{\Theta(0, 0)} = e^{-(\alpha t/L^2) B_0^2 L^2} \qquad (5\text{-}42)$$

The second part gives the relative spatial variation

$$\frac{\Theta(x, t)}{\Theta(0, t)} = \cos \left[B_0 L \left(\frac{x}{L} \right) \right] \qquad (5\text{-}43)$$

Because $B_0 L$ is determined by the Biot number, the dependence is parametric with the Biot number.

Charts encountered in the literature for time variation of center temperature, called Heisler charts (1947), may include an actual series expansion; that is,

$$\frac{\Theta(0, t)}{\Theta(0, 0)} = \frac{1}{\Theta_0} \sum_n A_n e^{-\text{Fo}(B_n L)^2} \qquad (5\text{-}44)$$

However to be used in conjunction with charts for spatial variation, Eq. (5-43) must be restricted to conditions such that only one term in the expansion is significant. These charts can be difficult to read with any accuracy.

When, as frequently is the case (according to a criterion to be discussed below), one term in the expansion is sufficient, the problem may be addressed as follows.

Step 1. Evaluate the Biot number.
Step 2. Set the Biot number equal to $B_0 L \tan B_0 L$ in Table 5-3-1 (or in Fig. 5-3-3) and obtain the corresponding value of $B_0 L$. (For example, if Bi is 0.240, $B_0 L$ is 0.471).
Step 3. Use Eq. (5-40) to evaluate A_0.
Step 4. Evaluate the Fourier modulus.
Step 5. Evaluate Eq. (5-4-1).

Note that from Fig. 5-3-4 that A_0/θ_0 actually can be greater than 1. Thus, there clearly will be times short enough (or Fo low enough) that one term in the expansion would not be sufficient.

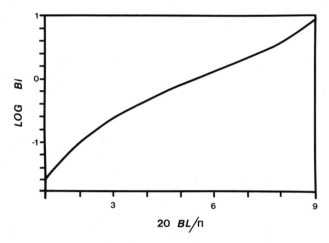

FIGURE 5-3-3. Relation between Biot number and BL in plane geometry.

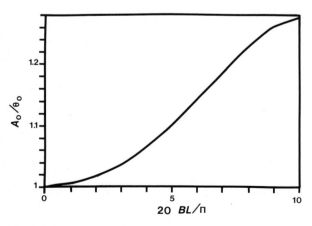

FIGURE 5-3-4. Leading term coefficient for plane geometry.

The Heisler charts used in the literature can be very difficult to read with any accuracy. They are mentioned here because they are so common. If the reader is restricted to a one-term analysis and wishes to use charts as aids, the graphs of Figs. 5-3-3 and 5-3-4 should yield reasonable estimates.

A limiting case occurs when the convection coefficient goes to infinity. This corresponds to specification of temperature at the wall. The tangent goes to infinity and

$$B_n L = (2n + 1)\frac{\pi}{2} \tag{5-45}$$

This analytical limit gives us an opportunity to estimate the validity of keeping only one term in the series.

We may specify that

$$\frac{e^{-(\alpha t/L^2)(B_1 L)^2}}{e^{-(\alpha t/L^2)(B_0 L)^2}} < \varepsilon \tag{5-46}$$

which leads to

$$\text{Fo} > \frac{-\ln \varepsilon}{(B_1 L)^2 - (B_0 L)^2} = \frac{-\ln \varepsilon}{8(\pi/2)^2} \tag{5-47}$$

If ε is 0.01, then Fo should exceed 0.23.

It is common to stipulate that Fo should exceed 0.2 for a one-term analysis to be valid, a reasonable assertion given the above analysis. It should be recognized, however, that the criterion in practical cases should depend on the accuracy requirements of particular situations. Thus, an engineer willing to tolerate a larger value of ε would be able to use only one term at a smaller value of Fo.

The same approach used to determine when only the first term in the series is adequate can be used to determine when two terms are adequate, and so on. If a computer code is prepared to evaluate Eq. (5-38), then it is useful to determine the appropriate number of terms to evaluate in the expansion. This evaluation is something the computer code itself may do. In other words, given the Fourier modulus (or the information from which the computer code may calculate the Fourier modulus), the computer code may determine the number of terms required and evaluate those terms. Extending Eqs. (5-46) and (5-47) to an estimate of the number of terms required is left as a problem at the end of this chapter.

For an engineer with a personal computer or with an office terminal, setting up a computer solution for an arbitrary number of terms in the expansion of Eq. (5-38) is a simple matter. The values of B_n that satisfy Eq. (5-36) would be obtained from the routine in Appendix C. From that point, a DO loop (or a FOR-NEXT loop in BASIC) would evaluate Eq. (5-38) using the coefficients of Eq. (5-40).

When preparing such a computer routine, you may observe that the effort involved is small and is essentially independent of the number of terms you take in the expansion (although the cost of the computer run can depend on the number of terms). Thus, in an environment where a computer or terminal is a handy tool, there is little incentive to make a potentially restrictive assumption of retaining only one term.

EXAMPLE 5-2. A block of glass, used for viewing a heating process, initially is at a temperature of $100°C$. The heating process is turned off and the air on both sides of the glass is then kept at $20°C$. Find the temperature at the center and surface of the 20 cm thick glass 1 hr later. Assume $h = 10 \text{ W/m}^2 \cdot °C$.

SOLUTION. First, let us examine whether a lumped parameter analysis would be adequate. In plane geometry

$$L_{eff} = \frac{V}{A_s} = \frac{A(2L)}{2A} = L$$

The total surface area is twice the area A of either face. The volume is A times the thickness $2L$. The Biot number is evaluated as

$$Bi = \frac{hL}{k} = \frac{(10)(0.1)}{1.4} = 0.71$$

This is not small enough to justify lumped parameter analysis.

Next, let us examine the Fourier modulus to see whether the first term in the expansion is sufficient.

$$Fo = \frac{\alpha t}{L^2} = \frac{(7.5 \times 10^{-7})(3600)}{(0.1)^2} = 0.27 > 0.23$$

Thus, one term should be sufficient.

The general solution at the center becomes

$$\frac{\Theta(0, t)}{\Theta(0, 0)} = \frac{A_0}{\Theta_0} e^{-Fo(B_0 L)^2} = \frac{2 \sin B_0 L}{B_0 L + \sin B_0 L \sin B_0 L} e^{-Fo(B_0 L)^2}$$

From Table 5-3-1, we obtain, with iteration, that for

$$Bi = BL \tan BL = 0.71$$

$$BL = 0.755$$

$$\frac{\Theta(0, t)}{\Theta(0, 0)} = 0.937$$

$$T(0, t) = 95°C$$

$$\Theta(L, t) = \Theta(0, t)\cos BL = 0.73\Theta(0, t) = 55°C$$

$$T(L, t) = 75°C$$

EXAMPLE 5-2 (Spreadsheet).   This problem may be treated with different levels of automation within the spreadsheet. Here we choose to use an intermediate level. We evaluate the Biot and Fourier numbers and find that one term is needed. We see where the Biot number lies in the $BL \tan BL$ column of Table 5-3-1 and obtain a more accurate value of BL by trial and error. Since the Biot number calculated in B7 is less than 0.785, we guess a slightly lower value, taken here to be 0.75. When the value of $BL \tan BL$ returned in B12 is too low, we increase the value guessed in B11. A small number of tries involving essentially no effort leads to an appropriate value in B11. This example illustrates the ease with which normally tedious iteration arithmetic can be handled. Inspecting the temperatures in B21 and B23 to see when changes in them become small provides guidance as to when the value of B11 is sufficiently accurate.

	A	B
1	TIME	DEPENDENT
2	SLAB	CONDUCTION
3	LENGTH	.2
4	LEFF	+B3 / 2
5	H	10.0
6	CONDUCTIVITY	1.4
7	BIOT	+ B4 * B5 / B6
8	ALPHA	7.5E - 7
9	TIME(SEC)	3.6E3
10	FOURIER NO	+ B8 * B9 / (B4 ∧ 2)
11	GUESS BL	+ .75
12	BL TAN BL	+ B11 * @ TAN (B11)
13	SIN	@ SIN (B11)
14	COS	@ COS (B11)
15	A0	2 * B13 / (B11 + B13 * B14)
16	EXPONENTIAL	@ EXP (-B10 * (B11 ∧ 2))
17	CENTER RATIO	+ B15 * B16
18	T0	100.0
19	TINFINITY	20.0
20	CENTER THETA	+ B17 * (B18 - B19)
21	CENTER TEMP	+ B20 + B19
22	SURFACE THETA	+ B20 * B14
23	SURFACE TEMP	+ B22 + B19

COMPUTER PROJECT 5-1.   Write a program to find the temperature in a plane wall at any position and time [i.e., Eq. (5-38)] for an initially uniform temperature. Your input should specify the number of terms to be taken in the series.

The restriction on the Fourier modulus is much more significant for thick planes than for thin ones. The restriction may be expressed as a restriction on time; that is,

$$t > \frac{0.23}{\alpha} L^2 \tag{5-48}$$

Thus, for a given material, the restriction on time is proportional to the square of the thickness. For a thin aluminum plate of 1 cm half thickness, the one-term solution would be valid after 0.25 sec. For a 10 cm half thickness, the solution would be valid only after 27 sec. In some steels, with thermal diffusivities an order of magnitude lower, waiting times would be 10 times longer.

A number of presentations of charts of results exist in the literature, which differ in presentation and assumptions from the charts discussed above. The reader may consult the references cited for an introduction to the literature. Thus, within the accuracy of reading parametrically plotted charts, the solutions of many problems of interest are available. However, the solutions for the principal geometries are sufficiently simple to code that the engineer with access to a terminal or personal computer may wish simply to program these solutions and have them available when needed.

5-4. CYLINDRICAL AND SPHERICAL GEOMETRIES

Separation of variables in cylindrical geometry leads to the Bessel equation

$$\frac{1}{r}\frac{d}{dr} r \frac{dG}{dr} = -B^2 G \tag{5-49}$$

which, once one recognizes the finite nature of the solution at the center of the cylinder, yields the solution

$$G(r) = C_1 J_0(Br) \tag{5-50}$$

The convection boundary condition

$$-k\frac{dG}{dr}(R) = hG(R) \tag{5-51}$$

yields

$$kBJ_1(BR) = hJ_0(BR) \tag{5-52}$$

which can be regrouped to

$$BR\frac{J_1(BR)}{J_0(BR)} = \frac{hR}{k} = 2\text{Bi} \tag{5-53}$$

Recall that

$$\text{Bi} = \frac{h}{k}\frac{V}{A} = \frac{h}{k}\frac{R}{2}$$

The left-hand side is tabulated in Table 5-4-1, so again, as in Section 5-3 we may obtain the value of BR corresponding to the Biot number. The Bessel functions are tabulated in Appendix B, which also provides convenient algorithms for their evaluation. With these algorithms used to evaluate the Bessel functions and with the routine in Appendix C for obtaining the solutions of Eq. (5-53) for B_n, it is a simple matter to code the evaluation of Eq. (5-56). Thus, as with the slab case, when there is access to a personal computer or a terminal, there is little incentive to rely on the potentially restrictive assumptions of the charts.

As in Section 5-3 the desired value of B can be obtained by a systematic search. A sample computer routine for conducting the search is given in Appendix C. As discussed in Appendix B, the J_0 function has properties analogous to a cosine. The first several zeros of the J_0 function (i.e., analogous to the values $\pi/2 + n\pi$ for the cosine) are listed in Appendix B. These zeros serve as the boundaries of the intervals in which successive solutions of Eq. (5-53) are sought.

The overall solution can be expressed as

$$\Theta(r, t) = \sum_{n=0}^{\infty} A_n J_0(B_n r) e^{-\alpha B_n^2 t} \tag{5-54}$$

TABLE 5-4-1
Function for Cylindrical Geometry
Separation Constant

BR	$BR\dfrac{J_1(BR)}{J_0(BR)}$
0.2	0.020
0.4	0.082
0.6	0.189
0.8	0.349
1.0	0.575
1.2	0.891
1.4	1.33
1.6	2.00
1.8	3.08
2.0	5.15
2.2	11.1
2.4048	∞

If the initial temperature distribution is uniform, the coefficients are given by

$$A_n = \Theta_0 \frac{\int_0^R dr\, rJ_0(B_n r)}{\int_0^R dr\, rJ_0^2(B_n r)} = \Theta_0 \frac{2B_n RJ_1(B_n R)}{(B_n R)^2 [J_0^2(B_n R) + J_1^2(B_n R)]} \quad (5\text{-}55)$$

and the series would be

$$\Theta(r,t) = \Theta_0 \sum_{n=0}^{\infty} \frac{2J_1(B_n R)}{B_n R[J_0^2(B_n R) + J_1^2(B_n R)]} J_n(B_n r) e^{-Fo(B_n R)^2/2} \quad (5\text{-}56)$$

For small $B_n R$, that is, for small-radius cylinders and low-order modes, the coefficients A_n are close to 1.

If, as per common practice, only one term in the expansion is considered we may write the solution as

$$\Theta(r,t) = A_0 J_0\left[B_0 R\left(\frac{r}{R}\right)\right] e^{-Fo(B_0 R|2)^2} \quad (5\text{-}57)$$

$$\frac{\Theta(0,t)}{\Theta(0,0)} = e^{-Fo(B_0 R^2/2)} = e^{-[\alpha t/(R/2)^2](B_0 R|2)^2} \quad (5\text{-}58)$$

and spatial variations are given by

$$\frac{\Theta(r,t)}{\Theta(0,t)} = J_0\left[B_0 R\left(\frac{r}{R}\right)\right] \quad (5\text{-}59)$$

When applying these equations on a personal computer, use may be made of compact convenient algorithms for evaluating the Bessel functions J_0 and

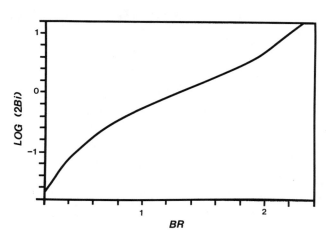

FIGURE 5-4-1. Relation between Biot number and BR in cylindrical geometry.

J_1, as was indicated in Chapter 3 to be the case for the modified Bessel functions I_0, I_1, K_0, and K_1. Such compact algorithms are provided in Appendix B.

When working with a calculator and keeping only one term in the series, the set of steps provided in plane geometry in Section 5-3 can be adapted directly (using appropriate equations for Bi, etc.). The information contained in Table 5-4-1 is plotted in Fig. 5-4-1 for convenience. The coefficient A_0 is plotted in Fig. 5-4-2.

For spherical geometry, analogous development leads to

$$\Theta(r, t) = \sum_{n=0}^{\infty} \left[\frac{2\Theta_0 (\sin B_n R - B_n R \cos B_n R)}{B_n R - \sin B_n R \cos B_n R} \right] \frac{\sin B_n r}{B_n r} e^{-\mathrm{Fo}(B_n R/3)^2} \quad (\textbf{5-60})$$

where the term in brackets is the coefficient A_n. The separation constant is obtained from

$$- BR \cot BR = -1 + 3\mathrm{Bi} \quad (\textbf{5-61})$$

The left-hand side is given in Table 5-4-2. Note that for large Biot numbers, BR will be in the range $(\pi/2, \pi)$ where the cotangent is negative. When only one term is significant, we obtain

$$\frac{\Theta(0, t)}{\Theta(0, 0)} = e^{-\mathrm{Fo}(B_0 R/3)^2} \quad (\textbf{5-62})$$

and

$$\frac{\Theta(r, t)}{\Theta(0, t)} = \frac{\sin[B_0 R(r/R)]}{B_0 R(r/R)} \quad (\textbf{5-63})$$

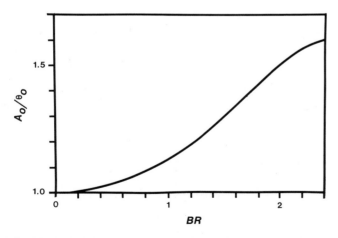

FIGURE 5-4-2. Leading term coefficient for cylindrical geometry.

102

Chapter Five

TABLE 5-4-2
Function for Spherical Geometry
Separation Constant

BR	− BR cot BR
0.314	− 0.967
0.628	− 0.865
0.942	− 0.685
1.26	− 0.408
1.57	− 0
1.88	0.612
2.20	1.60
2.51	3.46
2.83	8.70
3.14	∞

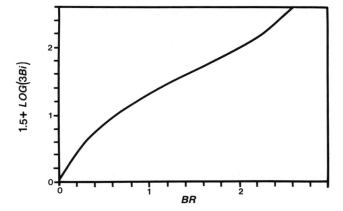

FIGURE 5-4-3.   Relation between Biot number and *BR* in spherical geometry.

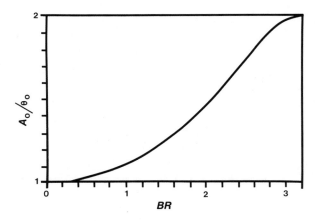

FIGURE 5-4-4.   Leading term coefficient for spherical geometry.

A systematic search for solution of Eq. (5-61) can be conducted in spherical geometry also, and a sample computer routine for doing so is given in Appendix C.

When using only one term, the set of steps of Section 5-3 can be adapted once again. The information in Table 5-4-2 is plotted in Fig. 5-4-3 and the coefficient A_0 is plotted in Fig. 5-4-4.

COMPUTER PROJECT 5-2.   Extend Computer Project 5-1 to include cylinders and spheres.

5-5. ANALYTICAL SOLUTION—LARGE MEDIA

The time-dependent heat conduction equation can be solved analytically in large media. The equation can be solved in different ways, including the use of Laplace transforms. We follow an approach that does not require transforms by making a change of variables that converts the partial differential equation to an ordinary differential equation.

Starting from the time-dependent equation for plane geometry

$$\rho c \frac{\partial T}{\partial t} = k \frac{\partial^2 T}{\partial x^2} \tag{5-64}$$

let us simplify by changing variables to

$$\tau = \frac{k}{\rho c} t = \alpha t \tag{5-65}$$

which leads to the equation

$$\frac{\partial T}{\partial \tau} = \frac{\partial^2 T}{\partial x^2} \tag{5-66}$$

Now define a new variable

$$u = \frac{x}{\sqrt{\tau}} \tag{5-67}$$

We may evaluate $\partial T/\partial \tau$ as

$$\frac{\partial T}{\partial \tau} = \frac{\partial T}{\partial u}\frac{\partial u}{\partial \tau} = -\frac{x}{2\sqrt{\tau}}\frac{1}{\tau}\frac{\partial T}{\partial u} = -\frac{1}{\tau}\frac{u}{2}\frac{\partial T}{\partial u} \tag{5-68}$$

We may evaluate $\partial^2 T/\partial x^2$ as

$$\frac{\partial^2 T}{\partial x^2} = \frac{\partial}{\partial u}\left(\frac{\partial T}{\partial x}\right)\frac{\partial u}{\partial x} = \frac{\partial}{\partial u}\left(\frac{\partial T}{\partial u}\frac{\partial u}{\partial x}\right)\frac{\partial u}{\partial x} = \frac{1}{\tau}\frac{\partial^2 T}{\partial u^2} \tag{5-69}$$

Equating $\partial T/\partial \tau$ and $\partial^2 T/\partial x^2$ yields

$$-\frac{u}{2}\frac{\partial T}{\partial u} = \frac{\partial^2 T}{\partial u^2} \tag{5-70}$$

Since there is dependence only on u, the partial derivatives become ordinary derivatives. We may proceed by solving a first-order differential equation for dT/du, which may be obtained simply as

$$\frac{dT}{du} = C_1 e^{-u^2/4} \tag{5-71}$$

If we integrate from $u = 0$ to some value u

$$T(u) = C_1 \int_0^u d\sigma \, e^{-\sigma^2/4} + T(u = 0) \tag{5-72}$$

We note that this integral is an error function. The error function is a commonly used integral in mathematics and the physical sciences. Its properties are discussed in Appendix B, where values of the function are tabulated. Appendix B also contains a convenient algorithm by which the error function can be evaluated.

$$\sqrt{\pi}\,\text{erf}\left(\frac{u}{2}\right) = \int_0^u d\sigma \, e^{-\sigma^2/4} \tag{5-73}$$

Our solution is thus

$$T(u) = C_1\sqrt{\pi}\,\text{erf}\left(\frac{u}{2}\right) + C_2 \tag{5-74}$$

This solution is useful in a general sense because it helps one to see that there is a tendency for transient heat flow to depend on the parameter $x/\sqrt{\tau} = x/2\sqrt{\alpha t}$. Whether the explicit solution will be useful depends on whether we can apply boundary conditions.

Suppose we have a semi-infinite block initially at temperature T_0. Suddenly, for time $t > 0$, we increase the temperature at $x = 0$ to T_1. Noting that $x = 0$ corresponds to $u = 0$, and $t = 0$ corresponds to $u = \infty$, we have boundary conditions

$$T(u = \infty) = T_0 \tag{5-75}$$

$$T(u = 0) = T_1 \tag{5-76}$$

This leads to

$$C_2 = T_1 \tag{5-77}$$

$$C_1\sqrt{\pi} = T_0 - C_2 = -(T_1 - T_0) \tag{5-78}$$

$$T - T_0 = (T_1 - T_0)\left[1 - \text{erf}\left(\frac{x}{2\sqrt{\alpha t}}\right)\right] \tag{5-79}$$

The heat flux is obtained from

$$q = -kA \frac{\partial T}{\partial x} \qquad (5\text{-}80)$$

Noting that

$$\frac{d}{dx}\left(\text{erf}\,\frac{u}{2}\right) = \frac{d}{du}\text{erf}\left(\frac{u}{2}\right)\frac{du}{dx} \qquad (5\text{-}81)$$

the heat flux, in general, is given by

$$q = \frac{C_1}{\sqrt{\alpha t}} e^{-x^2/4\alpha t} \qquad (5\text{-}82)$$

For the case examined, this becomes

$$q = \frac{T_1 - T_0}{\sqrt{\pi \alpha t}} e^{-x^2/4\alpha t} \qquad (5\text{-}83)$$

It may be observed that when $t = 0$, there is a singular character to the heat flux. The heat flux goes to infinity over a very narrow time interval in such a way that the integral over time is finite. Such behavior is called δ-function behavior and is discussed in the references. Other solutions, to be quoted below, also have this behavior.

The behavior comes about because we imposed a sudden change in temperature at the surface. A finite change in temperature in zero time implies an infinite heat flux. In reality, we cannot have an infinite heat flux, but we cannot have a truly sudden temperature increase either. We use such idealizations because they can be useful in practical application, for example, if the time interval over which the temperature is raised is short compared with the time scale of subsequent events.

Application of boundary conditions to other cases can be more complicated. We give the solutions here and leave their verification to problems at the end of the chapter. For the case of a uniform heat flux q_0 at the wall, the temperature distribution is given by

$$T - T_0 = \frac{2q_0}{kA}\sqrt{\frac{\alpha t}{\pi}}\, e^{-x^2/4\alpha t} - \frac{q_0 x}{kA}\left[1 - \text{erf}\left(\frac{x}{2\sqrt{\alpha t}}\right)\right] \qquad (5\text{-}84)$$

Note that the solution for this case does not depend only on the ratio $x/\sqrt{t}$. If we impose a condition of convective heat transfer on the surface

$$hA[T_\infty - T(0, t)] = -kA \frac{\partial T}{\partial x}(0, t) \qquad t > 0 \qquad (5\text{-}85)$$

then we obtain

$$\frac{T - T_0}{T_\infty - T_0} = 1 - \text{erf}\left(\frac{x}{2\sqrt{\alpha t}}\right) - e^{(h/k)[x + (h/k)\alpha t]}\left[1 - \text{erf}\left(\frac{x}{2\sqrt{\alpha t}} + \frac{h\sqrt{\alpha t}}{k}\right)\right]$$

$$(5\text{-}86)$$

The results for these cases are useful because there are practical situations that can be modeled as semi-infinite media. These cases also illustrate the difficulty in applying analytical methods. The analytical solution obtained by the change of variables $u = x/\sqrt{\tau}$ could not be used directly to match the boundary conditions of the imposed heat flux problem. Similarly, it would be difficult to match conditions at boundaries of finite media.

The types of boundary condition considered thus far pertain to step changes from one condition (prevailing for $t < 0$) to another (prevailing for $t > 0$). Another type of condition of interest is a localized burst of energy. At time $t = 0$, a pulse of energy is liberated. The solution to this problem is

$$T - T_0 = \frac{Q^{(n)}}{\rho c (4\pi\alpha t)^{n/2}} e^{-r^2/4\alpha t} \qquad (5\text{-}87)$$

where $n = 1$, 2, and 3 for plane, line, and point sources, respectively. The $Q^{(n)}$ are amounts of heat (in watt-second or joules) in spherical geometry and amounts of heat per meter and per square meter in cylindrical and plane geometry.

Note that more complex solutions can be built up as superpositions over these elementary solutions. For example, if a source $f(x, t)$ is used, then

$$T(x, t) - T_0 \int_{-\infty}^{\infty} dx' \int_0^t dt' \frac{f(x', t')}{\rho c [4\pi\alpha(t - t')]^{1/2}} e^{-(x-x')^2/4\alpha(t-t')} \qquad (5\text{-}88)$$

EXAMPLE 5-3.   A temperature change is suddenly imposed on the plane surface of a large mass of rock with $\alpha = 10^{-6} \, \text{m}^2/\text{sec}$. How long will it be before 84% of this temperature increase is seen at a distance of 1000 m into the rock?

SOLUTION

$$\frac{T - T_0}{T_i - T_0} = 0.84 = 1 - \text{erf}\left(\frac{x}{2\sqrt{\alpha t}}\right)$$

From the error function table (Appendix B)

$$\frac{x}{2\sqrt{\alpha t}} = 0.142 = \frac{1000}{2\sqrt{10^{-6}t}}$$

$$t = 1.24 \times 10^{13} \, \text{sec} \approx 32000 \text{ years}$$

This example illustrates the long time characteristics of heat-related activities performed deep underground.

The above example also illustrates the problem of evaluating a mathematical inverse, that is, of finding the argument given the function. For certain common functions, the inverse itself is a common function, so we evaluate it without thinking about it. For example, the inverse associated with an exponential is a logarithm. Logarithms and other common inverses (e.g., inverse sines and cosines) are sufficiently common that convenient algorithms exist for their evaluation. One of the computer projects below will call for evaluation of the inverse error function.

COMPUTER PROJECT 5-3.   Prepare a program to evaluate the temperature at a specified position and time within a semi-infinite block. Include, as options, the various boundary conditions considered in this section.

COMPUTER PROJECT 5-4.   Write a computer program to evaluate z given erf(z), so as to be able to treat cases such as Example 5-2. Use the approach of the computer routines of Appendix C of systematic search, that is, of establishing bounds and then shrinking the bounds. Because z can be infinite, you may wish to consider two regions ($z < z_0$ and $z > z_0$). For the large values of z, you can work with $1/z$ to avoid trying to find the midpoint of an infinite range.

5-6.   MULTIDIMENSIONAL PROBLEMS

Suppose we are concerned with a three-dimensional rectangular parallelepiped with dimensions $2L_x$, $2L_y$, $2L_z$. We assume that there is convection to a temperature T_∞ with coefficient h, and we let $\Theta = T - T_\infty$. Then

$$\frac{\partial^2 \Theta}{\partial x^2} + \frac{\partial^2 \Theta}{\partial y^2} + \frac{\partial^2 \Theta}{\partial z^2} = \frac{1}{\alpha}\frac{\partial \Theta}{\partial t} \tag{5-89}$$

Let us assume separability in the form

$$\Theta(x, y, z, t) = \Theta_x(x, t)\Theta_y(y, t)\Theta_z(z, t) \tag{5-90}$$

We then obtain

$$\frac{1}{\Theta_x}\left(\frac{\partial^2 \Theta_x}{\partial x^2} - \frac{1}{\alpha}\frac{\partial \Theta_x}{\partial t}\right) + \frac{1}{\Theta_y}\left(\frac{\partial^2 \Theta_y}{\partial y^2} - \frac{1}{\alpha}\frac{\partial \Theta_y}{\partial t}\right) + \frac{1}{\Theta_z}\left(\frac{\partial^2 \Theta_z}{\partial z^2} - \frac{1}{\alpha}\frac{\partial \Theta_t}{\partial t}\right) = 0$$

$$\tag{5-91}$$

If the one-space-dimension functions Θ_x, Θ_y, and Θ_z satisfy the respective time-dependent heat conduction problems, for example, if

$$\frac{\partial^2 \Theta_x}{\partial x^2} = \frac{1}{\alpha}\frac{\partial \Theta_x}{\partial t} \tag{5-92}$$

and if the boundary conditions on Θ_x, Θ_y, and Θ_z are such that the boundary

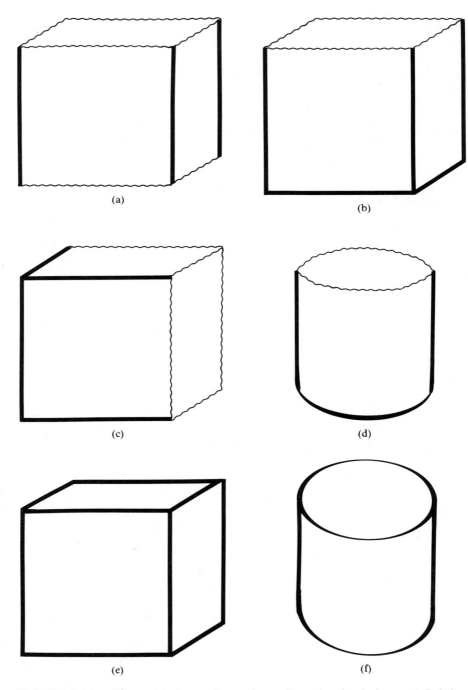

(a)

(b)

(c)

(d)

(e)

(f)

FIGURE 5-6-1. Geometries for products of one-dimensional solutions: (a) Infinite rectangular block; (b) Geometries for products of one-dimensional solutions; (c) Semi-Infinite block; (d) Semi-Infinite cylinder; (e) Rectangular block; (f) Finite cylinder.

conditions on Θ are satisfied, then the three-space dimension solution can be obtained by products of one-space-dimension solutions of the type already obtained.

While the discussions above have been for a three-dimensional rectangular parallelepiped, the discussion also applies to finite cylinders and to media that are semi-infinite in one-dimension but finite in another. The combinations are illustrated in Fig. 5-6-1.

EXAMPLE 5-4. A 304 stainless steel cylinder 10 cm in diameter and 22 cm long, initially at 80°C, is placed in water of 20°C (h = 100 W/m² · ° C). Assuming one spatial mode in each dimension is adequate, find the temperature at a radius of 5 cm at the midplane of the cylinder 1 hr later.

SOLUTION. The solution may be expressed as a product of two solutions, so

$$\Theta(r, z, t) = A_0^c J_0(B_0^c r) e^{-\alpha B_0^{c2} t} A_0^p \cos B_0^p x \left(e^{-\alpha B_0^{p2} t} \right)$$

where c and p superscripts refer to cylindrical and plane geometry values. The coefficients are evaluated from

$$A_0^c = \frac{2 J_1(B_0^c R)}{B_0^c R \left[J_0^2(B_0^c R) + J_1^2(B_0^c R) \right]}$$

$$A_0^p = \frac{2 \sin B_0^p L}{B_0^p L + \sin B_0^p L \cos B_0^p L}$$

For cylindrical geometry, we evaluate the Biot number

$$\text{Bi} = \frac{h L_{\text{eff}}}{k} = \frac{hV}{kA} = \frac{h\pi R^2 H}{k(2\pi RH)} = \frac{hR}{2k} = \frac{(100)(0.05)}{(2)(14)} = 0.175$$

Then we evaluate

$$B_0^c R \frac{J_1(B_0^c R)}{J_0(B_0^c R)} = \frac{hR}{k} = 2\text{Bi} = 0.35$$

From Table 5-4-1, we see that

$$B_0^c R = 0.8 \qquad B_0^c = \frac{0.8}{0.05} = 16$$

From Appendix B, we obtain

$$J_0(B_0^c R) = 0.846$$

$$J_1(B_0^c R) = 0.309$$

so that

$$A_0^c = \frac{2(0.309)}{0.8\left[(0.846)^2 + (0.309)^2 \right]} = 0.95$$

For plane geometry, we evaluate the Biot number

$$\text{Bi} = \frac{hL_{\text{eff}}}{k} = \frac{100}{14}\left(\frac{22}{2}\right) = 0.786$$

From Table 5-3-1, we take

$$B_0^p L = 0.785 \qquad B_0^p = \frac{0.785}{0.11} = 7.14$$

and obtain

$$\sin BL = \cos BL = 0.707$$

$$A_0^p = \frac{2(0.707)}{0.785 + (0.707)^2} = 1.10$$

At 1 hr (3600 sec) the time terms are

$$e^{-\alpha B_0^{p2}t} = e^{-(4\times10^{-6})(7.14)^2(3600)} = e^{-0.0734} = 0.480$$

$$e^{-\alpha B_0^{c2}t} = e^{-(4\times10^{-6})(16)^2(3600)} = 3.69 = 0.025$$

Therefore, temperature is evaluated from

$$\Theta(0.05, 0, 3600) = (0.95)(0.846)(0.025)(1.10)(1)(0.480) = 0.0106$$

$$T = 20 + (0.0106)(80 - 20) = 20.6°\text{C}$$

Since the above solution is a product of two solutions to one-dimensional problems, spreadsheet solution is especially convenient if one-dimensional spreadsheets have been prepared. A spreadsheet solution for Example 5-2, a plane geometry case, was provided in Section 5-3. Suppose that we also had available a spreadsheet for a cylindrical geometry problem. We could load the cylindrical geometry solution and insert columns twice at Column A. Columns A and B are now blank. We then load the spreadsheet for the plane geometry case (which required two columns). With some modest additional effort (e.g., we have to take a product of Θ terms before converting to temperature), we can develop the spreadsheet solution for the two-dimensional problem. Similarly, it is easy to prepare a spreadsheet solution for a three-dimensional problem from individual one-dimensional spreadsheets.

COMPUTER PROJECT 5-5. From your computer programs for time-dependent conduction, prepare a program that will treat multiple space dimensions.

5-7. FINITE DIFFERENCE APPROXIMATION

For problems too complicated to treat by the analytical or series methods, we again make use of finite difference approximations. We extend the procedure of constructing a difference approximation by integrating over time as well as over space.

Let us consider the case of one space dimension

$$\rho c \frac{\partial T}{\partial t} = \frac{1}{r^n} \frac{\partial}{\partial r}\left(kr^n \frac{\partial T}{\partial r}\right) \tag{5-93}$$

where again $n = 0$, 1, and 2 for plane, cylindrical, and spherical geometries, respectively. Let us integrate over time from t_{j-1} to t_j and over space from $r_{i-1/2}$ to $r_{i+1/2}$ after multiplying by r^n. This leads to (with $C_{i,i+1}$ as defined in Chapter 3)

$$\frac{r_{i+1/2}^{n+1} - r_{i-1/2}^{n+1}}{n+1}(\rho c)_i \left[T(r_i, t_{j+1}) - T(r_i, t_j)\right]$$

$$= k_i(t_{j+1} - t_j)\tfrac{1}{2}\left\{C_{i,i+1}\left[T(r_{i+1}, t_0) + T(r_{i+1}, t_{j+1})\right.\right.$$

$$\left. - T(r_i, t_j) - T(r_i, t_{j+1})\right]$$

$$- C_{i,i-1}\left[T(r_i, t_j) + T(r_i, t_{j+1}) - T(r_{i-1}, t_{j-1}) - T(r_i, t_{j-1})\right]\right\} \tag{5-94}$$

On the left-hand side of the equation, we have evaluated the spatial integral by using midpoint values, that is, values at spatial location r_i. On the right-hand side, we have evaluated the time integral by averaging values at ends of the time interval.

If $j = 0$, that is, if we are dealing with the first time interval, we would expect $T(r_i, t_j)$ to be known for all i from specified initial conditions. Thus, we must solve for $T(r_i, t_{j+1})$. We may rearrange the equation into the form

$$\frac{k_i}{2}(t_{j+1} - t_j)C_{i,i+1}T(r_{i+1}, t_{j+1})$$

$$- \left[\frac{k_i}{2}(t_{j+1} - t_j)(C_{i,i+1} + C_{i,i-1}) + \frac{r_{i+1/2}^{n+1} - r_{i-1/2}^{n+1}}{n+1}(\rho c)_i\right]T(r_i, t_{j+1})$$

$$+ \frac{k_i}{2}(t_{j+1} - t_j)C_{i,i-1}T(r_{i-1}, t_{j+1}) + S_{ij} = 0 \tag{5-95}$$

where S_{ij} has the appearance of a static source

$$S_{ij} = \frac{k_i}{2}(t_{j+1} - t_j)C_{i,i+1}T(r_{i+1}, t_j)$$

$$- \left[\frac{k_i}{2}(t_{j+1} - t_j)(C_{i,i+1} + C_{i,i-1}) + \frac{r_{i+1/2}^{n+1} - r_i^{n+1}}{n+1}(\rho c)_i\right]T(r_i, t_j)$$

$$+ \frac{k_i}{2}(t_{j+1} - t_j)C_{i,i+1}T(r_{i-1}, t_j) \tag{5-96}$$

This is a tridiagonal problem of the type encountered in Chapter 3 for static situations and thus can be solved by the method of forward substitution–backward elimination. Thus, each time step in a transient problem involves solution of a static problem. Therefore, one may use a computer routine developed for a one-dimensional static problem as a "subroutine" in a time-dependent problem to use at each step.

One may proceed in a similar manner for problems in two and three dimensions and obtain similar equations. The procedure is straightforward and is left for one of the problems at the end of the chapter. Again, an equivalent static problem is obtained at each time step. That problem may be solved by the iterative method discussed in Chapter 4. Thus, the computer code called for in a project in Chapter 4 can serve as a subroutine for a time-dependent code with multiple space dimensions.

There are alternate approaches for multiple space dimensions which may be more efficient but may also be more complicated. One example is the alternating-direction implicit method. This method is based on averaging differently the integrals over time of the spatial derivatives in successive time intervals. The result is that in each time interval, only a one-dimensional static problem (alternating particular dimensions in successive time intervals) need be solved. This procedure is discussed in Appendix B and in the references.

Different averaging in the time integration over the spatial terms can lead to different approximations. If, instead of averaging, we had evaluated the spatial derivatives at the beginning of the time interval, we would have obtained a very simple equation, because only one temperature on the left-hand side of the equation would have been evaluated at time t_{j+1}. While this equation can be solved very easily, it turns out that the solution of the equation will break down (not just get less accurate) unless very small time intervals are taken.

EXAMPLE 5-5.   Consider the situation examined in Example 3-7. Suppose that at time $t = 0$ the heat source is turned off and therefore the temperature declines toward an asymptotic condition where the temperature is uniform at the wall temperature. Set up a difference approximation to solve for the temperature distribution as a function of time. Density and specific heat are 2500 kg/m^3 and 800 J/kg$^2 \cdot °$ C.

SOLUTION.   Equation (5-95) is to be used with $n = 0$ (plane geometry) and with information from Example 3-7. Thus, we note that

$$-C_{11} = C_{12} = C_{21} = C_{23} = 18$$

$$C_{22} = C_{33} = -36$$

We have not specified a time mesh. Let us estimate a reasonable time step.

When the surface temperature is specified, we are, in effect, saying that the heat transfer coefficient is infinite, as noted in Section 5-2. Then we expect time behavior of the form

$$C - \frac{\alpha t}{L^2}(BL)^2$$

with BL equal to $\pi/2$. This means that the effective time constant of the problem is

$$\tau = \frac{L^2}{\alpha(\pi/2)^2} = \frac{L^2 \rho c}{k(\pi/2)^2} = \frac{(0.1)^2(2500)(800)}{(0.6)(\pi/2)^2} = 1.35 \times 10^4 \text{ sec}$$

We should select a time constant that is small compared to this characteristic time. Let us select

$$t_{j+1} - t_j = \frac{\tau}{100} = 135 \text{ sec}$$

The first equation corresponding to Eq. (5-95) becomes (using $\Theta = T - T_w$ instead of T as the variable), upon noting that

$$k(t_{j+1} - t_j)C_{21} = (0.6)(135)(18) = 1458$$

$$\Delta x \rho c = 6.667 \times 10^4$$

with similar expressions involving other C_{ij}

$$1458\Theta(x_2, t_{j+1}) - 6.813 \times 10^4 \Theta(x_1, t_{j+1}) = S_{1j}$$

$$1458\Theta(x_3, t_{j+1}) - 6.959 \times 10^4 \Theta(x_2, t_{j+1}) + 1458\Theta(x_1, t_{j+1}) = S_{2j}$$

$$- 7.104 \times 10^4 \Theta(x_3, t_{j+1}) + 1458\Theta(x_2, t_{j+1}) = S_{3j}$$

The S_{1j} are to be evaluated from Eq. (5-96) with Θ used in place of T. At the initial time step, the values of $\Theta(x_i, t_0)$ are the values obtained in the static problem of Example 3-7, that is, 33.3, 25.9, and 10.2. The numerical process proceeds in sequence as follows:

1. Evaluate the S_{i0}.
2. Evaluate the $\Theta(x_i, t_1)$ from the equations above by solving a one-dimensional "static" problem.
3. With these new values of Θ, evaluate the S_{i1}.
4. Evaluate the $\Theta(x_i, t_2)$.
5. Continue the process.

Since the transient approaches an asymptotic condition, in principle the problem does not end and an infinite number of time steps is required. In practice, we may be interested in following the transient over a time period comparable to the characteristic time constant τ, for example, 5τ.

COMPUTER PROJECT 5-6.   Using the static one-dimensional code developed as a project in Chapter 3 as a subroutine, prepare a one-dimensional time-dependent code.

To check whether your code is working, apply it to example problems of this chapter for which you know the solution.

COMPUTER PROJECT 5-7. Using the static two-dimensional code developed as a project in Chapter 4, prepare a two-dimensional time-dependent code. To check whether your code is working, apply it to example problems of this chapter for which you know the solution.

PROBLEMS

5-1. A long cylinder ($H \gg D$) is to be placed in an environment in which the heat transfer coefficient is 10 W/m^2 ·° K. Make a table or plot of the maximum diameter for which the lumped parameter model may be used (based on Bi $\leq$ 0.1) as a function of thermal conductivity for k in the range of 1–450 W/m ·° K. Identify locations in your table or plot for borosilicate glass, stainless steel, silicon, aluminum, copper, and silver with properties evaluated at 300° K.

5-2. Repeat Problem 5-1 using h of 100 and 1000 W/m^2 ·° K.

5-3. Evaluate the effective length to use in the Biot number for the following:
   a. A bar 20 cm long whose cross-section is an equilateral triangle with a 10 cm side.
   b. A conic section with an axis 20 cm long whose cross-section is circular, increasing in diameter linearly from a minimum of 5 cm to a maximum of 10 cm.

5-4. A cylinder block has a square cross-section with a 5 cm side and is 10 cm long.
   a. What is the effective length dimension for this block.
   b. We are interested in a time 30 min from the initial time. Tabulate or plot the Fourier number corresponding to this time versus thermal diffusivity α. The range of your table or plot should include borosilicate glass, stainless steel, silicon, aluminum, copper, and silver.

5-5. How long will it take the copper cylinder of Example 5-1 to cool to 80°C, 50°C, 25°C?

5-6. Repeat Problem 5-5 for cylinders of aluminum and stainless steel.

5-7. Prepare tables analogous to Table 5-3-1 that can be used to relate BL and $BL \tan BL$ for the second, third, and fourth terms in the solution of Eq. (5-39).

5-8. Refer to the second-level lumped parameter model of Eq. (5-15). Tabulate or plot the correction factor f as a function of Biot number, based on the information available in Table 5-3-1.

5-9. For a Biot number of 0.102, tabulate or plot the percent error in Θ_{av} as a function of Fourier number when the ordinary lumped parameter model is used.

5-10. Another possible criterion for validity of a one-term solution for Eq. (5-38) is a requirement that

$$A_0 e^{-Fo(B_0 L)^2} < 1.$$

   a. Why should this be considered a possible criterion?
   b. Evaluate the limit imposed in Fo by this criterion for the Bi $= BL \tan BL$ values in Table 5-3-1. Compare with the limit of 0.2 commonly suggested as a guideline.

5-11. Estimate a range of Fo for which you expect that one term would not be sufficient but that two terms would.

5-12. For Example 5-2, evaluate the second term in the expansion. Compare the contribution of the second term relative to that of the first term.

5-13. Repeat Problem 5-12 for times of 45 and 30 min.

5-14. Heat transfer coefficients obtained from correlated data may have uncertainties of $\pm 25\%$. What are the associated uncertainties in the cooling times calculated in Problems 5-5 and 5-6?

5-15. Derive an expression for the sensitivity of BL to small uncertainties in Bi in Eq. (5-36) (and therefore to small uncertainties in heat transfer coefficient).

5-16. Tabulate or plot, based on the solution to Problem 5-15, the percentage of uncertainty in $B_0 L$ due to a percentage uncertainty in heat transfer coefficient for $0 < h < \infty$. If you did not do Problem 5-15, evaluate the uncertainties numerically.

5-17. Find the range of possible temperatures called for in Example 5-2 if the heat transfer coefficient has an uncertainty of $\pm 25\%$.

5-18. An oven is preheated to $320°C$. An aluminum cylinder 10 cm in diameter and 10 cm long initially at $20°C$ is placed in the oven. If the heat transfer coefficient is 10 W/m² $\cdot°$ K, how long does it take for the cylinder to reach an average temperature of $300°C$? assume that radiation heat transfer can be neglected.

5-19. An uncertainty of $\pm 25\%$ exists in the heat transfer coefficient of Problem 5-18. How long must you wait before removing the cylinder to assure that the cylinder has reached a temperature of $300°C$.

5-20. A heat flux of 400 W/m² is suddenly turned on toward a plane surface of a large mass of rock with $\alpha = 10^{-6}$ m²/sec. There is a heat transfer coefficient of 10 W/m² $\cdot°$ K and the conductivity of the rock is 3 W/m $\cdot°$ K. The rock and the surrounding air are initially at a temperature of $20°C$. Tabulate or plot the temperature as a function of time over an 8 hr period at distances of 0, 10, and 20 cm into the rock assuming that the surrounding air is kept at $20°C$.

5-21. For the situation of Problem 5-20, tabulate or plot the heat transfer per unit area from the wall to the air as a function of time.

5-22. For the situation of Problem 5-20, determine what fraction of the total heat directed toward the surface over the 8 hr period is stored in the wall and what fraction is transferred to the air?

5-23. You wish to determine whether the criterion developed in Eq. (5-47) is overly restrictive when the heat transfer coefficient is finite. Make a table or plot of the limiting Fourier number as a function of the Biot number for $\varepsilon = 0.01$.

5-24. Suppose that, instead of Eq. (5-46), you applied the criterion

$$\left| \frac{A_1}{A_0} \right| \frac{e^{-Fo(B_1 L)^2}}{e^{-Fo(B_1 L)^2}} < \varepsilon$$

and evaluated $|A_1/A_0|$ based on the uniform initial condition [i.e., as given by Eq. (5-40). How is the limiting Fourier number affected?

5-25. A 304 stainless steel cube 22 cm on a side and initially at $80°C$ is placed in water of $20°C$ with $h = 200$ W/m² $\cdot°$ C. Find the temperature at the center of the cube 45 min later.

5-26. For the case of Example 5-3, how long will it take for the same temperature increase to be seen at 1, 10, and 100 m into the rock?

5-27. Consider the database of properties of metals and alloys in Appendix A.
a. Sort the listing of materials in rank order of thermal diffusivity ($\alpha = k/\rho C_p$);
b. Note specifically the substances with (1) the five highest values and (2) the five lowest values.
c. Compare the results with those for thermal conductivity (Problem 2-33).

REFERENCES

1. V. S. Arpaci, *Conduction Heat Transfer*, Addison-Wesley, Reading, MA, 1966.
2. H. S. Carslaw and J. C. Jaeger, *Conduction of Heat in Solids*, 2nd ed., Oxford University Press, London, 1959.
3. H. Grober, S. Erk, and U. Grigull, *Fundamentals of Heat Transfer*, McGraw-Hill, New York, 1961.
4. M. P. Heisler, "Temperature Charts for Induction and Constant Temperature Heating," *Trans. ASME*, **69**, 227 (1947).
5. S. S. Kutateladze, *Fundamentals of Heat Transfer*, Academic Press, New York, 1963.
6. S. S. Kutateladze, and V. M. Borishanskii, *A Concise Encyclopedia of Heat Transfer*, Pergamon, New York, 1966.
7. L. S. Lengston, "Heat Transfer from Multidimensional Objects Using One-Dimensional Solutions for Heat Loss," *Int. J. Heat Mass Transfer*, **25**, 149 (1982).
8. M. D. Mikhailov, and M. N. Ozisik, *Unified Analysis and Solutions of Heat and Mass Diffusion*, Wiley, New York, 1984.
9. M. N. Ozisik, *Heat Conduction*, Wiley, New York, 1980.
10. W. M. Rohsenow, and J. P. Hartnett, *Handbook of Heat Transfer*, McGraw-Hill, New York, 1973.
11. P. J. Schneider, *Conduction Heat Transfer*, Addison-Wesley, Reading, MA, 1955.
12. P. J. Schneider, *Temperature Response Charts*, Wiley, New York, 1963.

ELEMENTS OF CONVECTION— THE FLAT PLATE

6-1. INTRODUCTION

In dealing with conduction, we were concerned with finding a detailed temperature distribution; that is, we could find a temperature at any location. In dealing with convection, we usually deal with a process of sufficient complexity that we wish to avoid solving the detailed problem and express heat transfer in terms of a coefficient relating a wall temperature to an asymptotic or average fluid temperature. Generally, the coefficient is obtained on the basis of an empirical correlation.

There are some situations that are amenable to analysis, where we actually can make a reasonable direct prediction of the heat transfer coefficient. Such situations are important, because they help us to understand the processes involved in convection and guide us in the conduct of measurements for generating empirical correlations.

In this chapter, we analyze convection from a flat plate over which a fluid is flowing. This requires us to consider equations of fluid mechanics in addition to the equations we have considered thus far. We are especially concerned with viscosity (friction) and we develop the concepts of boundary layers for both fluid speed and fluid temperature.

We begin by developing the equations of fluid mechanics from a general conservation equation suitable for the conservation of any property. This general equation is inferred from the heat conduction equation obtained

117

previously and then applied to conservation of mass, momentum, and energy in a fluid. The resulting equations are then applied to heat transfer at a flat plate.

A useful concept in fluid mechanics and heat transfer is the concept of a boundary layer. The principal effects associated with both friction and heat transfer at a plate take place in a thin layer adjacent to the plate. We model these effects for laminar flow for which the equations can be solved. We find that heat transfer is determined by two principal dimensionless parameters, the Reynolds number and the Prandtl number.

In turbulent flow, which is characterized by statistical fluctuations about averages, it is not possible to write equations to determine precisely what is happening at every location at every instant. We attempt to infer the character of results to be expected in turbulent flow by extrapolating relationships developed for laminar flow.

In examining the various equations, we observe some important analogies. When the Prandtl number is 1, for example, we find that the momentum and energy equations become the same, with implications for understanding the relationship between friction and heat transfer. In addition, we observe strong similarities between the equations for mass transfer and the equations for heat transfer. Similarities are so strong that we are able to write down important mass transfer relations by direct analogy and to use techniques developed for heat transfer to treat mass transfer problems. This is of practical interest since some heat transfer situations involve mass transfer.

6-2. GENERAL CONSERVATION EQUATION

In dealing with heat conduction, we obtained the conservation equation

$$-\nabla \cdot k \nabla T + \rho c \frac{\partial T}{\partial \tau} = q''' \tag{6-1}$$

This equation states that the net flow of heat and the time rate of change of energy is equal to the source of heat. The principles behind this derivation apply to other quantities also. In this section, we introduce a general conservation equation and apply it to each of several specific properties.

For property p, we postulate the general conservation equation

$$\nabla \cdot \mathbf{J}_p + \frac{\partial N_p}{\partial t} = S_p \tag{6-2}$$

where $\mathbf{J}_p$ is the net current per unit area of property p, N_p is the amount of property p per unit volume, and S_p is the source of property p per unit volume. For the heat conduction equation of Chapter 3, we identify

$$J = -k\nabla T \tag{6-3}$$

$$N = \rho cT \tag{6-4}$$

$$S = q''' \tag{6-5}$$

For our treatment of fluid equations, we use this general conservation equation in steady state (no time derivatives) and apply it to conservation of mass, momentum, and energy.

For conservation of mass, we identify a material current as

$$\mathbf{J} = \rho\mathbf{V} \tag{6-6}$$

In two dimensions x and y (along and perpendicular to a flat plate), the velocity $\mathbf{V}$ is

$$\mathbf{V} = u\mathbf{i} + v\mathbf{j} \tag{6-7}$$

where u and v are the components of velocity in the x and y directions, respectively, and $\mathbf{i}$ and $\mathbf{j}$ are unit vectors. Since matter is neither created nor destroyed, there is no source and the continuity equation is

$$\nabla \cdot (\rho\mathbf{V}) = 0 \tag{6-8}$$

If we were to consider multiple constituents that could undergo transformations, for example, phase changes and chemical reactions, then we could have masses of individual constitutents being "created" and "destroyed." If the fluid were boiling water, for example, liquid could be destroyed and vapor could be created. In this section, we restrict ourselves to single-phase, single-component fluids.

In analyzing the flat plate problem, we consider that fluid properties are constant. Then the continuity equation becomes

$$\nabla \cdot \mathbf{V} = \frac{\partial u}{\partial x} + \frac{\partial v}{\partial y} = 0 \tag{6-9}$$

Momentum is a vector for which there are three components. Let us consider the x component (component along the plate). The current would be

$$\mathbf{J} = \rho u\mathbf{V} \tag{6-10}$$

and the momentum equation becomes (recall that force equals the rate of change of momentum)

$$\nabla \cdot (\rho u\mathbf{V}) = F_x \tag{6-11}$$

where F_x denotes components of force densities. This equation may be

rewritten

$$\frac{\partial}{\partial x}(u^2) + \frac{\partial}{\partial y}(uv) = \frac{F_x}{\rho} \qquad (6\text{-}12)$$

assuming constant properties.

There are two forces that we consider initially to cause momentum to change. One is the gradient of pressure in the x direction, as illustrated in Fig. 6-2-1. If pressure upstream is higher than pressure downstream, flow is accelerated.

The second force is illustrated in Fig. 6-2-2. If there is speed variation in the y direction, there is a stress

$$\tau = \mu \frac{\partial u}{\partial y} \qquad (6\text{-}13)$$

FIGURE 6-2-1. Upstream and downstream pressures p_u and p_d exert forces on ends of fluid element.

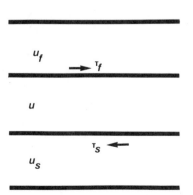

FIGURE 6-2-2. Layer of faster fluid (speed u_f) exerts accelerating stress τ_f on middle layer (speed u), while slower layer (speed u_s) exerts decelerating stress τ_s.

For a layer of fluid, as shown, the faster fluid on one side pulls the layer while the slower fluid on the other side drags the layer. The net force per unit area is

$$\mu \frac{\partial u}{\partial y}(y + dy) - \mu \frac{\partial u}{\partial y}(y) = \mu \frac{\partial^2 u}{\partial y^2} dy \qquad \text{(6-14)}$$

and the force density (force per unit volume) is $\mu \, \partial^2 u / \partial y^2$. The momentum equation is therefore

$$\frac{\partial}{\partial x}(u^2) + \frac{\partial}{\partial y}(uv) = \frac{\mu}{\rho} \frac{\partial^2 u}{\partial y^2} - \frac{1}{\rho} \frac{\partial p}{\partial x} \qquad \text{(6-15)}$$

In our analysis of the flat plate, we are concerned with a situation where the viscous force is important but the pressure gradient is not.

For conservation of energy, the current becomes

$$\mathbf{J} = -k \nabla T + \rho c T \mathbf{V} \qquad \text{(6-16)}$$

Heat flows from conduction in the medium and because the medium itself moves and carries energy along. In the absence of heat sources, the energy equation becomes

$$\frac{\partial}{\partial x}(uT) + \frac{\partial}{\partial y}(vT) = \frac{k}{\rho c} \left(\frac{\partial^2 T}{\partial x^2} + \frac{\partial^2 T}{\partial y^2} \right) \qquad \text{(6-17)}$$

Generally, when there is convection, conduction tends to be very small. However, conduction is required to transfer heat from the wall to the fluid. Accordingly, in analyzing the flat plate, we neglect only $\partial^2 T / \partial x^2$.

It may be observed that when the terms identified for neglect ($\partial p / \partial x$, $\partial^2 T / \partial x^2$) are dropped, the momentum and energy equations, Eqs. (6-15) and (6-17), become very similar. Indeed, they become identical if the coefficients on the right-hand side are in a certain ratio. This leads us to expect that there will be similarity between solutions for velocity and temperature distributions and that differences between these solutions will be characterized by the ratio of these right-hand side coefficients. (This ratio will be identified as the Prandtl number.) In addition, we expect relationships to exist between energy-related processes (heat transfer) and momentum-related processes (friction). We shall refer back to these similarities and relationships later in this chapter. That identical forms of solution are expected for heat transfer and friction when the Prandtl number is 1 is called the Reynolds analogy.

6-3. FLUID BOUNDARY LAYER

Consider the situation illustrated in Fig. 6-3-1. Fluid flows over a flat plate. Far from the plate, the fluid will not "know" that the plate is there. Consequently, we expect that far from the plate asymptotic conditions prevail.

FIGURE 6-3-1.

At the plate, viscous forces prevent motion of the fluid (i.e., u and v are zero). Therefore, a transition exists from the plate to the asymptotic condition away from the plate. We refer to this transition zone as the "boundary layer." As we shall see, this layer is generally thin compared to characteristic dimensions.

In reality, there is no sharply defined layer that ends at a well-defined location. Indeed, exact solutions of fluid mechanics equations indicate a transition that is felt from an infinite distance, although after a small distance, very little transition remains. A mathematical analogy is an exponential decay that gets very small very quickly but requires an infinite distance to reach zero.

For practical purposes, we may define a boundary layer with a finite thickness δ, as illustrated in Fig. 6-3-1. We assume that by this distance the influence of the plate will no longer be felt. We may express this assumption by the conditions

$$u(\delta) = u_\infty \tag{6-18}$$

$$\frac{\partial u}{\partial y}(\delta) = 0 \tag{6-19}$$

The first condition states that the speed has reached its asymptotic value. The second condition is required to assure that, beyond the boundary layer, the speed will not differ from the asymptotic value reached at δ. By δ, the speed has reached its final value and has stopped changing.

The analysis to obtain boundary layer thickness is complicated, so we list the steps to chart the path.

1. Integrate the momentum equation over the boundary layer.
2. Apply the boundary conditions to simplify.
3. To eliminate v (the y component of velocity), (a) integrate the continuity equation over the boundary layer and (b) substitute for v in the integrated momentum equation.
4. Apply the Leibnitz rule to interchange integration and differentiation.

These first four steps are performed in this section to obtain an integral equation for the boundary layer. In Section 6-4, we solve the equations by continuing the steps.

5. Assume a form for the velocity distribution with undetermined coefficients.
6. Choose the coefficients to match boundary conditions.
7. Substitute the velocity distribution into the boundary layer equation and carry out the integrations.
8. Solve the resulting differential equation for boundary layer thickness.

Keep the above steps in mind as you go through the analysis. Essentially, the same set of steps will be followed subsequently for the thermal boundary layer with the energy equation replacing the momentum equation.

Let us integrate the momentum equation obtained in Section 6-2 (neglecting the pressure gradient term) over the boundary layer.

$$\int_0^\delta dy \frac{\partial}{\partial x}(u^2) + u(\delta)v(\delta) - u(0)v(0) = \frac{\mu}{\rho}\frac{\partial u}{\partial y}(\delta) - \frac{\mu}{\rho}\frac{\partial u}{\partial y}(0) \quad (6\text{-}20)$$

Applying the known conditions, this reduces to

$$\int_0^\delta dy \frac{\partial}{\partial x}(u^2) + u_\infty v(\delta) = -\frac{\mu}{\rho}\frac{\partial u}{\partial y}(0) \quad (6\text{-}21)$$

we would like to evaluate $v(\delta)$ in this equation. To accomplish this, let us integrate the continuity equation

$$\int_0^\delta dy \frac{\partial u}{\partial x} + v(\delta) - v(0) = 0 \quad (6\text{-}22)$$

Noting that $v(0)$ is zero (no motion at the wall), we have obtained $v(\delta)$. Substitution into the integrated momentum equation yields

$$\int_0^\delta dy \frac{\partial}{\partial x}(u^2 - u_\infty u) = -\frac{\mu}{\rho}\frac{\partial u}{\partial y}(0) \quad (6\text{-}23)$$

We may apply the Leibnitz rule to interchange the processes of integration and differentiation. With a little rearrangement, we get the integral equation for the boundary layer

$$\frac{d}{dx}\int_0^\delta dy\, u(u_\infty - u) = \frac{\mu}{\rho}\frac{\partial u}{\partial y}(0) \quad (6\text{-}24)$$

Note that we could not simply move the derivative across the integral sign because δ is a function of x (Fig. 3-6-1). The Leibnitz rule provides that

$$\frac{d}{dx}\int_0^\delta dy\, u(u_\infty - u) = \int_0^\delta dy \frac{\partial}{\partial x}[u(u_\infty - u)] + u(\delta)[u_\infty - u(\delta)]\frac{d\delta}{dx} \quad (6\text{-}25)$$

Since $u(\delta)$ is equal to u_∞, the second term on the right-hand side vanishes even though δ varies with x.

Solving the integral equation for the boundary layer is not easy. We may generate a solution by assuming a form for the velocity profile that matches the boundary conditions. Let us assume a polynomial of the form

$$\frac{u}{u_\infty} = C_0 + C_1\frac{y}{\delta} + C_2\left(\frac{y}{\delta}\right)^2 + C_3\left(\frac{y}{\delta}\right)^3 \quad (6\text{-}26)$$

We already have specified three conditions on u that we repeat here for convenience:

$$u(\delta) = u_\infty \quad (6\text{-}27)$$

$$\frac{\partial u}{\partial y}(\delta) = 0 \quad (6\text{-}28)$$

$$u(0) = 0 \quad (6\text{-}29)$$

A fourth condition may be obtained by considering the conservation of momentum equation

$$\frac{\partial}{\partial x}(u^2) + \frac{\partial}{\partial y}(uv) = 2u\frac{\partial u}{\partial x} + u\frac{\partial v}{\partial y} + v\frac{\partial u}{\partial y} = \mu\frac{\partial^2 u}{\partial y^2} \quad (6\text{-}30)$$

Since the speeds vanish at the wall, the right-hand side of the equation also must vanish at the wall, yielding the fourth condition

$$\frac{\partial^2 u}{\partial y^2}(0) = 0 \quad (6\text{-}31)$$

Applying these four conditions to the polynomial yields four equations in four unknowns. The solution steps are left as a problem at the end of the chapter. The result is

$$\frac{u}{u_\infty} = \frac{3}{2}\frac{y}{\delta} - \frac{1}{2}\left(\frac{y}{\delta}\right)^3 \quad (6\text{-}32)$$

We now substitute this polynomial into the integral equation for the boundary layer, Eq. (6-24), and carry out the integrations. The result is

$$\frac{39}{280}u_\infty^2\frac{d\delta}{dx} = \frac{3}{2}\frac{u}{\rho}\frac{u_\infty}{\delta} \quad (6\text{-}33)$$

This is a first-order differential equation that may be integrated directly to yield

$$\delta^2 = \frac{280}{13} \frac{\mu x}{\rho u_\infty} \tag{6-34}$$

Note that the boundary layer thickness is zero at the leading edge of the plate and increases with distance along the plate.

It is convenient to recast the expression for δ into dimensionless form

$$\frac{\delta}{x} = \frac{4.64}{\sqrt{\mathrm{Re}_x}} \tag{6-35}$$

where Re_x is the Reynolds number

$$\mathrm{Re}_x = \frac{u_\infty x \rho}{\mu} = \frac{u_\infty x}{\nu} \tag{6-36}$$

having x as the length parameter.

As we noted earlier, the boundary layer as a sharply defined concept is artificial, but convenient. When the fluid equations are solved exactly, we find that the expression obtained corresponds to a condition where u/u_∞ is over 0.98. It is common to define the boundary layer thickness as the location at which u/u_∞ is 0.99. From precise solutions of the equations, it has been found that this definition implies that the coefficient 4.64 should be replaced by 4.92. Some authors recommend using 5.0.

Exact solutions for the velocity and temperature profiles have been obtained by Blasius (1968) and Pohlhausen (1921). The procedures are complicated and involve changes of variables and numerical solutions. At the introductory level, the text by Lienhard outlines how to make the variable changes and how to implement the numerical solutions. Advanced texts, such as that by Schlichting, tend to deal with these solutions in some depth.

We have emphasized an approximate approach leading to analytical solution rather than an exact approach leading to numerical solution. This is because the analytical solution provides insight as to the forms of dependence of heat transfer coefficients, friction coefficients, and boundary layer thicknesses on Reynolds and Prandtl numbers, and on the relationship between heat transfer and friction. The insight gained will influence the assumptions we make in dealing with turbulent flow.

It should be noted that our consideration of the fluid boundary layer has not been concerned with heat transfer; that is, we have assumed that the fluid behavior will determine heat transfer (as in subsequent sections) but that heat transfer will not perturb fluid behavior. In another situation that we consider in Chapter 8, natural convection, heat transfer will have a substantial influence on fluid motion.

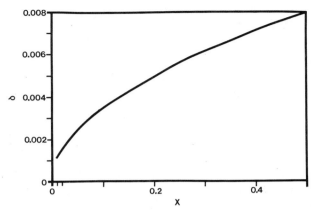

FIGURE 6-3-2.

EXAMPLE 6-1.   Air at atmospheric pressure and 300°K flows over a flat plate with a speed of 3 m/sec. What is the thickness of the boundary layer at distances of 0.01, 0.1, and 0.5 m from the leading edge?

SOLUTION.   We evaluate the boundary layer thickness by using the formula

$$\frac{\delta}{x} = \frac{4.92}{\sqrt{\text{Re}}}$$

that is, by using the more precise coefficient. From Appendix A, ν for air at 300°K is 1.57×10^{-5}. The results are obtained as follows:

x (m)	Re = ux/ν	$\sqrt{\text{Re}}$	δ (m)
0.01	1.91×10^3	43.7	1.13×10^{-3}
0.1	1.91×10^4	138	3.57×10^{-3}
0.5	9.55×10^4	309	7.96×10^{-3}

These results are plotted in Fig. 6-3-2. Thus, we see that boundary layer thickness is a small fraction of the distance from the leading edge of the plate.

EXAMPLE 6-1 (Spreadsheet).   This problem and several others in this chapter provide good examples of the use of spreadsheets for those who have access to personal computers because of repetitive evaluation of a single formula and because solutions to other problems will build on the solution to this problem. We proceed as follows:

	A	B
1	DISTANCE	.01
2	SPEED	3.0
3	NU	1.5 E − 5
4	REYNOLDS	+B1*B2 / B3
5	ROOT	+B4 ∧ .5
6	DELTA	4.92*B1 / B5

Once this is entered and the value of the boundary layer thickness is obtained from B6, change the value of B1 to .1. The other parameters adjust automatically and the new solution for boundary layer thickness appears in B6. Then change B1 to .5 for the last value of boundary layer thickness. The sequence can be saved for later use with other values.

If it is desired to record the sequence in this spreadsheet for each of the examples, then it is a simple matter to replicate the entries in A1–A6 twice into A7–A12 and A13–A18 and those in B1–B6 into B7–B12 and B13–B18. Then the entries for B7 and B13 would be changed to .1 and .5.

6-4. THERMAL BOUNDARY LAYER

Just as we postulated a transition zone for velocity, we may postulate a transition zone for temperature. However, we need not assume these zones to be the same size. We define the thickness of the thermal boundary layer to be δ_t, and we impose the conditions

$$T(\delta_t) = T_\infty \tag{6-37}$$

$$\frac{\partial T}{\partial y}(\delta_t) = 0 \tag{6-38}$$

Let us integrate the energy equation over y from 0 to δ_t:

$$\int_0^{\delta_t} \frac{\partial}{\partial x}(uT) + v(\delta_t)T(\delta_t) - v(0)T(0) = \frac{k}{\rho c}\left[\frac{\partial T}{\partial y}(\delta_t) - \frac{\partial T}{\partial y}(0)\right] \tag{6-39}$$

The known conditions on T and v permit reduction to

$$\int_0^{\delta_t} \frac{\partial}{\partial x}(uT) + v(\delta_t)T_\infty = -\frac{k}{\rho c}\frac{\partial T}{\partial y}(0) \tag{6-40}$$

The continuity equation may be integrated to yield

$$v(\delta_t) = -\int_0^{\delta_t} dy \frac{\partial u}{\partial x} \tag{6-41}$$

Substitution back into the energy equation and application of the Leibnitz rule yield the integral equation for the thermal boundary layer:

$$\frac{d}{dx}\int_0^{\delta_t} dy\, u(T - T_\infty) = -\frac{k}{\rho c}\frac{\partial T}{\partial y}(0) \tag{6-42}$$

Let us now assume a polynomial form for T that is consistent with boundary conditions. Two conditions already have been given for $y = \delta$. Conditions at the wall depend on the nature of the problem. Let us consider the case where the temperature at the wall is specified to be a constant T_w (a possible alternative is to specify heat flux as constant). In addition, we may evaluate the temperature equation at the wall and find that

$$\frac{\partial^2 T}{\partial y^2}(0) = 0 \tag{6-43}$$

For convenience, let us define a dimensionless variable

$$\phi = \frac{T - T_\infty}{T_w - T_\infty} \tag{6-44}$$

and expand in a polynomial

$$\phi = c_1 + c_2 \frac{y}{\delta_t} + c_3 \left(\frac{y}{\delta_t}\right)^2 + c_4 \left(\frac{y}{\delta_t}\right)^3 \tag{6-45}$$

Application of the four boundary conditions yields

$$\phi = 1 - \frac{3}{2}\frac{y}{\delta_t} + \frac{1}{2}\left(\frac{y}{\delta_t}\right)^3 \tag{6-46}$$

We now substitute the polynomials for ϕ and for u/u_∞ into the integral equation for the thermal boundary layer. We assume that δ_t is less than δ. If this were not the case, then the polynomial representation for u would not apply to the entire integral. We also assume that the ratio of boundary layer thickness $\zeta = \delta_t/\delta$ does not vary with x. Carrying out the resulting integrations yields

$$\frac{d}{dx}\left(\delta_t^2\right) = \frac{3k}{\rho c u_\infty \left(\frac{3}{20}\zeta - \frac{3}{280}\zeta^3\right)} \tag{6-47}$$

Noting that $k/\rho c = \alpha$, we integrate to obtain

$$\delta_t = \frac{\sqrt{\dfrac{3\alpha x}{u_\infty}}}{\sqrt{\frac{3}{20}\zeta - \frac{3}{280}\zeta^3}} \tag{6-48}$$

Let the left-hand side be divided by δ to yield ζ and the right-hand side be divided by its value $4.64x/\sqrt{\mathrm{Re}_x}$. After some algebra, we obtain

$$\zeta = \frac{1}{1.025}\mathrm{Pr}^{-1/3}\left(1 - \frac{\zeta^2}{14}\right)^{-1/3} \tag{6-49}$$

with the Prandtl number Pr given by

$$Pr = \frac{c\mu}{k} \qquad (6\text{-}50)$$

If, as assumed, ζ is less than 1, it is reasonable to neglect $\zeta^2/14$ and to obtain

$$\zeta = \frac{Pr^{-1/3}}{1.025} \qquad (6\text{-}51)$$

This result, for uniform properties, is consistent with the earlier assumption that ζ would be independent of position.

The assumption that ζ is constant is useful in the evaluation of integrals and in obtaining and solving Eq. (6-47). That the assumption turns out to be valid should not be surprising. As discussed in Section 6-2, if we refer back to Eqs. (6-15) and (6-17) for momentum and energy (neglecting the $\partial p/\partial x$ and $\partial^2 T/\partial x^2$ terms), we may note that these equations are of the same form. Indeed, if we replace T in Eq. (6-17) by u, and if we replace $k/\rho c$ in Eq. (6-17) by μ/ρ, we obtain Eq. (6-15). The ratio of the combinations μ/ρ and $k/\rho c$ is the Prandtl number which determines the ratio of boundary layer thicknesses. In other words, since the equations are of the same form, it is not surprising that the solutions are of the same form. We shall refer to this similarity again when we relate heat transfer and friction in Section 6-6.

In our discussion so far, we have assumed fluid properties to be uniform. Actually, properties will vary with the state of the fluid, particularly with respect to temperature. In dealing with problems over flat plates, it is advisable to evaluate properties at a "film temperature," which is defined as the average of the wall and asymptotic temperatures. Thus, if air at 400°K is sent over a plate that is maintained at 300°K, then the properties of the air should be evaluated at 350°K.

EXAMPLE 6-2.   For air of 400°K traveling over a plate at 300°K, what is the ratio of thermal and velocity boundary layers?

SOLUTION.   At 350°K, we see from Appendix A that Pr = 0.706. Then we evaluate

$$\zeta = \frac{Pr^{-1/3}}{1.025} = 1.09$$

Thus, the thermal boundary layer in this case is actually greater than the fluid boundary. This point will be discussed below.

In the derivation, it was assumed that δ_t is less than δ. From the expression obtained for $\zeta = \delta_t/\delta$, we see that this assumption implies that $Pr^{1/3}$ should be greater than 0.975. This condition tends to be satisfied for most liquids (except for liquid metals which we consider separately), as may be

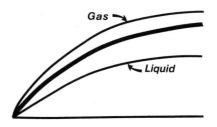

FIGURE 6-4-1. Thermal boundary layer (thin curves) can be larger than velocity boundary layer (thick curve) for gas and smaller for liquid.

seen from the data in Appendix A and the illustration of Fig. 6-4-1. For many gases, as Appendix A also shows, the Prandtl number is in the vicinity of 0.7, which corresponds to $Pr^{1/3}$ in the vicinity of 0.9.

While the assumption is violated for gases, the degree of violation is not serious for most purposes. In principle, the polynomial velocity profile should be used for the first 90% of the thermal boundary layer in the expression

$$\frac{d}{dx} \int_0^{\delta_t} u(T - T_\infty)\, dy$$

In the final 10%, u_∞ should be used in the integrand. The assumption does not create a basic difficulty because its violation occurs only in 10% of the y interval. The integrand is very small in this interval, since T is close to T_∞ near the end of the thermal boundary layer.

6-5. HEAT TRANSFER COEFFICIENT

The heat transfer coefficient is defined in the relationship

$$q = hA(T_w - T_\infty) \tag{6-52}$$

and is also obtained by the condition

$$q = -kA \frac{\partial T}{\partial y}(0) = -kA(T_w - T_\infty)\frac{\partial \phi}{\partial y}(0) \tag{6-53}$$

Combining Eqs. (6-52) and (6-53) and obtaining the derivative of Eq. (6-46) yield

$$h = \frac{q}{A(T_w - T_\infty)} = \frac{3}{2}\frac{k}{\delta_t} = \frac{3}{2}\frac{k}{\zeta\delta} \tag{6-54}$$

Substituting for ζ and δ yields

$$h = \frac{3}{2}k\frac{\sqrt{\mathrm{Re}_x}}{4.64x}1.025\,\mathrm{Pr}^{1/3} = \frac{0.33k}{x}\mathrm{Re}^{1/2}\mathrm{Pr}^{1/3} \qquad (6\text{-}55)$$

This provides us with an expression for the convection heat transfer coefficient.

The heat transfer coefficient may be placed in a dimensionless grouping called the Nusselt number

$$\mathrm{Nu}_x = \frac{hx}{k} = 0.33\,\mathrm{Re}^{1/2}\mathrm{Pr}^{1/3} \qquad (6\text{-}56)$$

A more precise value for the coefficient is 0.332. Note that the Nusselt number is the same type of grouping as the Biot number. However, in the Nusselt number, the thermal conductivity used is that of the fluid, while in the Biot number the conductivity was that of the solid body. The Nusselt number is a measure of relative importance of conduction (via conductivity over length) as a component of the overall heat transfer by convection (via the convection coefficient). Usually, in convection, fluid motion is more important than conduction and Nusselt numbers are typically larger than 1.

The Nusselt number for a Prandtl number of unity is plotted against the logarithm of the Reynolds number in Fig. 6-5-1. The Nusselt number varies from about 10 when the Reynolds number is 1000 (log Re = 3) to over 200 by the time the Reynolds number gets to 5×10^5. In contrast with the order of magnitude variation with Reynolds number, the variation with Prandtl num-

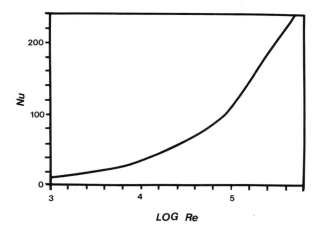

FIGURE 6-5-1. The Nusselt number for laminar flow over a flat plane when the Prandtl number is One.

ber is more modest, as may be seen from Fig. 6-5-2, which includes a range of Prandtl numbers characteristic of common gases and liquids. (More extreme low and high Prandtl numbers are encountered for liquid metals and some organic fluids.)

A more appropriate length dimension to characterize this physical interpretation for the flat plate would be a dimension normal to the plate, like the boundary layer thickness. The Nusselt number defined in Eq. (6-56) could be expressed as $(h\delta/k)(x/\delta)$ with the first term providing the physical interpretation we assigned. In other situations, for example, flow in pipes where the length dimension is the diameter (a dimension normal to the wall), the interpretation applies directly without applying an additional ratio of lengths.

We may observe that since $\text{Re}_x^{1/2}$ varies as the square root of x, the heat transfer coefficient varies inversely with $\sqrt{x}$; that is,

$$h = \frac{A}{\sqrt{x}} \qquad (6\text{-}57)$$

where A is a constant. We therefore conclude that the heat transfer coefficient will be much larger near the leading edge of the plate than farther along the plate.

It is reasonable to expect the heat transfer coefficient to decline with distance along the plate. Since the boundary layer gets thicker, the temperature derivative declines and the heat flux declines. At the leading edge where the boundary layer thickness goes to zero, a finite change in temperature (from T_w to T_∞) must take place in a zero distance, so the heat transfer coefficient must be infinite. Again, as with some solutions obtained with time-dependent heat conduction, idealizations in modeling have led to the appearance of an infinite parameter prevailing over an infinitesimal distance, although with a finite integral (i.e., the heat transferred over any length of plate is finite).

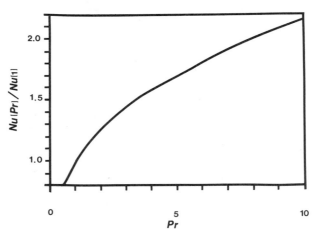

FIGURE 6-5-2.

Frequently, we are interested in the average heat transfer coefficient for an entire surface. For a flat plate, we define the average according to

$$\bar{h} = \frac{1}{L} \int_0^L h(x)\, dx \qquad (6\text{-}58)$$

With h varying as $A/\sqrt{x}$, we obtain

$$\bar{h} - \frac{1}{L} 2A\sqrt{x}\Big|_0^L = \frac{2A}{\sqrt{L}} = 2h(L) \qquad (6\text{-}59)$$

Thus, we may evaluate the heat transfer coefficient at the end of the plate and double the value to get the average.

Although we have assumed that properties are constant in deriving our equations, in reality properties will vary. As noted earlier, it is generally reasonable to evaluate fluid properties at the "film temperature," which is defined by

$$T_f = \tfrac{1}{2}(T_w + T_\infty) \qquad (6\text{-}60)$$

This procedure provides a mechanism for estimating an effective average temperature.

EXAMPLE 6-3.   Air at atmospheric pressure and 400°K flows over a plate at 300°K at 3 m/sec. Evaluate the heat transfer coefficient at points 0.01, 0.1, and 0.5 m from the leading edge.

SOLUTION.   We evaluate the Nusselt number from

$$\mathrm{Nu} = 0.332\,\mathrm{Re}^{1/2}\,\mathrm{Pr}^{1/3}$$

and the heat transfer coefficient from the Nusselt number by

$$h = \frac{k}{x}\,\mathrm{Nu}$$

We evaluate properties at the film temperature $\tfrac{1}{2}(300°\mathrm{K} + 400°\mathrm{K}) = 350°\mathrm{K}$. The properties we need are

$$\nu = 2.06 \times 10^{-5} \quad \text{(for the Reynolds number)}$$

$$\mathrm{Pr} = 0.706$$

$$k = 0.0297 \quad \text{(for the heat transfer coefficient)}$$

We may construct the following table:

x	$\mathrm{Re} = \dfrac{ux}{\nu}$	$\mathrm{Re}^{1/2}$	$0.332\,\mathrm{Pr}^{1/3}$	Nu	h
0.01	1.46×10^3	38.2	0.296	11.3	33.5
0.1	1.46×10^4	121	0.296	35.7	10.6
0.5	7.30×10^4	270	0.296	80.0	4.8

These results are plotted in Fig. 6-5-3.

EXAMPLE 6-3 (Spreadsheet).   By reference back to the solution of Example 6-1, we can see how to build on the earlier solution. In that problem, we already evaluated the square root of the Reynolds number. We thus enter the "old" spreadsheet and change the value of the viscosity since we are now dealing with a different temperature. We proceed as follows:

	A	B
1	DISTANCE	.01
2	SPEED	3.0
3	NU	2.06 E − 5
4	REYNOLDS	+B1*B2 / B3
5	ROOT	+B4 ∧ .5
6	DELTA	4.92*B1 / B5
7	PRANDTL	.706
8	CUBE ROOT	+B7 ∧ .3333
9	NUSSELT	.332*B5*B8
10	CONDUCTIVITY	.0297
11	H	+B9*B10 / B1

To obtain the heat transfer coefficients for the other cases, we enter 0.1 and 0.5 in B1 and obtain the results from B11. Note that we do not need DELTA, the boundary

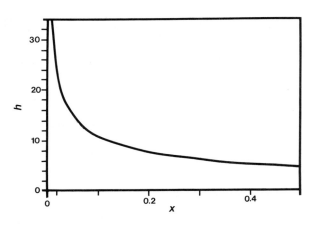

FIGURE 6-5-3.

layer thickness, and thus would not have included it had we been starting this spreadsheet from scratch.

6-6. RELATION TO FRICTION

In fluid mechanics, a friction coefficient is defined relating shear stress at the wall to asymptotic kinetic energy density

$$C_f = \frac{\tau_w}{\frac{1}{2}\rho u_\infty^2} \tag{6-61}$$

Using the polynomial obtained for velocity profile in the boundary layer, the stress at the wall is

$$\tau_w = \mu \frac{\partial u}{\partial y}(0) = \frac{3}{2}\frac{\mu u_\infty}{\delta} \tag{6-62}$$

Using the expression obtained for boundary layer thickness, we obtain

$$C_f = \frac{0.647}{\sqrt{\mathrm{Re}_x}} \tag{6-63}$$

A more precise solution for the derivative of the velocity profile yields

$$C_f = \frac{0.664}{\sqrt{\mathrm{Re}_x}} \tag{6-64}$$

It may be observed that convective heat transfer and fluid friction both depend on the square root of the Reynolds number. This has design implications for the engineer. Measures to improve heat transfer by increasing flow rate, for example, lead to greater friction that must be overcome by pumping.

The relationship between heat transfer and friction may be expressed via regrouping of dimensionless groups. Let us define the Stanton number as

$$\mathrm{St} = \frac{\mathrm{Nu}}{\mathrm{Re}\,\mathrm{Pr}} = \frac{h}{\rho c u_\infty} \tag{6-65}$$

which for flow over a flat plate yields

$$\mathrm{St}\,\mathrm{Pr}^{2/3} = 0.332\,\mathrm{Re}^{-1/2} = \tfrac{1}{2}C_f \tag{6-66}$$

The Stanton number is a measure of the total heat transfer by convection (the h in the numerator) to the rate at which heat is carried away (the $\rho c u_\infty$ in the denominator). This interpretation of the Stanton number becomes more evident if the numerator and denominator are multiplied by temperatures, so $\rho c T$ appears as heat content and u_∞ as a speed at which this heat is carried. Relation (6-66) implies a possibility of inferring information about heat transfer from knowledge about friction and is referred to as the Reynolds–Colburn analogy.

We make considerable use of the relationship of heat transfer and friction, particularly in situations involving turbulent flow. Such relationships are important either because it may be easier to generate empirical information about friction than about heat transfer or because such information may be available from fluid mechanics without heat transfer.

It is reasonable to expect the two phenomena to be related once we recognize that heat transfer depends on viscosity. Viscosity is fluid friction, so a friction-related pressure drop should depend on viscosity. Two phenomena dependent on the same quantity should be related to one another. That the phenomena should be governed by formulas very similar in form is not surprising. As already noted in Sections 6-2 and 6-4, the energy and momentum equations are very similar in form; that is, the momentum equation form is obtained if T in the energy equation is replaced by u. In addition, the functional dependence of heat conduction (depending on $\partial T/\partial y$) is similar to that of viscous stress (depending on $\partial u/\partial y$). The quantity $St\,Pr^{2/3}$ sometimes is referred to as the Colburn j factor.

6-7. LIQUID METALS

For liquid metals, the thermal conductivity is very large and the Prandtl number is very small. Thus, if the Prandtl number is a guide to the ratio of fluid and thermal boundary layer, we would expect thermal boundary layer to be much larger than fluid boundary layer. Therefore, we expect the assumption, used in Section 6-4, that the thermal boundary layer is the smaller of the two to lead to significant contradictions.

As indicated in Section 6-4, we could apply the polynomial for the fluid boundary layer up to $y = \delta$ and then use $u = u_\infty$. However, since we expect the fluid boundary layer to be very small compared to the thermal boundary layer, we simply assume that $u = u_\infty$ for the purpose of integrating over the thermal boundary layer. In this respect, the case of liquid metal heat transfer is simpler than the other cases considered thus far.

The integral equation for the thermal boundary layer becomes, with $u = u_\infty$ and with the same polynomial for temperature profile,

$$\frac{d}{dx} \int_0^{\delta_t} u_\infty \left[1 - \frac{3}{2}\frac{y}{\delta_t} + \frac{1}{2}\left(\frac{y}{\delta_t}\right)^3 \right] dy = \frac{3\alpha}{2\delta_t} \tag{6-67}$$

Performing the integrations leads to

$$\frac{3}{8} u_\infty \frac{d\delta_t}{dx} = \frac{3}{2}\frac{\alpha}{\delta_t} \tag{6-68}$$

which may be integrated to yield

$$\delta_t = \sqrt{\frac{8\alpha x}{u_\infty}} \qquad (6\text{-}69)$$

As before, we obtain the heat transfer coefficient by

$$h = \frac{q}{A(T_w - T_\infty)} = \frac{3}{2}\frac{k}{\delta_t} \qquad (6\text{-}70)$$

Again, h varies inversely with the square root of the distance along the plate.

As usual, we may rearrange parameters in terms of dimensionless groups. For this case, we obtain

$$\text{Nu} = \frac{hx}{k} = 0.53\sqrt{\text{Pe}_x} \qquad (6\text{-}71)$$

where the Peclet number Pe_x is defined by

$$\text{Pe}_x = \frac{u_\infty x}{\alpha} \qquad (6\text{-}72)$$

A more precise analysis yields a coefficient of 0.564 instead of 0.53. We may observe that the Peclet number is the product of the Reynolds and Prandtl numbers.

We also may observe that the form of dependence of Nusselt number on Reynolds number (i.e., varying as the square root) is the same for fluids in general. For liquid metals, however, variation with Prandtl number goes as the square root rather than as the cube root. Churchill and Zoe (1973) introduced the correlation

$$\text{Nu}_x = \frac{0.3387\,\text{Re}_x^{1/2}\text{Pr}^{1/3}}{\left[1 + (0.0468/\text{Pr})^{2/3}\right]^{1/4}} \qquad \text{Pe}_x > 100 \qquad (6\text{-}73)$$

to apply over the full range of Prandtl numbers. It is left as a problem at the end of the chapter to show that Eq. (6-73) yields the proper limit as the Prandtl number goes toward zero. At very high Prandtl numbers, corresponding to very viscous liquids, this formula leads to a limiting coefficient of 0.3387, again consistent with exact solution of fluid mechanics equations.

EXAMPLE 6-4. Evaluate the heat transfer coefficient 0.5 cm from the leading edge by the flow of liquid sodium at 400°K flowing at a speed of 2 m/sec over a plate that has a uniform temperature of 600°K.

SOLUTION. The heat transfer coefficient is given by

$$h = \frac{k}{x} \text{Nu}$$

$$\text{Nu} = 0.564\sqrt{\text{Pe}}$$

The Peclet number is evaluated at the film temperature of 400°K by

$$\text{Pe} = \frac{u_\infty x}{\alpha} = \frac{(2)(0.05)}{6.8 \times 10^{-5}} = 1.47 \times 10^3$$

The Nusselt number is then 21.6 and

$$h = \frac{k}{x}\text{Nu} = \frac{48}{0.05}(21.6) = 20{,}736$$

6-8. TURBULENCE

Let us consider the ratio of kinetic energy density and viscous force density. The kinetic energy (KE) may be expressed as

$$\text{KE} = \tfrac{1}{2}\rho u_\infty^2 \qquad\qquad (6\text{-}74)$$

The viscous force per unit area is

$$\text{VF} = \mu \frac{\partial u}{\partial y}(0) = \frac{3\mu u_\infty}{2\delta} \qquad\qquad (6\text{-}75)$$

The ratio of these quantities is

$$\frac{\text{KE}}{\text{VF}} = \frac{\rho u_\infty \delta}{3\mu} = \frac{1}{3}\frac{\rho u_\infty x}{u}\frac{\delta}{x} = \frac{1}{3}\text{Re}_x \frac{\delta}{x} \qquad (6\text{-}76)$$

The boundary layer thickness depends on Reynolds number, so the ratio may be expressed as

$$\frac{\text{KE}}{\text{VF}} = \frac{4.92}{3}\sqrt{\text{Re}_x} \qquad\qquad (6\text{-}77)$$

Flow tends to be layered in a well-defined velocity distribution (laminar flow occurs) when viscous forces are large enough to restrain the kinetic energy of the fluid. As the ratio of kinetic energy to viscous force increases, we approach a situation where kinetic energy no longer can be channeled into neat layers by viscosity, and we obtain turbulent flow. For flow over a flat plate, we observe from the above equations that the Reynolds number characterizes this ratio. Empirically, we find that a transition from laminar to turbulent flow occurs when the Reynolds number approaches

$$\text{Re}_x > 5 \times 10^5 \qquad (6\text{-}78)$$

The precise value for transition will depend on conditions specific to the situation, for example, roughness of plate surface.

When we speak of a transition to turbulence of the boundary layer, we are making a simplification in terminology. Even when there is turbulence in the boundary layer, flow at the surface of the plate goes to zero. We therefore obtain a laminar sublayer in the immediate vicinity of the plate. We then would encounter a buffer layer between the laminar sublayer and the fully turbulent portion of the boundary layer. The reader interested in further study of the fluid mechanics of the components of a turbulent boundary layer should consult the references at the end of this chapter.

Since the Reynolds number increases with distance along the plate, eventually, if the plate is long enough, a transition to turbulence will take place as illustrated in Fig. 6-8-1. We thus must be concerned with evaluating heat transfer in the turbulent region.

Turbulence involves substantial fluctuations about average behavior, associated with motion of macroscopic amounts of fluid. Thus, we may expect to find an eddy of fluid cross a "layer," as in Fig. 6-8-2. Let us represent speeds by average and fluctuating quantities:

$$u = \bar{u} + u' \qquad (6\text{-}79)$$

$$v = \bar{v} + v' \qquad (6\text{-}80)$$

The average values of u' and v' are zero. The x momentum being transferred in the y direction at any time is $\rho u'v'$. On the average, we obtain $\overline{\rho u'v'}$. Note

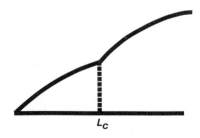

FIGURE 6-8-1. Boundary layer with transition to turbulence at L_c.

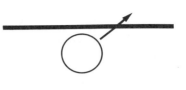

FIGURE 6-8-2. Macroscopic amount of fluid crosses "layer" in turbulent flow. (Average motion is to the right.)

that while the average values of u' and v' are zero, the average value of their product need not be zero.

Shear stress is the rate of transfer of x-momentum density in the y direction (force is the rate of change of momentum), so we may identify the momentum from this fluctuating motion as a turbulent shear stress

$$\tau_t = -\rho\overline{u'v'} \tag{6-81}$$

The negative sign may be explained as follows. When the fluctuating fluid moves in the positive y direction ($v' > 0$), it is moving to where fluid speed ($\bar{u}$) is greater, so this transfer of momentum provides a retarding force.

To place relationships in the same form as used in laminar flow, let us express stress as

$$\tau = \rho(\nu + \varepsilon_m)\frac{\partial u}{\partial y} \tag{6-82}$$

where ν is the kinematic viscosity and ε_m is defined as the eddy diffusivity (or eddy viscosity). This means that

$$\rho\varepsilon_m\frac{\partial u}{\partial y} = \rho\overline{u'v'} \tag{6-83}$$

Similarly, we may have heat carried in the eddy as it crosses a layer. Let us make an analogous definition for heat flow:

$$q = -\rho c(\alpha + \varepsilon_H)\frac{\partial T}{\partial y} \tag{6-84}$$

where ε_H is the thermal eddy diffusivity. In turbulent flow, we expect to have

$$\varepsilon_m \gg \nu \qquad \text{(6-85)}$$

$$\varepsilon_H \gg \alpha \qquad \text{(6-86)}$$

although in laminar flow, we expect ε_m and ε_H to go to zero.

While it is difficult to evaluate ε_m and ε_H explicitly, we may assume from our discussion that in turbulent flow, as in laminar flow, the processes of heat transfer and friction occur by related mechanisms. The relationship between these two phenomena in laminar flow is expressed in the Prandtl number

$$\text{Pr} = \frac{\mu/\rho}{k/\rho c} = \frac{\nu}{\alpha} \qquad \text{(6-87)}$$

We assume that for turbulent flow, the relationship between these mechanisms is still equal to the Prandtl number:

$$\text{Pr} = \frac{\nu}{\alpha} = \frac{\nu + \varepsilon_m}{\alpha + \varepsilon_H} \qquad \text{(6-88)}$$

Accordingly, we assume that the relationship developed earlier between heat transfer and friction,

$$\text{St Pr}^{2/3} = \frac{C_f}{2} \qquad \text{(6-89)}$$

applies; that is, the Reynolds–Colburn analogy applies for turbulent flow.

On this basis, we can obtain correlations for friction coefficients in turbulent flow and use these to gain information about heat transfer. For example, Schlichting (1979) recommends

$$C_f = 0.0592\,\text{Re}_x^{-1/5} \qquad 5 \times 10^5 < \text{Re}_x < 10^7 \qquad \text{(6-90)}$$

Thus, we may infer that

$$\text{St Pr}^{2/3} = 0.0296\,\text{Re}_x^{-1/5} \qquad 5 \times 10^5 < \text{Re}_x < 10^7 \qquad \text{(6-91)}$$

Similarly, we may use the friction formula of Schultz-Grunow (1940)

$$C_f = 0.370\left[\log(\text{Re}_x)\right]^{-2.584} \qquad 10^7 < \text{Re}_x < 10^9 \qquad \text{(6-92)}$$

for obtaining heat transfer coefficients at higher Reynolds numbers via

$$\text{St Pr}^{2/3} = 0.185\left[\log(\text{Re}_x)\right]^{-2.584} \qquad 10^7 < \text{Re}_x < 10^9 \qquad \text{(6-93)}$$

If we wish to obtain the average heat transfer coefficient for the plate, we may define an average Stanton number by

$$\overline{\text{St}}\,\text{Pr}^{2/3} = \frac{1}{L}\int_0^L \tfrac{1}{2}C_f(x)\,dx \qquad (6\text{-}94)$$

Let us assume that Re_L is between 5×10^5 and 10^7. We would then break the integral into two parts for the laminar (recall that the leading portion of the plate always will be characterized by laminar flow) and turbulent portions. Then

$$\overline{\text{St}}\,\text{Pr}^{2/3} = \frac{1}{L}\left[\int_0^{L_c} 0.332\,\text{Re}_x^{-1/2}\,dx + \int_{L_c}^L 0.0296\,\text{Re}_x^{-1/5}\,dx\right] \qquad (6\text{-}95)$$

This may be integrated to

$$\overline{\text{St}}\,\text{Pr}^{2/3} = \frac{L_c}{L}\frac{0.664}{\sqrt{\text{Re}_{L_c}}} + \frac{5}{4}(0.0296)\,\text{Re}_L^{-1/5} - \frac{5}{4}(0.0296)\frac{L_c}{L}\text{Re}_{L_c}^{-1/5} \quad (6\text{-}96)$$

We note that Re_{L_c} is 5×10^5 and that

$$\frac{L_c}{L} = \frac{\text{Re}_{L_c}}{\text{Re}_L} = \frac{5 \times 10^5}{\text{Re}_L} \qquad (6\text{-}97)$$

The result is

$$\overline{\text{St}}\,\text{Pr}^{2/3} = 0.037\,\text{Re}_L^{-1/5} - 871\,\text{Re}_L^{-1} \qquad (6\text{-}98)$$

It is significant that averaging over a plate involves averaging over both laminar and turbulent portions. We are concerned again with the laminar and turbulent portions when dealing with high-speed flow.

The average Nusselt number may be obtained from the average Stanton number by

$$\overline{\text{Nu}}_L = \overline{\text{St}}\,\text{Re}_L\text{Pr} = \text{Pr}^{1/3}\left(0.037\,\text{Re}_L^{0.8} - 871\right) \qquad (6\text{-}99)$$

We shall encounter $\text{Re}^{0.8}$ dependence again when we deal with forced convection in pipes. Note that if the transition Reynolds number were to differ from 5×10^5, then the appropriate alternate value should be used in evaluating the average.

The average Nusselt number for plates with turbulent flow is plotted in Fig. 6-8-3. Values for turbulent flow are substantially higher than values for laminar flow (see the local Nusselt numbers plotted in Fig. 6-5-1).

Note also that we have assumed that there is a clear and sharp break between laminar and turbulent flow which occurs at location L_c. This implies a sudden change in Nusselt number by a factor of 4.6. In effect, we are assuming that the transition occurs over a relatively short distance.

In this section, we have encountered two factors that influence how we deal with heat transfer problems involving convection. One is the fact that flow may be either laminar or turbulent. The other is that we may have to use correlated empirical data, because the problem configuration is too complicated to yield a solution directly from first principles.

As a result of these factors, we must ask two questions in approaching problems:

1. What type of problem is this?
2. What correlation is appropriate to the problem?

To answer the first question, we may evaluate the Reynolds number. It is logical to evaluate the Reynolds number for the trailing edge of the plate. If that indicates laminar flow, then the entire plate sees laminar flow. If that indicates turbulent flow, we know there is a transition point corresponding to the critical Reynolds number. To answer the second question, we consult the available correlations and select one appropriate to the conditions of the problem. Recall that different correlations apply for turbulent flow, depending on the Reynolds number.

As we proceed with our study of convection, we shall encounter additional factors that determine the nature of the problem. High-speed flow considerations may apply. For flow in pipes (Chapter 7), we shall be concerned with whether the pipe is long enough to be characterized by fully developed flow. Improper answers to the two basic questions account for a large fraction of the errors this author has observed in dealing with convection heat transfer.

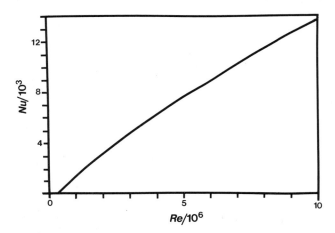

FIGURE 6-8-3. Average Nusselt number for turbulent flow over a flat plate (Prandtl number in One).

Since there is a need to deal with empirical correlations, we shall find, as we proceed, that more than one correlation may apply to a given situation. The engineer in a specific situation must make a specific choice. There is not necessarily an obvious "right" or "wrong" choice. Sometimes the alternate formulas will be of equivalent accuracy, but of slightly different form. On the other hand, some correlations may attempt to represent effects neglected or simplified in other correlations. Such effects could include Prandtl number dependence and temperature variation of properties (particularly viscosity). The degrees to which the engineer understands the physical phenomena associated with a situation and the representation of such phenomena in particular correlations provide the basis for good judgment in selecting correlations to use.

The requirement to use correlations introduces uncertainty into predictions. Correlations that do not account for all physical phenomena introduce errors beyond just the error associated with experimental measurement. A correlation is a force-fit of a formula to a situation. It tends to average over phenomena not represented explicitly. Specific cases are not averages and thus can deviate considerably from the correlations. Correlations of simple form covering a broad range of situations may have substantial uncertainty, say $\pm 25\%$. Correlations of elaborate form can have lower uncertainty. Similarly, correlations applying to limited ranges of conditions can have lower uncertainty. The issue of uncertainty is emphasized here because students frequently come to believe that formulas in courses, by definition, are exact. This may not be the case, even when formulas are obtained by theoretical derivation because the real physical situation may not correspond exactly to the idealized conditions used to derive the formula. It is especially not the case when correlations are involved.

EXAMPLE 6-5.   Air at atmospheric pressure and 400°K flows at a speed of 3 m/sec over a 10 m long plate that is at 300°K. Evaluate the average heat transfer coefficient for the plate.

SOLUTION.   Let us first determine what type of problem we have by evaluating the Reynolds number at the trailing edge. Evaluating properties at the film temperature of 350°K, we have

$$\mathrm{Re}_L = \frac{uL}{\nu} = \frac{(3)(10)}{2.06 \times 10^{-5}} = 1.46 \times 10^6$$

This is greater than 5×10^5 so turbulent flow exists. The average Nusselt number, from which we obtain the average heat transfer coefficient, is

$$\overline{\mathrm{Nu}}_L = \mathrm{Pr}^{1/3}\left(0.037\,\mathrm{Re}_L^{0.8} - 871\right)$$

$$= (0.706)^{1/3}\left[0.037(1.46 \times 10^6)^{0.8} - 871\right] = 2038$$

Then the average heat transfer coefficient is given by

$$\bar{h} = \frac{k}{L}\overline{\mathrm{Nu}}_L = \frac{0.0297}{10}(2038) = 6.05$$

EXAMPLE 6-5 (Spreadsheet).   This problem illustrates how choices are made in a spreadsheet. In this problem, we have to choose whether flow is laminar or turbulent and evaluate the heat transfer coefficients. Consider the following spreadsheet:

	A	B
1	LENGTH	10.0
2	SPEED	3.0
3	NU	2.06 E – 5
4	REYNOLDS	+B1*B2 / B3
5	PRANDTL	.706
6	PR ROOT	+B5 ∧ .3333
7	LAMINAR	OPTION
8	RE POWER	+B4 ∧ .5
9	NUSSELT	.664*B5*B8
10	TURBULENT	OPTION
11	RE POWER	+B4 ∧ .8
12	NUSSELT	+B6*(.037*B11 – 871)
13	STANDARD	B4 > 5E5
14	TEST	@IF (B13, 1,2)
15	CHOICE	@CHOOSE (B14, B12, B9)
16	CONDUCTIVITY	.0297
17	H	+B15*B16 / B1

This spreadsheet starts out similarly to the one for Example 6-3. Provision is made to calculate the Nusselt number for both laminar (B9) and turbulent (B12) situations. Line 13 is the standard that determines whether turbulence exists. Entry B13 is an assertion that the standard is satisfied, which yields a TRUE or FALSE response. Entry B14 checks on this response, making B14 either 1 or 2 depending on whether B13 is TRUE or FALSE. Entry B15 chooses the turbulent or laminar Nusselt number depending on whether B14 is 1 or 2. From the Nusselt number, the heat transfer coefficient then is evaluated. This spreadsheet can be made an ingredient of more elaborate spreadsheets involving heat transfer from flat plates.

For those not using spreadsheets and trying to construct standard programs following the pattern set in Appendix D with Examples 2-1 and 2-2, some variation in procedure comes about in dealing with certain functions such as IF and CHOOSE. Appendix D provides demonstration programs for converting the above spreadsheet.

6-9. HIGH-SPEED FLOW

When dealing with high-speed flow, as in aircraft problems, some effects not considered important in our discussions in previous sections become important. Two effects in particular are of concern—the increase in temperature resulting from slowing down of a high-speed fluid and the heating effect of viscosity.

When high-speed flowing fluid is brought reversibly to rest, the kinetic energy is converted to internal energy. The temperature T_0 of the stagnant fluid is related to the temperature T_∞ of the free stream by the energy balance

$$\rho c_p (T_0 - T_\infty) = \frac{1}{2g_c} \rho u_\infty^2 \qquad (6\text{-}100)$$

The constant g_c is unity in SI units but has the value 32.16 ft · lbm/lbf · sec^2 in English units. We therefore observe that if fluid is brought to rest at a flat plate, the temperature of the fluid at the wall will be higher than the free-stream temperature. Whether heat will be transferred from the plate to the fluid will depend on whether the temperature of the wall of the plate T_w exceeds T_0, not whether it exceeds T_∞.

Actually, the fluid is not brought to rest reversibly. Viscosity leads to irreversibility, and the actual temperature of the fluid at the wall, the adiabatic wall temperature T_{aw}, is given by

$$r = \frac{T_{aw} - T_\infty}{T_0 - T_\infty} \qquad (6\text{-}101)$$

where r is called the recovery factor. For gases with Prandtl numbers near 1,

$$r = \begin{cases} \mathrm{Pr}^{1/2} & \text{laminar flow} \\ \mathrm{Pr}^{1/3} & \text{turbulent flow} \end{cases} \qquad (6\text{-}102)$$

When dealing with high-speed flow, speeds often are expressed in terms of the Mach number

$$\mathrm{M}_\infty = \frac{u_\infty}{u_s} \qquad (6\text{-}103)$$

where u_s is the speed of sound

$$u_s = \sqrt{\gamma g_c R T_\infty} \qquad (6\text{-}104)$$

Noting that

$$\gamma = \frac{C_p}{C_v} \qquad (6\text{-}105)$$

the ratio of specific heats at constant pressure and volume, and that

$$R = C_p - C_v \qquad (6\text{-}106)$$

it follows that

$$T_0 = T_\infty\left(1 + \frac{\gamma - 1}{2}M_\infty^2\right) \quad (6\text{-}107)$$

Since the transfer of heat depends on the adiabatic wall temperature, we may express convection by

$$q = hA(T_w - T_{aw}) \quad (6\text{-}108)$$

With this relation, it is possible to use correlations already obtained for flat plates. However, there is some uncertainty regarding the temperature at which to evaluate properties of the fluid to calculate dimensionless groups, since the temperature differences can be large. Eckert (1956) has recommended that properties be evaluated at a reference temperature

$$T^* = \tfrac{1}{2}(T_w + T_\infty) + 0.22(T_{aw} - T_\infty) \quad (6\text{-}109)$$

At low speeds, the adiabatic wall temperature reduces to T_∞ and T^* becomes the film temperature.

We now may evaluate heat transfer coefficients via

$$\mathrm{St}_x^*\mathrm{Pr}^{*2/3} = 0.332\,\mathrm{Re}_x^{*-1/2} \qquad \mathrm{Re}_x^* < 5 \times 10^5 \qquad (6\text{-}110)$$

$$\mathrm{St}_x^*\mathrm{Pr}^{*2/3} = 0.0296\,\mathrm{Re}_x^{*-1/5} \qquad 5 \times 10^5 < \mathrm{Re}_x^* < 10^7 \qquad (6\text{-}111)$$

$$\mathrm{St}_x^*\mathrm{Pr}^{*2/3} = 0.185(\log\mathrm{Re}_x^*)^{-2.584} \qquad 10^7 < \mathrm{Re}_x^* < 10^9 \qquad (6\text{-}112)$$

In principle, the process for evaluating heat transfer coefficients in high-speed flow is iterative. The iteration is usually rapidly convergent but can be tedious for hand calculations. The procedure can be programmed easily on a personal computer.

To evaluate properties at the reference temperature T^*, we first must obtain the adiabatic wall temperature T_{aw}. To obtain T_{aw}, we must evaluate the Prandtl number. However, the Prandtl number should be evaluated at T^*, which is not known yet. We thus follow the sequence:

1. Estimate Pr^*, r.
2. Evaluate T^*.
3. Reevaluate Pr^*, r.
4. Reevaluate T^*.
5. Check to see whether the T^* of Step 4 is within a specified percentage of the last evaluation.
6. (a) If so, select the latest value and proceed. (b) If not, return to Step 3.

Because the recovery factor is a root of the Prandtl number, a reasonable initial guess of the Prandtl number can lead to a good estimate of the recovery factor and rapid convergence. For modest requirements on accuracy, the first estimate may be sufficient.

For total heat transferred from a plate of length L and width w in high-speed flow involving laminar flow near the leading edge and turbulent flow further downstream, we should evaluate

$$q = \int_0^{L_c} dx\, hW(T_w - T_{aw}) + \int_{L_c}^{L} dx\, hW(T_w - T_{aw}) \qquad (6\text{-}113)$$

where L_c is the location at which the Reynolds number reaches the transition value of 5×10^5.

In the laminar region, we again find that

$$\int_0^{L_c} dx\, h(x) = 2L_c h^*(L_c) \qquad (6\text{-}114)$$

Evaluating $h^*(L_c)$ from the laminar flow equation

$$\int_0^{L_c} dx\, h(x) = 0.664 \rho u_\infty C_p \Pr^{-2/3} \mathrm{Re}_{L_c}^{-1/2} \qquad (6\text{-}115)$$

In the turbulent region, for $\mathrm{Re}_L^* < 10^7$, we obtain

$$\int_{L_c}^{L} dx\, h(x) = 0.0296 \rho u C_p \Pr^{-2/3} \left(\frac{u_\infty \rho}{\mu} \right)^{-115} \int_{L_c}^{L} \frac{dx}{x^{1/5}} \qquad (6\text{-}116)$$

which integrates to

$$\int_{L_c}^{L} dx\, h(x) = 0.037 \rho u C_p \Pr^{-2/3} \left[L\,\mathrm{Re}_L^{-1/5} - L_c\,\mathrm{Re}_{L_c}^{-1/5} \right] \qquad (6\text{-}117)$$

We must exercise caution in combining the two integrals. For the integral over the laminar portion, properties are evaluated at T^* for laminar flow (using a laminar recovery factor) while for turbulent flow properties are evaluated at T^* for turbulent flow. Thus, if we use $\mathrm{Re}_{L_c} = 5 \times 10^5$ in the laminar region to determine the transition to turbulent flow, the value of Re_{L_c} in the evaluation of the turbulent integral will be slightly different.

For rough approximation purposes, since hand calculations can become tedious, one may sometimes use one recovery factor (preferably turbulent, since the majority of heat transfer usually is turbulent) to evaluate one set of properties. The high-speed flow formulas provide another example where coding for a personal computer is easy to implement and can save a significant amount of labor.

As noted previously, an important concern in convection heat transfer is recognition of the character of the problem. Let us consider when we would have to consider a problem to be a high-speed flow problem. We do this by attempting to find when treating the problem as an ordinary (low-speed) problem leads to excessive error.

The heat transfer in low-speed flow is

$$q_{LS} = h(T_w - T_\infty) \qquad \textbf{(6-118)}$$

The heat transfer in high-speed flow (assuming the same heat transfer coefficient applies, that is, assuming we can use the film temperature to evaluate properties) is

$$q_{HS} = h(T_w - T_{aw}) \qquad \textbf{(6-119)}$$

The error is

$$E = q_{LS} - q_{HS} = h(T_{aw} - T_\infty) \qquad \textbf{(6-120)}$$

The ratio R_E of error to low-speed flow heat transfer is

$$R_E = \frac{E}{q_{LS}} = \frac{T_{aw} - T_\infty}{T_w - T_\infty} \qquad \textbf{(6-121)}$$

The relative error depends on how much the adiabatic wall temperature differs from T_∞ compared with how much the wall temperature differs from T_∞. Thus, the closer T_∞ and T_w, the lower the fluid speed needed to make the flow high speed. Expressed in terms of Mach number [using Eqs. (6-102) and (6-108)], the ratio is

$$R_E = r\frac{\gamma - 1}{2}\left(\frac{M_\infty^2}{T_w/T_\infty - 1}\right) \qquad \textbf{(6-122)}$$

This ratio should be small if low-speed flow analysis is used.

We often think of high-speed flow as involving flow at supersonic speeds. However, it is possible, when T_w and T_∞ are fairly close, for high-speed considerations to become important at more modest speeds.

High-speed flow presents us with a possibility that at first may seem like a paradox. It is possible for heat to be transferred from a cold high-speed fluid to a warmer wall. This is because it is possible to have

$$T_{aw} - T_w = T_\infty - T_w + r\frac{\gamma - 1}{2}M_\infty^2 T_\infty > 0 \qquad \textbf{(6-123)}$$

even though

$$T_\infty - T_w < 0 \qquad \textbf{(6-124)}$$

Since heat transfer is determined by temperature difference at the wall, it is possible for the cold fluid to give heat to the warm wall.

EXAMPLE 6-6. A high-speed train travels at 45 m/sec (about 100 mi/hr). The outside air temperature is 40°C (95°F). Analysis of air-conditioning and insulation requirements for passenger cars based on low-speed flow indicates that the outside wall temperature of the passenger car under these conditions will be 35°C while the inside of the car is at 25°C. Assuming that the outside wall temperature is kept at 35°C, estimate the significance of high-speed flow considerations.

SOLUTION.   For estimating purposes, let us consider the train to be a plate. Then

$$R_E = r \frac{\gamma - 1}{2} \left(\frac{M_\infty^2}{T_w/T_\infty - 1} \right)$$

The temperature ratio term is small, that is,

$$\frac{T_w}{T_\infty} - 1 = \frac{273 + 35}{273 + 40} - 1 = 0.016$$

The Mach number term is

$$M_\infty^2 = \frac{u_\infty^2}{\gamma g_c RT} = \frac{(45)^2}{(1.4)(1)(287)(273 + 40)} = 0.016$$

which is about the same as the temperature term. Evaluating the recovery factor based on turbulent flow, we obtain

$$R_E = (0.71)^{1/3} \frac{1.4 - 1}{2} \left(\frac{0.016}{0.016} \right) = 0.18$$

High-speed flow considerations imply a need for about an 18% correction, a significant effect even though the Mach number is low. Admittedly, this example was contrived to obtain a case with a small difference $T_w - T_\infty$. However, the point should be clear that high-speed considerations need not be restricted to problems involving supersonic aircraft and rockets.

COMPUTER PROJECT 6-1.   Write a computer program to calculate the heat transfer coefficient at any specified location along a flat plate and averaged over the length of the plate. Your program should analyze the known conditions of the problem to determine the appropriate formula to apply. In particular, it should check

1. Whether there is turbulence.
2. Whether the fluid is a gas or liquid.
3. If a gas, whether the high-speed formulas are needed.
4. If a liquid, whether (via the Prandtl number) liquid metal formulations are needed.

6-10.   ANALOGIES TO MASS TRANSFER

Just as there was found to be similarity between the equations governing heat transfer and friction leading to useful analogies and similar results, there is similarity between the equations of mass transfer and heat transfer leading to useful analogies and similar results. The study of mass transfer in convection thus leads to similar equations and correlations, with analogous useful dimensionless constants.

The basic governing equation for mass transfer is Fick's Law of diffusion, which is a direct analogue of Fourier's law of heat conduction. In a stagnant fluid, a constituent A will flow according to

$$J = -D_A \frac{\partial C}{\partial x} \tag{6-125}$$

or, in more general geometry,

$$J = -D_A \nabla C \tag{6-126}$$

where J is the current (amount per unit area per unit time), C is concentration (amount per unit volume), and D_A is a diffusion coefficient (the subscript A identifies the constituent and also distinguishes notation from diameter). In a flowing fluid, the current would be analogous to Eq. (6-16),

$$\mathbf{J} = -D_A \nabla C + C \mathbf{V} \tag{6-127}$$

and the balance equation for concentration near a plate would be analogous to Eq. (6-17),

$$\frac{\partial}{\partial x}(uC) + \frac{\partial}{\partial y}(vC) = D_A \left(\frac{\partial^2 C}{\partial x^2} + \frac{\partial^2 C}{\partial y^2} \right) \tag{6-128}$$

This differs from Eq. (6-17) in that $k/\rho c$ is replaced by D_A. If we have similar boundary conditions at the plate for concentration and temperature, for example, if there is a specified concentration at the surface of the plate, and the constituent may enter the flow stream at the surface (as may be the case if the plate is a porous wall through which the constituent has diffused to get to the surface), then an entirely parallel development may be pursued, defining a concentration boundary layer instead of a temperature boundary layer.

In a mass transfer problem, analogous to a heat transfer problem, we assume that the velocity distribution is established independently. Thus, the velocity boundary layer will be determined by the Reynolds number as before. The ratio of the concentration boundary layer to that of the velocity boundary layer will be analogous to Eq. (6-51),

$$\frac{\delta c}{\delta} = \frac{1}{1.026} \mathrm{Sc}^{-1/3} \tag{6-129}$$

where the Schmidt number Sc is given by

$$\mathrm{Sc} = \frac{\nu}{D_A} \tag{6-130}$$

The Schmidt number is analogous to the Prandtl number with thermal diffusivity α replaced by diffusion coefficient D_A.

Analogous to the heat transfer case, we may define a convective coefficient h_m such that

$$J = -D_A \frac{\partial C}{\partial y}(0) = h_m(C_w - C_\infty) \tag{6-131}$$

Since Fourier's Law is replaced by Fick's Law, we may define a Sherwood number Sh analogous to the Nusselt number with D_A replacing k:

$$Sh = \frac{h_m L}{D_A} \tag{6-132}$$

Since all other considerations are directly analogous, we infer that the local Sherwood number is, analogous to Eq. (6-57),

$$Sh = \frac{h_m x}{D_A} = 0.332\,Re^{1/2}Sc^{1/3} \tag{6-133}$$

and that the average over the plate is double the value at the trailing edge.

When we deal with turbulence, we again make analogies between processes in laminar flow and processes in turbulent flow. We may infer an eddy mass transfer diffusion coefficient ε_D and assume that

$$Sc = \frac{\nu}{D} = \frac{\nu + \varepsilon_m}{\nu + \varepsilon_D} \tag{6-134}$$

We may introduce a friction mass transfer relation

$$St_M Sc^{2/3} = \tfrac{1}{2}C_f \tag{6-135}$$

with the mass transfer Stanton number St_M defined by

$$St_M = \frac{Sh\,Re}{Sc} \tag{6-136}$$

Then, analogous to Eq. (6-91), we infer

$$St_M Sc^{2/3} = 0.0296\,Re^{-1/5} \tag{6-137}$$

The parallels between heat and mass transfer extend to other geometric configurations as long as the boundary conditions associated with the mass transfer and the heat transfer problems are the same. Thus, the student learning heat transfer at the same time learns many important techniques for analyzing mass transfer.

From a problem-solving point of view, the student who has developed a set of spreadsheets and computer routines for treating heat transfer problems will find that the set also applies to mass transfer problems.

EXAMPLE 6-7.   A swimming pool is 25 m long. A breeze blows in the lengthwise direction of the pool at a rate of 3 m/sec. The air temperature is 300°K (about 80°F) and the relative humidity is 50%. Estimate the mass transfer coefficient 1 m from the leading edge of the pool if the diffusion coefficient of water vapor in air is 2.6×10^{-5} m^2/sec.

SOLUTION.   We evaluate properties at 300°K and the Reynolds number is

$$Re = \frac{u_\infty x}{\nu} = \frac{(3)(1)}{1.59 \times 10^{-5}} = 1.89 \times 10^5$$

so laminar flow still applies. We evaluate the Sherwood number from

$$Sh = 0.332 \, Re^{1/2} Sc^{1/3}$$

The Schmidt number is

$$Sc = \frac{\nu}{D_A} = \frac{1.59 \times 10^{-5}}{2.6 \times 10^{-5}} = 0.61$$

The Sherwood number is

$$Sh = 0.332(1.89 \times 10^5)^{1/2}(0.61)^{1/3} = (0.332)(435)(0.848) = 122$$

The mass transfer coefficient is

$$h_m = Sh \frac{D_A}{x} = (122)\frac{2.6 \times 10^{-5}}{1} = 3.17 \times 10^{-3}$$

This example is a direct parallel with Example 6-3. Thus, the spreadsheet solution provided for that problem applies if we associate Schmidt number with Prandtl number, Sherwood number with Nusselt number, and diffusion coefficient with conductivity.

To obtain the rate of mass transfer we take the vapor density at the pool surface to be the saturation density at that temperature, 0.0256/m^3. The density in the air at 50% relative humidity is half that. Thus, the mass flow rate at that location is

$$J = h_m A(\rho_w - \rho_\infty)$$

$$\frac{J}{A} = (3.17 \times 10^{-3})(0.0256 - 0.0128) = 4.06 \times 10^{-5} \text{ kg/sec} \cdot \text{m}^2$$

Note that if we wish to calculate the mass transfer from the entire pool, we must evaluate the Reynolds number at the trailing edge, to determine whether turbulent flow exists, and select the proper formula for average mass transfer. This is left as an exercise.

6-1. Suppose, as an alternative to Eq. (6-26), that we assume a velocity

$$\frac{u}{u_\infty} = \sin\frac{y}{\delta}$$

    a. Does this profile satisfy the boundary conditions?
    b. Use this profile in Eq. (6-24) and obtain an equation analogous to Eq. (6-33).
    c. Solve for the boundary layer thickness.

6-2. A fluid flows over a flat plate at a speed of 3 m/sec. For a distance of 0.01 m from the leading edge, make a table or plot of boundary layer thickness versus $\nu = \mu/\rho$. The range of ν should include values expected for most gases and liquids (other than liquid metals). The spreadsheet of Example 6-1 may be convenient for this problem. Identify values associated with water, engine oil, and helium at 300°K. Note for which types of fluid boundary layers are thinnest and thickest. Do not use values of ν that would make the Reynolds number exceed 5×10^5.

6-3. Extend the spreadsheet of Example 6-1 so that it also calculates thermal boundary layer thickness.

6-4. Properties of water vary significantly with temperature. Tabulate or plot the variation of velocity and thermal boundary layer thickness for water as a function of temperature for a speed of 3 m/sec at a distance of 0.01 m from the leading edge for temperatures from 300 to 600°K.

6-5. For the conditions and fluids stipulated in Problem 6-2, evaluate the local heat transfer coefficients.

6-6. A fluid at 400°K flows over a plate at 300°K. The fluid speed and the location at the plate are such that the flow is laminar. Your task is to determine the significance of evaluating fluid properties at the film temperature. For air, helium, water, and engine oil, find the ratios of the heat transfer coefficients evaluated at T_∞ and the wall temperature to the heat transfer coefficient evaluated at the film temperature. Examine your results and determine what is the principal contributor to the deviation of the ratios from 1.

6-7. A plate over which fluid flow is laminar has a friction coefficient that varies according to Eq. (6-64). If the plate is L meters long (in the direction of flow) and W meters wide, obtain an expression for the total force exerted by the fluid on one surface of the plate.

6-8. If the plate of Example 6-3 is 0.5 m long and 0.5 m wide, what is the total force on one surface of the plate.

6-9. Air and water flow over plates 10 cm long and 10 cm wide. Fluid temperature is 400°K and plate temperature is 300°K. Asymptotic fluid speed is 0.5 m/sec. Find the following for each fluid:
    a. The total heat transfer from a plate surface
    b. The total force on a plate surface.
    c. The ratio of heat transfer and force.

6-10. Extend the spreadsheet of Example 6-3 so that it calculates local friction factors and local Stanton numbers.

6-11. For the conditions of Example 6-3, tabulate or plot the Stanton number and the friction factor versus position.

6-12. For the conditions of Example 6-4, tabulate or plot the fluid boundary layer thickness, the thermal boundary layer thickness, and the heat transfer coefficients versus position from the leading edge to 0.5 cm. Note how your results tend to compare with boundary layer thicknesses and heat transfer coefficients for other fluids.

6-13. Adapt the spreadsheet of Example 6-3 to one that will treat liquid metals.

6-14. A single empirical correlation for laminar flow local heat transfer, intended to be valid for a wide variety of fluids from liquid metals to very viscous liquids, is Eq. (6-73)

$$\mathrm{Nu} = \frac{0.3387\,\mathrm{Re}^{1/2}\,\mathrm{Pr}^{1/3}}{\left[1 + (0.0468/\mathrm{Pr})^{2/3}\right]^{1/4}} \qquad \mathrm{Re}\,\mathrm{Pr} > 100$$

a. Find a limiting form for this formula as the Prandtl number goes to zero. Compare with the form found in Section 6-7.

b. Tabulate or plot the term $0.3387/[1 + 0.0468/\mathrm{Pr})^{2/3}]^{1/4}$ as a function of the Prandtl number. Note at what value of Pr this is equal to 0.332 and what the value is when the Prandtl number is 1.

c. Exact solution of the equations at a flat plate has led to the tabulated result

$$\mathrm{Nu} = 0.332\,\mathrm{Re}^{1/2}\,\mathrm{Pr}^{1/3} f(\mathrm{Pr})$$

where $f(\mathrm{Pr})$ is given by the table

Pr	0.01	0.1	0.7	1	10	100
$f(\mathrm{Pr})$	0.7194	0.9092	0.9922	1.0	1.0179	1.0199

What are the percentage errors associated with using the correlation at these Prandtl numbers.

6-15. For a plate 0.5 m long, tabulate or plot the speed at which the transition to turbulence occurs as a function of $\nu = \mu/\rho$. Identify common specific fluids—air, water, helium, engine oil, and liquid sodium—at 400°K in your table or plot.

6-16. For a fluid speed of 1 m/sec, tabulate or plot versus $\nu = \mu/\rho$ the plate length above which turbulent flow is to be expected. Identify air, water, helium, engine oil, and liquid sodium at 400°K in your table or plot.

6-17. Tabulate or plot the local friction factor and local heat Nusselt number versus Reynolds number for Pr = 0.7 for $0 < \mathrm{Re} < 10^9$.

6-18. Refer to the conditions of Example 6-5.
a. Over what fraction of the plate is the flow laminar?
b. What fraction of the heat transfer comes from the laminar flow?
c. What is the minimum heat transfer coefficient and when does it occur?

6-19. Some buildings, like the Flatiron Building in New York City, have been built such that they are very thin. Assume such a building can be represented as a flat plate for the purpose of estimating heat transfer coefficients. Estimate the heat transfer coefficient (making a table or plots) as a function of distance from the leading edge and as a function of wind speed. Assume the building length to be 100 m and the range of wind speeds to be from 10 to 60 mi/hr.

6-20. When the heat flux (not the temperature) is constant on the surface of a plate it can be shown that the local Nusselt number for laminar flow is given by

$$\mathrm{Nu} = \frac{hx}{k} = 0.453\,\mathrm{Re}^{1/2}\,\mathrm{Pr}^{1/3}$$

a. Find an expression for the wall temperature as a function of distance along the plate.
b. Find an expression for the average temperature on the plate.
c. Accounting for spatial variation of both heat transfer coefficient and wall temperature, show that the Nusselt number to be used with the average plate temperature is

$$\overline{\mathrm{Nu}} = \frac{\bar{h}L}{k} = 0.68\,\mathrm{Re}^{1/2}\,\mathrm{Pr}^{1/3}$$

d. Suggest an appropriate film temperature at which to evaluate properties for the average correlation of part c.

6-21. An electronic device has a back surface 1 cm^2 from which a heat flux of 1 W/cm^2 is to be dissipated. Room temperature (20°C) air is available to be blown over the surface. You would like to avoid turbulence and you would like to keep the average surface temperature from exceeding 65°C. Can you accomplish the objective of heat removal? Note that this is a constant heat flux problem as per Problem 6-20.

6-22. If the answer to Problem 6-21 is yes, then characterize the range of design parameters under which it will be possible to remove the heat.

6-23. If, within the allowed constraints for Problems 6-21 and 6-22, you have a range of design parameters that will permit you to achieve your heat transfer objective, suggest a criterion according to which you might select a design condition. Evaluate the design condition on the basis of this criterion.

6-24. For the design condition you picked in Problem 6-23, tabulate or plot the surface temperature as a function of position. Compare the maximum and minimum local temperatures with the average temperature.

6-25. Let the design specifications of Problem 6-21 be modified so that the maximum surface temperature does not exceed 65°C. Repeat the sequence of questions asked in Problems 6-21–6-24.

6-26. Suppose the designer wished to maintain, in Problem 6-21, the surface temperature to be uniformly equal to 65°C. Repeat Problems 6-21–6-23 with this objective. Tabulate or plot the heat flux as a function of position.

6-27. Assume that the turbulent flow heat transfer correlations apply to either isothermal wall or constant heat flux conditions. Derive an expression for the average wall temperature when the trailing edge Reynolds number is in the range $5 \times 10^5 < \mathrm{Re} < 10^7$.

6-28. Suppose the specifications of Problem 6-21 were changed to permit turbulent flow, but that the Reynolds number is restricted to be below 10^7 out of concern for potential vibration. Repeat Problems 6-21–6-23.

6-29. Air at atmospheric pressure and 20°C is blown at various speeds over a flat plate 1 m long. The plate surface is initially at 20°C. If the wind is turned on essentially instantaneously, what is the initial heat transfer to the plate as a function of speed for air speeds ranging from 100 mi/hr (45 m/sec) to 1000 mi/hr (450 m/sec)?

6-30. Repeat Problem 6-29 if the air is at 0.1 atmospheric pressure.

6-31. We wish to see the significance of using the recommended procedure for evaluating the effective temperature for property evaluation in high-speed flow. Repeat Problem 6-29 using the film temperature and compare results.

6-32. It has been found that if a flat plate consists of two sections, an unheated length L_u starting at the leading edge, followed by a heated (or cooled) length L_h, then the heat transfer coefficient is given by

$$\mathrm{Nu} = \frac{0.332 \, \mathrm{Re}^{1/2} \, \mathrm{Pr}^{1/3}}{1 - \left(L_u/x\right)^{3/4}} \qquad x > L_u$$

For the conditions of Example 6-3, plot or tabulate the heat transfer coefficient versus position for the first 10 cm of the heated section for unheated lengths of 0, 10, 20, 30, and 40 cm.

6-33. To treat the case of the temperature boundary layer for the case of uniform heat flux, it is suggested that instead of Eq. (6-44) a dimensionless temperature difference be defined as

$$\psi = \frac{k\left(T - T_\infty\right)}{\delta_t q''}$$

where q'' is the known uniform heat flux and ψ can be expanded in a polynomial

$$\psi = C_0 + C_1\left(\frac{y}{\delta_t}\right) + C_2\left(\frac{y}{\delta_t}\right)^2 + C_3\left(\frac{y}{\delta_t}\right)^3$$

a. Determine a set of conditions for evaluating the coefficients.
b. Evaluate the coefficients.

6-34. Show that, for laminar flow over a flat plate for ordinary fluids, heat transfer depends on fluid properties according to

$$h \sim \rho^{1/2}C_p^{1/3}k^{2/3}\mu^{-1/6}$$

6-35. Show that for general fluids (including liquid metals and highly viscous fluids) a more general dependence than that given in Problem 6-34 is

$$h \sim \frac{\rho^{1/2}C_p^{1/3}k^{2/3}\mu^{-1/6}}{\left[1 + (0.466/\mathrm{Pr})^{2/3}\right]^{1/4}}$$

6-36. Define a new property

$$Y = \frac{\rho^{1/2}C_p^{1/3}k^{2/3}\mu^{-1/6}}{\left[1 + (0.466/\mathrm{Pr})^{2/3}\right]^{1/4}}$$

that provides a measure of the merit of a fluid as a heat transfer medium in laminar flow.
a. If you have a computerized database, add this property to your database.
b. Compare the values of this property at 300°K for air, water, engine oil, and mercury.

6-37. Tabulate or plot the variation of property Y of Problem 6-36 versus temperature for air, water, engine oil, and mercury. Note for which fluids the tendency for heat transfer improves with temperature.

6-38. How would the result of Problem 6-34 be changed if there were turbulent flow?

REFERENCES

1. V. S. Arpaci and P. S. Larsen, *Convection Heat Transfer*, Prentice-Hall, Englewood Cliffs, NJ, 1984.
2. L. C. Barmeister, *Convective Heat Transfer*, Wiley, New York, 1983.
3. A. Bejan, *Convective Heat Transfer*, Wiley, New York, 1984.
4. H. Blasius, "Grenzschleten in Flussigkeiten mit Kleiner Reibung." *Z. Math. Phys.*, **56**, 1 (1908).
5. T. Cebeci and P. Bradshaw, *Momentum Transfer in Boundary Layers*, McGraw-Hill, New York, 1977.
6. T. Cebeci and P. Bradshaw, *Physical and Computational Aspects of Convective Heat Transfer*, Springer, New York, 1984.
7. S. W. Churchill and H. Zoe, "Correlations for Laminar Forced Convection over an Isothermal Flat Plate and in Developing and Fully Developed Flow in an Isothermal Tube," *J. Heat Transfer*, **95**, 78 (1973).
8. E. R. G. Eckert, "Engineering Relations for Heat Transfer and Friction in High-Velocity Laminar and Turbulent Boundary Layer Flow over a Surface with Constant Pressure and Temperature," *Trans. ASME*, **78**, 1273 (1956).

9. R. W. Fox and A. T. McDonald, *Introduction to Fluid Mechanics*, 3rd ed., Wiley, New York, 1985.

10. W. M. Kays and M. E. Crawford, *Convective Heat and Mass Transfer*, McGraw-Hill, New York, 1980.

11. J. H. Lienhard, *A Heat Transfer Textbook*, Prentice-Hall, Englewood Cliffs, NJ, 1981.

12. E. Pohlhausen, Der Wärmeaustausch zwischen festen Körperm und Flüssigkeiten mit kleiner Reibung und kleiner Wärm leitung, *Z. Angew. Math. Mech.*, **1**, 115 (1921).

13. A. Prasum, *Fundamentals of Fluid Mechanics*, Prentice-Hall, New York, 1980.

14. R. H. Sabersky, A J. Acosta and E. G. Hauptmann, *Fluid Flow*, 2nd ed., MacMillan, New York, 1971.

15. H. Schlichting, *Boundary Layer Theory*, 7th ed., McGraw-Hill, New York, 1979.

16. F. Schultz-Grunow, "Newes Widerstands gerety fur glatte Platten," *Luftfahrt Forsch.*, **17**, 239 (1940).

17. F. M. White, *Fluid Mechanics*, McGraw-Hill, New York, 1979.

Chapter Seven

FORCED CONVECTION

7-1. INTRODUCTION

The flat plate analyses of Chapter 6 provide us with a basis for proceeding more generally with analysis of forced convection. In this chapter we consider forced convection inside pipes and over bluff bodies. Considerable emphasis is placed on the use of empirical correlations. The forms of these correlations are motivated, at least in part, by the forms of solutions obtained analytically for simple cases, for example, those considered in Chapter 6. It should be recognized that considerable uncertainty is associated with the application of a general correlation in a specific situation. Reasonable practice would be to assign an uncertainty of $\pm 25\%$ to the use of information from a general correlation. We emphasize this point, since there is a tendency to assume that numbers from formulas are perfect.

Two categories of convection are considered—internal flow and external flow. Internal flow corresponds to flow in pipes, tubes, or channels. External flow implies flow across a body. Internal flow includes situations where the flow is channeled within a well-defined cross-sectional area. External flow includes situations where the body is an obstruction that causes a change in flow pattern. There may not be a well-defined flow cross-section, as with the flat plate.

We begin with internal flow and again treat laminar flow first. This is an interesting case since the parallel viscous flow results in a situation where heat transfer is determined by conduction through the fluid. In turbulent flow, however, we again have a situation where the Reynolds–Colburn analogy can be invoked. We thus again relate friction and heat transfer. We also consider the influence of the roughness of the pipe surface on both friction and heat transfer.

In external flow, we find that in addition to the viscous drag associated with boundary layer effects, there is form drag associated with pressure distribution about the body. This was not a concern with flat plates in Chapter 6 which were assumed to be extremely thin. For bodies with finite frontal area, for example, a cylinder with finite diameter, form drag becomes significant. We thus no longer have a convenient friction–heat transfer analogy, and heat transfer relationships must be developed independently. It is still possible to correlate in terms of the Reynolds and Prandtl numbers. A special case of interest is heat transfer from flow over an array of bodies (a bank of tubes).

7-2. LAMINAR FLOW IN A LONG TUBE

As was the case for the plate, we begin our discussion with laminar flow. In this section, we assume that the tube is very long. In Section 7-3, we consider the significance of entrance effects in pipes of finite length.

Since we must be concerned both with momentum and with energy, as was the case for the flat plate, we again must be concerned with velocity and temperature profiles. The velocity profile is developed assuming constant properties and assuming that it influences but is not influenced by the temperature profile. These assumptions are similar to those made in dealing with the flat plate in Chapter 6.

In dealing with the flat plate, however, we neglected pressure gradient terms. There was assumed to be an infinite flow (e.g., the dimension normal to the plate was infinite) sustained by forces larger than those encountered in regard to friction at the plate surface. In a pipe of finite diameter, the flow rate is finite, and the pressure gradient becomes an essential feature of the situation.

Steady flow can be maintained in a pipe if there is a pressure differential sufficient to balance the viscous resistance. Consider a fluid element of radius r (less than radius R of the pipe) and thickness dx as illustrated in Fig. 7-2-1. The viscous and pressure forces are balanced if

$$\left(\mu \frac{du}{dr}\right)(2\pi r\, dx) = (dp)(\pi r^2) \tag{7-1}$$

If we assume that the pressure gradient is constant in the pipe, we may obtain the simple first-order equation

$$\frac{du}{dr} = \frac{1}{2\mu}\frac{dp}{dx}r \tag{7-2}$$

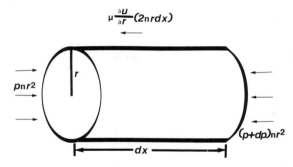

FIGURE 7-2-1. Force balance on cylindrical element of fluid.

Since the velocity must be zero at the pipe wall (i.e., at $r = R$), we have a boundary condition and can integrate the equation to obtain

$$u = u_{max} \frac{(R^2 - r^2)}{R^2} \tag{7-3}$$

$$u_{max} = -\frac{R^2}{4\mu} \frac{dp}{dx} \tag{7-4}$$

Note that the pressure gradient should be negative (i.e., fluid flows from locations of high pressure to locations of low pressure).

Let us now consider the temperature profile. Heat is conducted radially from the wall of the pipe toward the center of the fluid. Each annulus (cylindrical shell) of thickness dr contains flowing fluid. In the region of thickness dx, fluid is heated and carries away the energy acquired, as illustrated in Fig. 7-2-2. The equation describing the process is

$$\frac{1}{r} \frac{\partial}{\partial r}\left(kr\frac{\partial T}{\partial r}\right) = \rho c_p u \frac{\partial T}{\partial x} \tag{7-5}$$

If we were traveling in the fluid at speed u, the fluid would appear stationary (at least at our radius) and there would be heating given by $\rho c_p \, \partial T/\partial t$. Since the speed is $u = \partial x/\partial t$, we transform back to the frame of reference that is not moving by replacing $1/\partial t$ with $u(1/\partial x)$.

Let us consider the case of uniform heat flux along the wall. Then we can take $\partial T/\partial x$ as a constant (since the fluid heating is uniform along the length of the pipe). The result is of the form of a heat conduction problem with a source depending on radius (since we have found that the velocity profile depends on radius), a type of problem treated in Chapter 3.

$$\frac{1}{r} \frac{\partial}{\partial r}\left(r\frac{\partial T}{\partial r}\right) = \frac{u_{max}}{R^2 \alpha} \frac{\partial T}{\partial x}(R^2 - r^2) \tag{7-6}$$

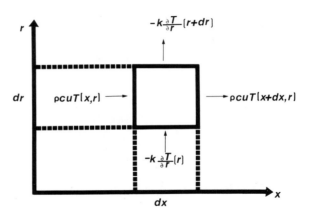

FIGURE 7-2-2. Energy balance with in fluid.

The equation is solved by following the set of steps given in Chapter 3. Multiplying by r, integrating from 0 to r, and noting that $r\partial T/\partial r(0)$ is zero, we obtain

$$r\frac{\partial T}{\partial r}(r) = \frac{u_{max}}{\alpha R^2}\frac{\partial T}{\partial x}\left(\frac{R^2 r^2}{2} - \frac{r^4}{4}\right) \tag{7-7}$$

Dividing by r, integrating from r to R, and defining $T(R) = T_w$, we obtain

$$T_w - T(r) = \frac{u_{max}}{\alpha R^2}\frac{\partial T}{\partial x}\left(\frac{R^4}{4} - \frac{R^2 r^2}{4} - \frac{R^4}{16} + \frac{r^4}{16}\right) \tag{7-8}$$

In tubes, we define a heat transfer coefficient in terms of a bulk temperature

$$q = hA(T_w - T_b) \tag{7-9}$$

where the bulk temperature is an average taken so as to yield the total energy carried by the fluid. Thus,

$$T_b = \frac{\int_0^R 2\pi r\, dr\, \rho u c_p T(r)}{\int_0^R 2\pi r\, dr\, \rho u c_p} \tag{7-10}$$

Note that in a tube, unlike over a flat plate, there is no asymptotic temperature. Evaluating the integrals yields

$$T_w - T_b = \frac{11}{96}\frac{u_{max}}{\alpha}\frac{\partial T}{\partial x} \tag{7-11}$$

The heat transfer coefficient is obtained from

$$h = \frac{q}{A(T_w - T_b)} = \frac{k(\partial T/\partial r)(R)}{T_w - T_b} \qquad \text{(7-12)}$$

We may evaluate

$$\frac{\partial T}{\partial r}(R) = \frac{u_{max}R}{4\alpha} \frac{\partial T}{\partial x} \qquad \text{(7-13)}$$

which leads to

$$h = \frac{24}{11}\frac{k}{R} = \frac{48}{11}\frac{k}{D} \qquad \text{(7-14)}$$

where D is the diameter. The Nusselt number is

$$\text{Nu} = \frac{hD}{k} = \frac{48}{11} = 4.364 \qquad \text{(7-15)}$$

At first glance, this result, which does not depend on Reynolds number or Prandtl number as did the Nusselt number for the flat plate, may seem surprising. However, we note that for the long tube, flow is perfectly parallel and heat is transferred radially only by conduction. In the flat case, the boundary layer thickness varied with position and there was flow normal to the plate. Thus, the case of laminar flow in a long tube is a convection problem in the sense that it involves a moving medium that carries away heat, but not in the sense of the transfer of heat from the surface into the medium.

The result of this section serves to illustrate the physical interpretation that may be assigned to the Nusselt number. The total heat transfer to the fluid is $hA(T_w - T_b)$. The heat transfer by conduction alone may be approximated as $kA(T_w - T_b)/R$ or $2kA(T_w - T_b)/D$. The ratio of these expressions is proportional to the Nusselt number, so the Nusselt number may be interpreted as a measure of the ratio of total heat transfer to heat transfer when only conduction is applicable. The constant of proportion may be expected to depend on geometry. When all the heat transfer is by conduction, the ratio and therefore the Nusselt number will be independent of parameters not related to conduction (e.g., flow rate, viscosity).

When fluid flows through a channel of noncircular cross-section, a length dimension used to characterized the channel in dimensionless parameters such as the Nusselt number is the hydraulic diameter

$$D_h = \frac{4A}{P} \qquad \text{(7-16)}$$

where A is the channel cross-section area and P is the wetted perimeter. Nusselt numbers for constant wall temperature and uniform heat flux are

TABLE 7-2-1
Laminar Flow Nusselt Numers

Channel geometry cross-section	Constant heat flux	Uniform temperature
Circle	4.36	3.66
Square	3.61	2.98
Equilateral triangle	3.11	2.47
Isosceles right triangle	2.98	2.34
Hexagon	4.00	3.34
Rectangle,[a] $R = 2$	4.12	3.39
Rectangle,[a] $R = 4$	5.33	4.44
Rectangle,[a] $R \to \infty$	8.24	7.54
Oval,[b] $R = 2$	4.56	3.74
Oval,[b] $R = 4$	4.88	3.79

[a] R = Ratio of sides.
[b] R = Ratio of maximum to minimum diameter.

given in Table 7-2-1 for several geometries for both uniform heat flux and uniform wall temperature. The text of Shah and London (1979) provides many such solutions.

Generally, when evaluating properties for flow in a tube, it is recommended that the bulk temperature be used. In a long tube, we would use the average of the inlet and outlet bulk temperatures. However, we may not know both temperatures in advance. For example, we may be asked to find the heat transferred to a fluid initially at a given temperature when it goes through a pipe having a constant wall temperature. Until we know the heat transfer, we do not know the outlet temperature, and thus we cannot evaluate the average bulk temperature.

This type of problem is iterative. We begin by assuming a temperature at which to evaluate fluid properties. One possibility is to use the known inlet temperature. A second possibility is to estimate in advance the temperature change in the pipe to get an estimated average bulk temperature. A third possibility is to use a temperature different from the inlet temperature (higher if the fluid is being heated, lower if it is being cooled) that is given explicitly in the available tables of data (Appendix A in this text), so that interpolation is avoided when making a first estimate.

For laminar flow in a long tube, there is only one property—thermal conductivity—that is needed to obtain the heat transfer coefficient. In short tubes and in turbulent flow, however, more properties are needed, and the issue of interpolation can be significant.

After the first estimate of heat transfer, temperatures can be reevaluated. Iteration on heat transfer can take place until the desired level of accuracy.

In this chapter, we consider heat transfer to be used on the average heat transfer coefficient and the difference between average wall temperature and average bulk temperature. More elaborate averaging may turn out to be appropriate, but we defer those considerations to our discussion of heat exchangers in Chapter 11.

Consider two parallel flat plates, as shown in Fig. 7-3-1. At the leading edges, boundary layers are formed and increase in thickness with distance along the plates. Near the leading edges, the boundary layers can be thin by comparison with distance between the plates, so heat transfer from each plate may be independent of one another. We thus would expect heat transfer near the leading edge to vary as

$$h \sim \frac{1}{\sqrt{x}} \qquad (\text{7-17})$$

that is, to be high near the leading edge and to decline with distance.

As boundary layers become thicker, a point is reached when they no longer can be considered independent of one another as Fig. 7-3-1 illustrates. Far from the leading edges, the boundary layers merge and a condition of fully developed flow is reached. The heat transfer coefficient should then approach an asymptotic value consistent with fully developed flow as illustrated in Fig. 7-3-2.

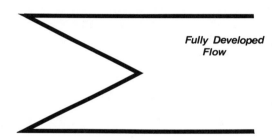

FIGURE 7-3-1. Boundary layer growth between parallel plates.

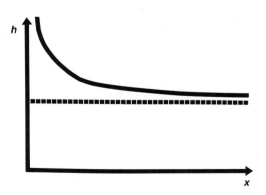

FIGURE 7-3-2. Heat transfer coefficient variation with distance into a pipe.

It is possible to correlate the influence of entrance effects of a pipe of finite length with laminar flow in terms of the Graetz number Gz,

$$Gz = Re\,Pr\frac{D}{L} \tag{7-18}$$

where the Reynolds number is evaluated with diameter as the length dimension. For a tube with constant surface temperature, for example, we may use

$$Nu = 3.66 + \frac{0.0668\,Gz}{1 + 0.04\,Gz^{2/3}} \tag{7-19a}$$

$$Nu = 1.86\,Gz^{1/3}\left(\frac{\mu}{\mu_s}\right)^{0.14} \tag{7-19b}$$

Equation (7-19a) given by Hausen applies to cases where heating begins well beyond the entrance to the tube. For such cases, the velocity profile has become fully developed by the location at which heating begins, and "entrance" effects apply one to temperature. Equation (7-19b) given by Seider and Tate applies to cases where entrance effects apply to both velocity and temperature. For long tubes, Eq. (7-19a) approaches an asymptotic value, Eq. (7-19b), however, does not. Equation (7-19b) should be used when it predicts a Nusselt number greater than 3.66. Otherwise, the Nusselt number should be taken as 3.66. Equation (7-19b) with $\mu/\mu_s = 1$ is plotted in Fig. 7-3-3. The reciprocal of the Graetz number is used as the horizontal axis so that the coordinate increases as pipe length increases.

It is of interest to compare the Nusselt number obtained here with that obtained for a flat plate. The tube Nusselt number increases as the length of the tube decreases. The plate Nusselt number declined with decreasing length.

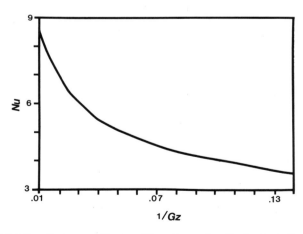

FIGURE 7-3-3. Entrance effect on Nusselt number for laminar flow in a tube.

The plate heat transfer coefficient increased with decreasing length because the Nusselt number was divided by length to obtain the heat transfer coefficient. Since the tube Nusselt number is divided by diameter to get heat transfer coefficient, any decline in heat transfer must be exhibited directly in the Nusselt number. For short tubes in laminar flow, the Nusselt number increases with fluid speed and density and decreases with viscosity as with the plate, another indication of boundary layer effects prior to achievement of fully developed flow where conductivity is the only significant fluid property.

EXAMPLE 7-1.   Atmospheric air at 0°C in a long tube of 1 cm diameter enters a 10 cm long heated section with a speed of 2 m/sec. The wall of the tube in this section is maintained at a temperature of 125°C. How much heat is transferred to the air?

SOLUTION.   Since the 10 cm section is in a long tube, Eq. (7-19a) applies, so

$$Nu = 3.66 + \frac{0.0668\,Gz}{1 + 0.04\,Gz^{2/3}}$$

To evaluate the Graetz number, we need the average bulk temperature

$$T_{b,av} = \tfrac{1}{2}(T_{b,in} + T_{b,out})$$

$$T_{b,in} = 273°K + 0 = 273°K$$

$T_{b,out}$ is not known. Since the fluid is being heated, let us evaluate properties at the temperature above 273°K that appears in the table of Appendix A, that is, at 300°K.

$$Re = \frac{VD}{\nu} = \frac{(2)(0.01)}{1.59 \times 10^{-5}} = 1258$$

$$Pr = 0.707$$

$$Gz = Re\,Pr\frac{D}{L} = (1258)(0.707)\frac{0.01}{0.1} = 88.9$$

$$Nu = 3.66 + \frac{(0.0668)(88.9)}{1 + 0.04(88.9)^{2/3}} = 6.97$$

$$h = Nu\frac{k}{D} = \frac{6.97(0.026)}{0.01} = 18.1$$

$$q = hA(T_w - T_{b,av}) = hA\left[T_w - \tfrac{1}{2}(T_{b,in} + T_{b,out})\right]$$

We also know that (with A_c the tube cross-section and area)

$$q = \rho v A_c C_p(T_{b,out} - T_{b,in})$$

that is, the heat transferred to the fluid increases the temperature of the fluid. Equating the expressions for q,

$$T_{b,\text{out}} = \frac{hA\left(T_w - \frac{1}{2}T_{b,\text{in}}\right) + \rho v A_c C_p T_{b,\text{in}}}{\frac{1}{2}hA + \rho v A_c C_p}$$

$$hA = (h\pi DL) = (18.1)(\pi)(0.01)(0.1) = 0.0569$$

$$\rho v A_c C_p = (1.16)(2)\frac{\pi}{4}(0.01)^2(1007) = 0.183$$

$$T_{b,\text{out}} = \frac{(0.569)\left[125°C - \frac{1}{2}(0)\right] + (0.183)(0)}{\frac{1}{2}(0.569) + 0.183} = 34°C$$

This yields $T_{b,\text{av}} = 290°K$. Let us reevaluate parameters by interpolating

$$\nu = \left[1.14 + 0.8(1.59 - 1.14)\right] \times 10^{-5} = 1.50 \times 10^{-5}$$

$$\text{Re} = \frac{\nu D}{\nu} \quad \frac{(2)(0.01)}{1.50 \times 10^{-5}} = 1333$$

$$\text{Pr} = 0.720 - 0.8(0.720 - 0.707) = 0.710$$

$$\text{Gz} = \text{Re}\,\text{Pr}\frac{D}{L} = (1333)(0.710)\frac{0.01}{0.1} = 94.6$$

$$\text{Nu} = 3.66 + \frac{(0.0668)(94.6)}{1 + (0.04)(94.6)^{2/3}} = 7.11$$

$$k = 0.022 + 0.6(0.028 - 0.022) = 0.025$$

$$h = \text{Nu}\frac{k}{D} = 17.8$$

The heat transfer coefficient is not changed substantially (compared with the accuracy of the heat transfer correlation).

$$\rho = 1.39 - 0.6(1.39 - 1.16) = 1.21$$

$$C_p = 1007$$

$$hA = 0.0556$$

$$\rho v A_c C_p = 0.191$$

$$T_{b,\text{out}} = \frac{(0.0556)(125)}{\frac{1}{2}(0.0556) + 0.191} = 32°C$$

This yields $T_{b,\text{av}} = 16°C$ or $289°K$. This is close enough that properties need not be reevaluated. The total heat transfer is then

$$q = hA(T_w - T_{b,\mathrm{av}}) = (0.0556)(125 - 16) = 6.06 \text{ W}$$

This problem is a good candidate for spreadsheet application because of the iteration on temperature. A substantial amount of arithmetic is involved in obtaining $T_{b,\mathrm{out}}$. Upon reevaluation of parameters at the new $T_{b,\mathrm{av}}$, recalculation of $T_{b,\mathrm{out}}$ conventionally requires repeating a task of substantial magnitude. With the spreadsheet, evaluation of the new value of $T_{b,\mathrm{out}}$ involves very little additional effort. Preparation of a spreadsheet for this problem is left an exercise.

COMPUTER PROJECT 7-1.   In view of the potential amount of interpolation that may be required in dealing with convection problems, prepare an interpolation program that will evaluate properties at specified conditions. Put into your program data for the commonly used heat transfer fluids air and water. Add other fluids as you find the need.

7-4. TURBULENT FLOW IN TUBES

When we dealt with turbulent flow over a flat plate, we invoked the analogy between friction and heat transfer and applied the relationship

$$\mathrm{St}\,\mathrm{Pr}^{2/3} = \tfrac{1}{2}C_f \tag{7-20}$$

The friction coefficient for flow in a tube is defined somewhat differently from friction coefficients for flow over a flat plate. Let us therefore write the equation in terms of the physical quantities involved

$$\mathrm{St}\,\mathrm{Pr}^{2/3} = \frac{\tau}{\rho u^2} \tag{7-21}$$

We postulate that the relationship expressed in this way will apply to turbulent flow in tubes. This postulate is intuitive based on our experience with the flat plate. It will be justified to the extent that its implications are consistent with empirical relationships.

While the friction coefficient for flow over a flat plate is defined directly in terms of viscous stress, the friction coefficient for flow in a tube, referred to in the literature as the Darcy, Darcy–Weisbach, or Moody friction factor, is defined in terms of pressure drop, such that

$$\Delta p = f \frac{L}{D} \frac{\rho u_m^2}{2} \tag{7-22}$$

where u_m is the mean (bulk) velocity in the tube (in the literature, one may find $4f$ used to define a Fanning friction coefficient instead of f). In steady flow, pressure and viscous stress effects must balance, so

$$\tau \pi D L = \Delta p \frac{\pi}{4} D^2 \qquad (7\text{-}23)$$

The last two equations may be combined to yield

$$\tau = \frac{f}{8} \rho u_m^2 \qquad (7\text{-}24)$$

Thus, our heat transfer correlation for tubes may be expressed as

$$\text{St} \, \text{Pr}^{2/3} = \frac{f}{8} \qquad (7\text{-}25)$$

One empirical expression for the friction factor in smooth pipes at high Reynolds number is

$$f = 0.184 \, \text{Re}^{-1/5} \qquad (7\text{-}26)$$

This would lead, upon use with Eq. (7-25), to a Nusselt number of

$$\text{Nu} = 0.023 \, \text{Re}^{0.8} \text{Pr}^{1/3} \qquad (7\text{-}27)$$

This equation is referred to as the Colburn equation. Direct measurements of heat transfer in smooth pipes have led to the correlation known as the Dittus–Boelter equation

$$\text{Nu} = 0.023 \, \text{Re}^{0.8} \text{Pr}^n \qquad (7\text{-}28)$$

where n is 0.4 and 0.3 for heating and cooling, respectively. Thus, while explicit correlation of heat transfer yields somewhat different coefficients, the friction–heat transfer analogy appears to provide a reasonable basis for understanding the mechanism of heat transfer. As we shall find below, the variation of the friction factor with Reynolds number is actually more elaborate than Eq. (7-26), so we may expect to find more elaborate heat transfer correlations than Eq. (7-28). However, these more elaborate correlations are consistent with an understanding based on a close relationship between friction and heat transfer.

The exponent on the Prandtl number can be different in heating and cooling because properties such as viscosity vary with temperature. This property variation upsets the symmetry between heating and cooling expected from a simple analysis, particularly for liquids. In other formulas below, alternate means are provided to account for this asymmetry.

Entrance effects are expected to be of significance for turbulent flow also. The average heat transfer in a pipe of finite length may be obtained from the Nusselt number

$$\text{Nu} = 0.036 \, \text{Re}^{0.8} \text{Pr}^{1/3} \left(\frac{D}{L} \right)^{0.055} \qquad 10 < \frac{L}{D} < 400 \qquad (7\text{-}29)$$

Note that when $L/D = 400$, we have

$$0.036 \left(\frac{D}{L} \right)^{0.055} = 0.026 \qquad (7\text{-}30)$$

which is close to the coefficient in the long-tube formula. Some authors recommend a smaller value of L/D for transition to the use of full developed flow relations, for example, $L/D = 60$. Note from Fig. 7-4-1 that evaluation of Eq. (7-29) at L/D of 60 and 400 yields results that differ by about 10%. It may be argued that the correction may be comparable to the accuracy of the correlation. In this and other matters in the literature, different people may make alternate plausible choices on the basis of available information.

It should be noted that, in principle, it takes longer (in distance) to establish an asymptotic average heat transfer coefficient than an average local heat transfer coefficient. Thus, at the point that flow becomes fully developed, the average heat transfer coefficient (averaging from the entrance to that point) will be higher than the fully developed coefficient (see Fig. 7-3-1).

To aid in estimating when fully developed conditions may be considered to prevail, we invoke the concept of entrance length L_E, that is, the length into a pipe by which the local heat transfer coefficient comes within 1% of its asymptotic value. For air, this entrance length in dimensionless form (L_E/D)

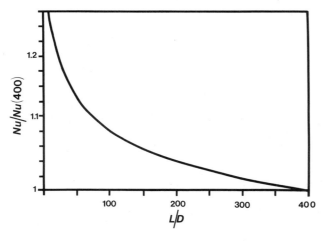

FIGURE 7-4-1. Entrance effect on Nusselt number for turbulent flow in a pipe.

is given by $2.4\,\mathrm{Re}^{-0.2}$. Entrance lengths calculated with this expression are about 15–30 diameters. If we say that to achieve an asymptotic average we need a distance double that to achieve a local asymptotic value, then an $L/D = 60$ will constitute fully developed flow. At higher Prandtl numbers, that is, for liquids, entrance lengths tend to be shorter.

Since entrance lengths tend to be shorter for liquids than for gases, and since they also depend on Reynolds number, it is to be expected that the true entrance correction should be more elaborate than that given by Eq. (7-29). The reader is referred to the text in convection by Kays and Crawford (1980) for more details.

When dealing with liquids for which viscosity may decline substantially with temperature (for gases, viscosity tends to increase with temperature), the following correlation of Seider and Tate (1936) may be used

$$\mathrm{Nu} = 0.027\,\mathrm{Re}^{0.8}\mathrm{Pr}^{1/3}\left(\frac{\mu}{\mu_w}\right)^{0.14} \tag{7-31}$$

where μ_w is evaluated at the wall temperature and all other parameters are evaluated at the bulk temperature. Because of the 0.14 power dependence (i.e., about the 7th root) the viscosity correction will not be important unless there is a major difference between μ and μ_w. In Eq. (7-31), the temperature variation of properties is treated in the viscosity ratio. Asymmetry between heating and cooling enters through this ratio. Accordingly, a single power of the Prandtl number is used, and that is the $1/3$ that would be expected.

Sometimes, we will have explicit information about the friction factor in a tube for which, because of atypical degree of roughness, we may expect a general heat transfer correlation not to be applicable. Then we may use the friction–heat transfer analogy in the form

$$\mathrm{St}_b\mathrm{Pr}_f^{2/3} = \frac{f}{8} \tag{7-32}$$

where b and f denote bulk and film temperatures (here film implies average between wall and bulk temperatures). The use of the film temperature for Prandtl number does not usually make much difference for gases but introduces an averaging process for liquids. Figure 7-4-2 shows friction factor variation for a smooth pipe, and Fig. 7-4-3 shows the impact of pipe roughness.

A convenient formula for evaluating the friction factor with reasonable accuracy was introduced by Haaland (1983):

$$f = 4C_f = \frac{1}{3.24\log_{10}^2\left[(\varepsilon/3.7D)^{1.11} + 6.9/\mathrm{Re}\right]} \tag{7-33}$$

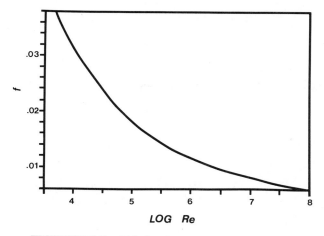

FIGURE 7-4-2. Friction factor in a smooth pipe.

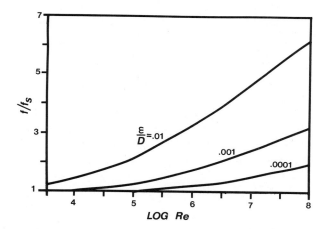

FIGURE 7-4-3. Impact of pipe roughness on friction factor.

Actual correlation of data has led to

$$\frac{1}{\sqrt{f}} = -2\log_{10}\left(\frac{\varepsilon}{3.7D} + \frac{2.51}{\mathrm{Re}\sqrt{f}}\right) \tag{7-34}$$

Evaluation of the latter formula involves iteration. The simpler formula which is easy to program into a personal computer or into a spreadsheet and which is within about 1.5% of Eq. (7-34), should be appropriate for most circumstances.

To obtain the heat transfer coefficient given the friction factor, one approach used in the literature has been to generalize a relationship developed originally for smooth pipes by Petukhov (1970),

$$\text{Nu} = \frac{(f/8)\,\text{Re}\,\text{Pr}}{1.07 + 12.7\sqrt{f/8}\,(\text{Pr}^{2/3} - 1)}\left(\frac{\mu}{\mu_w}\right)^n$$

$$n = \begin{cases} 0.11 & T_w > T_b \\ 0.25 & T_w < T_b \\ 0 & \text{constant heat flux} \end{cases} \qquad (7\text{-}35)$$

by using the actual (nonsmooth) friction factor in the equation. However, this may lead to errors for substantial roughness. Consequently, one may evaluate the Nusselt number for a smooth pipe and correct according to Norris (1970) by

$$\frac{\text{Nu}}{\text{Nu}_{\text{smooth}}} = \left[\min\left(\frac{f}{f_{\text{smooth}}}, 3\right)\right]^{0.68\,\text{Pr}^{0.215}} \qquad (7\text{-}36)$$

which holds for Prandtl numbers between 0.7 and 6 (i.e., it does not apply for liquid metals). Equation (7-36) applies best for sand-grain roughness. For other types of roughness, for example, squared-edged ribs introduced to augment heat transfer, Norris found similar qualitative behavior (Nusselt ratio varying as a power of friction factor ratio up to a maximum value) with different parameters.

The ratio of actual and smooth pipe friction factors needed in Eq. (7-36) is plotted in Fig. 7-4-3 for several degrees of roughness. (A perfectly smooth pipe would have $\varepsilon/D = 0$.) Note that the ratio increases with Reynolds number. This behavior may be understood as follows. For a smooth pipe, as per Fig. 7-4-2, the friction factor declines with the Reynolds number. For a rough pipe (finite ε/D), an asymptotic value of f in Eq. (7-33) is approached as the Reynolds number declines in influence. Thus, the ratio increases with the Reynolds number.

The ratio of actual and smooth pipe Nusselt numbers is plotted in Fig. 7-4-4 for values of the Prandtl number defining the range of applicability of Eq. (7-36). Increase of factors of 2–3 (depending on the fluid, i.e., on the Prandtl number) in Nusselt number are possible due to roughness.

It should be recognized that formulas for roughness effects tend to apply most reliably in situations where the roughness has been designed in explicitly as a means of augmenting heat transfer. Natural roughness, which is likely to involve substantial irregularity in ε within the pipe even for a given average roughness, may not be represented as accurately. The handbook by Rohsenow and Hartnett (1973) discusses this point and others related to augmenting heat transfer. We note that while roughness is an important factor in heat transfer in turbulent flow, it is not an important factor in connection with laminar flow.

We note that Eq. (7-35) is more complicated than Eq. (7-31). There are many situations in the literature where correlations of an elaborate nature

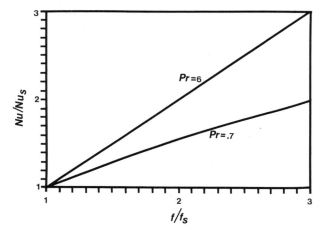

FIGURE 7-4-4. Influence of roughness on Nusselt number.

have been developed to provide improved accuracy relative to simpler correlations. The student or practicing engineer with access to a personal computer or desk top terminal need not be as concerned as those of an earlier generation with the added complexity of a more elaborate correlation, since a simple code or spreadsheet can serve to evaluate the formula for all future applications.

The more elaborate representation can be considered to be appropriate. Even for smooth pipes, Eq. (7-33) or (7-34) for friction factor versus Reynolds number would not yield a straight line on a log–log plot over the full range of Reynolds numbers as implied by Eq. (7-26). Thus, use of a heat transfer formula with the actual friction factor, as may be evaluated with Eq. (7-33), should be more general. In addition, the simple friction–heat transfer analogy tends to be most nearly applicable when the Prandtl number is near 1. Equation (7-35) includes in the denominator a term that provides for corrections for viscous liquids. There is a theoretical basis for the form of this correction, as discussed in the text by White, derived from explicit consideration of the forms of temperature and velocity profiles in the vicinity of a wall in turbulent flow. The factors contributing to differences between Eqs. (7-31) and (7-35) will be checked in Example 7-3.

The transition from laminar to turbulent flow in tubes occurs at a Reynolds number of about 2300. This transition value is much lower than that for flow over flat plates (5×10^5). Consider again the ratio of kinetic energy to viscous force

$$\frac{\text{KE}}{\text{VF}} = \frac{\frac{1}{2}\rho u_m^2}{-\mu(\partial u/\partial r)(R)} \tag{7-37}$$

The negative sign occurs because we enter the tube in the negative r direction.

For laminar flow in a tube, we find

$$\frac{\partial u}{\partial r} = \frac{2u_{\max}}{R} = -\frac{4u_{\max}}{D} \qquad (7\text{-}38)$$

It is easily shown (as in a problem at the end of the chapter) that

$$u_m = \tfrac{1}{2}u_{\max} \qquad (7\text{-}39)$$

The ratio of kinetic energy to viscous stress is

$$\frac{\mathrm{KE}}{\mathrm{VF}} = \frac{\rho u_m D}{16\mu} = \frac{\mathrm{Re}}{16} \qquad (7\text{-}40)$$

Thus, for a tube, we see that the ratio varies with the Reynolds number, whereas for the flat plate it varies with the square root of the Reynolds number. Therefore, it is reasonable to expect the transition to turbulent flow to occur at a lower Reynolds number in a tube.

Empirical correlations have been developed for liquid metals in tubes with the form

$$\mathrm{Nu} = A + B\,\mathrm{Pe}^n \qquad (7\text{-}41)$$

for fully developed turbulent flow ($L/D > 60$) in the range $100 < \mathrm{Pe} < 10^4$. These yield an A of 5 and 4.82 for constant wall temperature (Seban and Shimazuki, 1951) and constant heat flux (Skupinshi, Tortel, and Vautrey, 1965), a B of 0.025 and 0.0185, and an n of 0.8 and 0.827.

In dealing with flow in tubes, we again see that we must ask the two basic questions posed in Chapter 6. First, we have to determine the nature of the problem. Evaluation of the Reynolds number tells whether flow is laminar or turbulent. Evaluation of the length-to-diameter ratio for turbulent flow or the Graetz number for laminar flow tells us whether entrance effects are significant. Second, we use the answer to the first question (the nature of the problem) to select the proper heat transfer correlation.

EXAMPLE 7-2. Air at 3 atm pressure rises from 500 to 600°K as it passes through a 1.5 m long, 3 cm diameter tube with constant wall temperature of 650°K. How fast is the air flowing?

SOLUTION. We know the inlet and outlet temperatures. We can express the total heat transfer as

$$q = \rho v A_c C_p (T_{b,\,\mathrm{out}} - T_{b,\,\mathrm{in}})$$

We also can express the heat transfer as

$$q = hA(T_w - T_{b,\,\mathrm{av}})$$

We do not know yet whether flow is turbulent or laminar because we do not know fluid speed. Let us assume turbulence. If we are wrong, we will obtain a contradiction. We see that

$$\frac{L}{D} = \frac{1 \text{ m}}{0.03 \text{ m}} = 33 < 400$$

We may evaluate the heat transfer coefficient from

$$h = \frac{k}{D}(0.036)\text{Re}^{0.8}\text{Pr}^{1/3}\left(\frac{D}{L}\right)^{0.055}$$

where the Reynolds number involves velocity

$$\text{Re} = \frac{vD\rho}{\mu}$$

Equating the two formulas for heat transfer

$$\rho v A_c(T_{b,\text{out}} - T_{b,\text{in}}) = \frac{k}{D}0.036v^{0.8}\left(\frac{D\rho}{\mu}\right)^{0.8}\text{Pr}^{1/3}\left(\frac{D}{L}\right)^{0.055}A(T_w - T_{b,\text{av}})$$

All quantities may be evaluated from tables or are known except for velocity. Evaluating properties at the average bulk temperature of 550°K leads to (noting that density at 3 atm may be taken as three times density at 1 atm)

$$\rho = 3(0.6329) = 1.90 \text{ kg/m}^3$$

$$C_p = 1040 \text{ J/kg} \cdot °\text{K}$$

$$k = 0.0439 \text{ W/m} \cdot °\text{K}$$

$$\text{Pr} = 0.683$$

$$\mu = 2.88 \times 10^{-5} \text{ N} \cdot \text{sec/m}^2$$

$$A_c = \frac{\pi}{4}D^2 = \frac{\pi}{4}(0.03)^2 = 7.069 \times 10^{-4} \text{ m}^2$$

$$A = \pi D L = \pi(0.03)(1) = 0.0942 \text{ m}^2$$

$$(1.90)v(7.069 \times 10^{-4})(1040)(600 - 100)$$

$$= \frac{0.0439}{0.03}0.036v^{0.8}\left[\frac{(0.03)(1.90)}{2.88 \times 10^{-5}}\right]^{0.8}(0.7)^{1/3}\left(\frac{0.03}{1}\right)^{1055}(0.0942)(650 - 550)$$

$$v = 1.83 \text{ m/sec}$$

We now check to see if our assumption of turbulence was correct

$$\text{Re} = \frac{vD\rho}{\mu} = \frac{(1.83)(0.03)(1.90)}{2.88 \times 10^{-5}} = 3619 > 2300$$

The turbulent flow formula was appropriate.

EXAMPLE 7-3. One kg/sec of water with an average temperature of 60°C flows through a tube 2 cm in diameter and 10 m long. A constant heat flux is to be imposed on the surface of the tube, which has a relative roughness (ε/D) of 0.002. Find the heat transfer coefficient.

SOLUTION. We will consider several options for dealing with this problem, but first let us establish the type of problem we have. The Reynolds number is

$$\text{Re} = \frac{u_m}{\mu} D\rho = \frac{\rho u_m \pi D^2}{4}\left(\frac{4}{\pi \mu D}\right) = 1\frac{4}{\pi(4.7 \times 10^{-4})(0.02)}$$

$$\text{Re} = 1.35 \times 10^5$$

so the flow is turbulent

$$\frac{1}{D} = \frac{10}{0.02} = 500$$

so the flow is fully developed. Let us evaluate the heat transfer coefficient for a smooth pipe by

$$\text{Nu}_{\text{smooth}} = 0.023\,\text{Re}^{0.8}\text{Pr}^{0.4} = (0.023)(1.35 \times 10^5)^{0.8}(3.01)^{0.4} = 454$$

Now let us evaluate the ratio of friction by using Eq. (7-33) (with $\varepsilon = 0$ for the smooth case):

$$\frac{f}{f_{\text{smooth}}} = \frac{3.24\log_{10}^2(6.9/\text{Re})}{3.24\log_{10}^2\left[(\varepsilon/3.7D)^{1.11} + 6.9/\text{Re}\right]} = \frac{59.67}{40.64} = 1.47$$

We correct for roughness using Eq. (7-36):

$$\text{Nu} = \text{Nu}_{\text{smooth}}(1.47)^{0.68\,\text{Pr}^{0.215}} = 1.39\,\text{Nu}_{\text{smooth}} = 633$$

As an alternative, let us evaluate Eq. (7-35) with the rough f value of $1/40.64 = 0.0246$, noting that $n = 0$ because of uniform heat flux

$$\text{Nu} = \frac{(f/8)\text{Re}\,\text{Pr}}{1.07 + 12.7\sqrt{(f/8)}\,(\text{Pr}^{2/3} - 1)} = \frac{(0.0246/8)(1.35 \times 10^5)(3.01)}{1.07 + 12.7\sqrt{(0.0246/8)}\left[(3.01)^{2/3} - 1\right]}$$

$$\text{Nu} = 681$$

which is slightly higher than the first value.

Let us compare the smooth heat transfer coefficient already obtained with that which would be obtained from Eq. (7-35) using the smooth $f = 1/59.67 = 0.0168$. We

$$\mathrm{Nu}_{smooth} = \frac{(0.0168/8)(1.35 \times 10^5)(3.01)}{1.07 + 12.7\sqrt{(0.0168/8)}\left[(3.01)^{2/3} - 1\right]} = 502$$

If we apply the friction factor correction calculated above to be 1.39, we get

$$\mathrm{Nu} = 698$$

From the above, we see that the biggest discrepancy is in the evaluation of the smooth heat transfer coefficients, that is, in using Eq. (7-35) relative to Eq. (7-28). The difference of 10% is not surprising given the accuracy of correlations. Given a difference, the value from the more elaborate Eq. (7-35) is preferred. Note that the main difference between the two correlations arises from the Prandtl number correction term for viscous liquids in the denominator of Eq. (7-35). Had we used the form of simple correlation directed more specifically at liquids, Eq. (7-31), we would have obtained (assuming the viscosity correction for water is 1)

$$\mathrm{Nu}_{smooth} = 0.027(1.35 \times 10^5)^{0.8}(3.01)^{1/3} = 495$$

which is close to the value of 502 obtained with the more elaborate formula above. The higher coefficient (0.027 vs. 0.023) compensates for the smaller degree of Prandtl number dependence in formulas like Eqs. (7-28) and (7-31) compared with Eq. (7-35). The friction coefficient implied by Eq. (7-26) is 0.0173, close to the smooth pipe value of 0.0168 obtained from Eq. (7-33). On the other hand, the difference between the values of 681 and 698 obtained using the alternative mechanisms for accounting for roughness is small (about 2%). We will select the 698 value to obtain

$$h = \mathrm{Nu}\frac{k}{D} = 698\left(\frac{0.654}{0.02}\right) = 22,800$$

Note that the roughness has a substantial effect on heat transfer.

EXAMPLE 7-3 (Spreadsheet).   This problem is a good candidate for a spreadsheet because it places a complicated formula, Eq. (7-35), in a situation where it can be evaluated again and because a particular formula, for friction factor, Eq. (7-33), can be replicated. In this spreadsheet, as in others, we have taken more steps than are necessary in order to provide for display of intermediate information. For example, in the friction factor development, we made separate steps to evaluate the roughness and Reynolds terms in the argument of the logarithm. Note that the friction factor evaluation for smooth tubing (B6–B11 with $\varepsilon/D = 0$) is replicated in B16–B21 with $\varepsilon/D = .002$.

In this example, the mass flow rate is specified, and the Reynolds number is calculated accordingly. In other problems, the fluid speed may be specified. Under those circumstances, the first few lines of the spreadsheet could be modified leaving the remainder intact. Similarly, this problem involved a uniform heat flux. If this were not the case, then rows could be inserted after Row 14 to evaluate the appropriate viscosity ratio correction.

Finally, if there is a desire to evaluate the Nusselt number directly with the roughness friction factor, then one simply has to change the entry in B7 from 0.0 to .002 and examine the Nusselt number in B15. To get the heat transfer coefficient, change B24 to B15 in B26.

	A	B
1	ROUGHNESS	EFFECT
2	FLOW RATE	1.0
3	VISCOSITY	4.17E − 4
4	DIAMETER	.02
5	REYNOLDS	4.0*B2 / (@PI*B3*B4)
6	FRICTION	FACTOR
7	EPSD	0.0
8	RUFF TERM	(B7 / 3.7) ∧ 1.11
9	RE TERM	6.9 / B5
10	LOGARITHM	@ LOG (B8 − B9)
11	FRICTION	1./ (3.24*(B10 ∧ 2))
12	PRANDTL	3.01
13	NUMERATOR	+B11*B5*B12 / 8.0
14	PR TERM	+B12 ∧ (2 / 3) − 1
15	SQRT (F / 8)	(B11 / 8) ∧ .5
16	DENOMINATOR	12.7*B14*B15 + 1.07
17	SMOOTH NU	+B13 / B16
18	FRICTION	FACTOR
19	RELATIVE ROUGHNESS	.002
20	ROUGHNESS TERM	(B11 / 3.7) ∧ 1.11
21	REYNOLDS TERM	6.9 / B5
22	LOGARITHM	@ LOG (B20 + B21)
23	FRICTION	1./ 3.24*(B22 ∧ 2)
24	RATIO	+B23 / B11
25	RATIO LIMIT	@ MIN (B24, 3)
26	NU FACTOR	1.47 ∧ (.68*(B12 ∧ .215))
24	NUSSELT	+B15*B23
25	CONDUCTIVITY	.654
26	H	+B24*B25 / B4

EXAMPLE 7-4.   Air at atmospheric pressure and an average temperature of 350°K flows at 5 m/sec through a long insulated square box 7 cm on a side, at the center of which is a metal pipe 5 cm O.D. with a surface temperature of 400°K. Evaluate the heat transfer coefficient.

SOLUTION.   For this problem, the hydraulic diameter must be evaluated in order to obtain a Reynolds number. The hydraulic diameter is

$$D_H = \frac{4A}{P}$$

The area A is

$$A = L^2 - \tfrac{1}{4}\pi D^2 = (0.07)^2 - \tfrac{1}{4}\pi(0.05)^2 = 0.00294$$

The perimeter P is

$$P = 4L + \pi D = 4(0.07) + \pi(0.05) = 0.437$$

so the hydraulic diameter is

$$D_H = \frac{4A}{P} = 0.0269$$

The Reynolds number is

$$\text{Re} = \frac{\mu D}{\nu} = \frac{(5)(0.0269)}{2.09 \times 10^{-5}} = 6.44 \times 10^3$$

so flow is turbulent. The Nusselt number may be obtained from

$$\text{Nu} = 0.023\,\text{Re}^{0.8}\,\text{Pr}^{0.4} = (0.023)(6440)^{0.8}(0.697)^{0.4} = 22.2$$

$$h = \text{Nu}\frac{k}{D} = 22.2\left(\frac{0.03}{0.0269}\right) = 24.8$$

Example 7-4 serves two purposes. One is to illustrate the procedure for dealing with flow through noncircular channels. The other is to provide a basis for comparison with a cross-flow problem in Section 7-5 in which many conditions (temperature, speed) are similar. The comparison will be important with respect to motivation of heat exchanger design.

COMPUTER PROJECT 7-2. Extend your computer program for evaluating heat transfer coefficients to include tubes. Analogous to the case for plates, include tests to determine the geometry, type of flow, and treatment of entrance effects.

7-5. FLOW ACROSS BLUFF BODIES

Frequently, we must consider flow across the outside of a tube. In some heat exchangers, for example, one fluid may flow inside a tube and another across the outside of the tube. Flow across a bluff body can be complicated. Considerable variation can exist in the flow pattern around the body. Indeed, total separation of flow from the body can occur, as illustrated in Fig. 7-5-1 for a cylinder.

A bluff body results in a retarding or drag force, which is represented by

$$F = C_D A \frac{\rho u_\infty^2}{2g_c} \tag{7-42}$$

where C_D is the drag coefficient and A is the frontal area (LD for a cylinder of length L and diameter D). This drag force is made up of two parts. One is

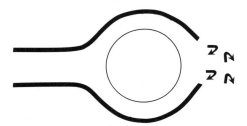

FIGURE 7-5-1. Flow across a bluff body leading to separated flow.

viscous drag, which we have considered in dealing with flow over flat plates and inside tubes. The other is form drag, which results from the pressure distribution about the body. We cannot invoke the Reynolds–Colburn analogy, and we cannot use the total drag force to infer the heat transfer coefficient (although the friction–heat transfer analogy has been applied over portions of bluff bodies where viscous effects predominate).

While the drag coefficient may not be of direct use for evaluating heat transfer, it is of interest, since the friction portion of the drag is related to heat transfer. Actions taken to promote heat transfer thus influence the drag and therefore, the cost of pumping fluid. The drag coefficient is plotted in Fig. 7-5-2. The drag coefficient may be estimated by

$$C_D = \begin{cases} 11.3 \, \mathrm{Re}^{-0.623} & 0.1 < \mathrm{Re} < 8 \\ 5.2 \, \mathrm{Re}^{-0.239} & 8 < \mathrm{Re} < 10^3 \\ 0.7 + 0.1 \log \mathrm{Re} & 10^3 < \mathrm{Re} < 10^5 \end{cases} \qquad (7\text{-}43)$$

This analytical estimate may be used in spreadsheets or in computer programs to evaluate drag force. The drag coefficient drops sharply at a Reynolds number of about 3×10^5 when the boundary layer becomes turbulent and the separation point shifts.

While we may be interested in an overall average heat transfer coefficient, we should recognize that heat transfer can vary considerably around the surface of the bluff body, as may be observed from Fig. 7-5-3 for cylinders. At low Reynolds numbers, the heat transfer coefficient starts out relatively high and then declines. This behavior can be understood by analogy to the decline of the heat transfer coefficient from the leading edge of a plate. A minimum occurs in the vicinity of the point of flow separation. Beyond the separation point, turbulent eddies come in contact with the cylinder, leading to an increase in heat transfer.

At higher Reynolds numbers, after the initial decline in heat transfer occurs, a sudden increase is encountered corresponding to transition from laminar to turbulent boundary layer. Heat transfer subsequently declines, with a second minimum occurring when flow separates. Turbulent eddy motion then leads to an increase in heat transfer.

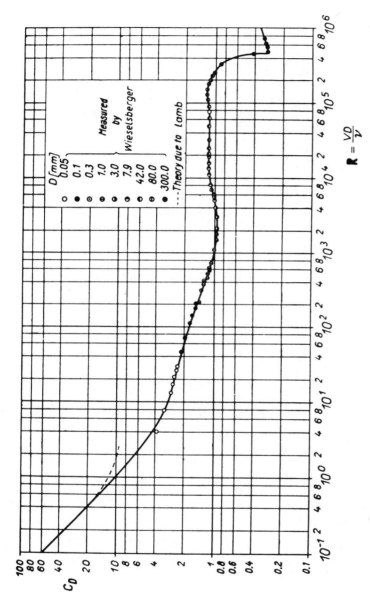

FIGURE 7-5-2. Drag coefficient vs. Reynolds number for flow across a cylinder. From Schlichting, BOUNDARY LAYER THEORY, McGraw-Hill, 1979. Reproduced with permission.

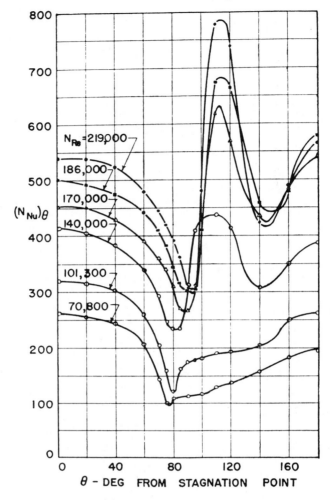

FIGURE 7-5-3. Variation of Nusselt number around a cylinder in cross flow. From W. H. Giedt, ASME Trans. 71, 375 (1949). Reproduced with permission.

The highly nonuniform heat transfer need not have nonuniform implications in the cylinder itself. The cylinder may be a tube containing a well-mixed fluid. The tubing material may be a good conductor of heat, which would also minimize nonuniformity in temperature.

Average heat transfer coefficients tend to be correlated in terms of Reynolds and Prandtl numbers. A commonly used expression is

$$\mathrm{Nu} = C\,\mathrm{Re}^n\mathrm{Pr}^{1/3} \tag{7-44}$$

where C and n are different in different ranges of Reynolds number, as given

TABLE 7-5-1
Values of C and n for Flow over Circular Cylinders

Re	C	n
0.4–4	0.919	0.330
4–40	0.911	0.385
40–4000	0.683	0.466
4,000–40,000	0.193	0.618
40,000–400,000	0.0266	0.805

TABLE 7-5-2
Values of C and n for Flow over Noncircular Cylinders

Geometry	Flow incidence	Re range	C	n
Square	Corner	$5 \times 10^3 – 10^5$	0.246	0.588
Square	Side	$5 \times 10^3 – 10^5$	0.102	0.675
Hexagon	Corner	$5 \times 10^3 – 10^5$	0.153	0.638
Hexagon	Side	$5 \times 10^5 – 1.95 \times 10^4$	0.160	0.650
		$1.95 \times 10^4 – 10^5$	0.0385	0.782
Vertical plate	Normal	$4 \times 10^3 – 1.5 \times 10^4$	0.228	0.731

by Holman (1981) in Table 7-5-1. Properties are evaluated at the film temperature. Values of C and n to use with noncircular cylinders, given in the text by Jakob (1949), are listed in Table 7-5-2. In evaluating the Reynolds numbers for cross-flow over the noncircular cylinders, the length dimension to use is the cross-section dimensional normal to the flow. For example, when flow is incident on a side of a square cylinder, the length dimension is the side of the square. When flow is incident on a corner of a square cylinder, the length dimension is the diagonal of the square.

EXAMPLE 7-5. Air at atmospheric pressure and 300°K blows at 5 m/sec across a metal pipe 5 cm O.D. containing a mixture of saturated water and steam. The flow is well mixed, such that the pipe is at a uniform temperature of 400°K. How much heat is transferred to the air per meter of pipe?

SOLUTION. We choose to apply the correlation

$$\text{Nu} = C\,\text{Re}^n\,\text{Pr}^{1/3}$$

The film temperature is $\frac{1}{2}(300°K + 400°K) = 350°K$. The Reynolds number is

$$\text{Re} = \frac{uD}{\nu} = \frac{(5\text{ m/sec})(0.05\text{ m})}{2.09 \times 10^{-5}\text{ m}^2/\text{sec}} = 1.2 \times 10^4$$

The proper formula from Table 7-5-1 is

$$\text{Nu} = 0.193\,\text{Re}^{0.618}\text{Pr}^{1/3} = 0.193(1.2 \times 10^4)^{0.618}(0.697)^{1/3} = 56.7$$

The heat transfer coefficient is

$$h = \text{Nu}\frac{k}{D} = (56.7)\left(\frac{0.03}{0.05}\right) = 34$$

The heat transfer per meter of length is

$$q = h\pi D(1)\Delta T = (34)(\pi)(0.05)(1)(100) = 534 \text{ W}$$

EXAMPLE 7-5 (Spreadsheet).   This problem may be treated by looking up a value in a short table containing values of C and n for appropriate Reynolds numbers. The spreadsheet is as follows

	A	B	C	D
1	FLOW	ACROSS	CYLINDER	
2	FLUID TEMP	300.0		
3	PIPE TEMP	400.0		
4	FILM TEMP	.5*(B2+B3)		
5	DIAMETER	.05		
6	SPEED	5.0		
7	VISCOSITY	2.09 E-5		
8	REYNOLDS	+B6*B5/B7		
9	TABLE			
10	REYNOLDS	C	REYNOLDS	N
11	.4	.989	.4	.330
12	4	.911	4	.385
13	40	.683	40	.466
14	4000	.193	4000	.618
15	40000	.0266	40000	.805
16	400000		400000	
17			LOOKUP C	@ LOOKUP (B8, A11... A16)
18			LOOKUP N	@ LOOKUP (B8, C11 ··· C16)
19			PRANDTL	.697
20			NUSSELT	+D17*(B8 ∧ D18)* (D19 ∧ .3333)
21			CONDUCTIVITY	.03
22			HEAT	+D20*D21/B5
23			H	+D22*@PI*B5* (B3 - B2)

Note that the LOOKUP function seeks a value higher than the reference value $B8 = 1.12 \times 10^4$ so it finds 40000. It then goes back to the previous value (4000) and returns the value in this row in the next column (.193 for C and .618 for N).

The LOOKUP function is one for which there is no corresponding single command in standard programming. A basic routine is provided in Appendix D which is equivalent to this spreadsheet and which illustrates how to provide for the LOOKUP function.

It is instructive to compare the results of Examples 7-4 and 7-5. The heat transfer coefficient in cross-flow was substantially higher than the heat transfer coefficient in axial flow along the cylinder. This enhanced heat transfer in cross-flow is one reason why baffles are used in certain types of heat exchanger to promote cross-flow in what would ordinarily be axial flow situations.

It is useful to observe that, while in both cases, air flowed on the outside of a pipe, in one case (Example 7-4) the flow was treated as flow inside a channel (of an effective hydraulic diameter) while in the other case (Example 7-5) the flow was treated as flow over a bluff body.

A single general correlation for cylinders has been developed by Churchill and Bernstein (1977) in the form

$$Nu = 0.3 + \frac{0.62\,Re^{1/2}\,Pr^{1/3}}{\left[1 + (0.4/Pr)^{2/3}\right]^{1/4}} \left[1 + \left(\frac{Re}{282{,}000}\right)^{5/8}\right]^{4/5} \qquad (7\text{-}45)$$

For Reynolds numbers between 20,000 and 400,000, this correlation tends to underpredict the Nusselt number, and a slightly modified formula

$$Nu = 0.03 + \frac{0.62\,Re^{1/2}\,Pr^{1/3}}{\left[1 + (0.4/Pr)^{2/3}\right]^{1/4}} \left[1 + \left(\frac{Re}{282{,}000}\right)^{1/2}\right] \qquad (7\text{-}46)$$

has been suggested. At low Reynolds numbers (below 4000) the last part of the formula does not contribute, and, for hand calculations, it should be sufficient to evaluate

$$Nu = 0.3 + \frac{0.62\,Re^{1/2}\,Pr^{1/3}}{\left[1 + (0.4/Pr)^{2/3}\right]^{1/4}} \qquad (7\text{-}47)$$

In all these formulas, properties should be evaluated at the film temperature. The above three equations apply for liquid metals (i.e., for low Prandtl numbers), but for very small Peclet numbers (below 0.2) one may use the formula of Nakai and Okazaki (1975)

$$Nu = \frac{1}{0.8237 - 0.5\ln Pe} \qquad (7\text{-}48)$$

It may be observed that for laminar flow (low Reynolds numbers) of liquid metals (low Prandtl numbers), Eqs. (7-45)–(7-47) reduce to dependence on the square root of the Peclet number. The proof is left as an exercise at the end of the chapter.

EXAMPLE 7-6.   Repeat Example 7-5 with the formula of Eq. (7-45).

SOLUTION

$$\text{Nu} = 0.3 + \frac{0.62\sqrt{1.2 \times 10^4}\,(0.697)^{1/3}}{\left[1 + (0.4/0.697)^{2/3}\right]^{1/4}}\left[1 + \left(\frac{1.2 \times 10^4}{2.82 \times 10^5}\right)^{5/8}\right]^{4/5}$$

$$\text{Nu} = 0.3 + \left(\frac{60.2}{1.14}\right)1.11 = 58.9$$

This is close to the result from Example 7-5. The formulas of Eqs. (7-45) and (7-46) have the advantage of including Prandtl number dependence that does not appear in Eq. (7-44).

For flow past spheres, we may use an equation due to Whitaker (1972)

$$\text{Nu} = 2 + (0.4\,\text{Re}^{1/2} + 0.06\,\text{Re}^{2/3})\text{Pr}^{0.4}\left(\frac{\mu_\infty}{\mu_w}\right)^{1/4} \qquad \begin{array}{l} 3.5 < \text{Re} < 8 \times 10^4 \\ 0.7 < \text{Pr} < 380 \end{array}$$

for gases and liquids. Properties are evaluated at T_∞ (except for μ_w). Diameter is the length dimension used in the Reynolds number. The sphere formula is identical to another cylinder correlation except for the addition of the number 2. One view toward sphere analysis could then be to add 2 to a cylinder correlation.

There is a theoretical basis to expect a limiting value of 2 for the sphere. This value is characteristic of a pure conduction problem from a sphere into a large pool of stagnant liquid. An analogous limiting value does not exist for the cylinder. A finite sphere will provide a finite heat source, so as the distance from the sphere increases, the influence of this heat source is reduced and an asymptotic temperature is approached, providing the conditions from which a Nusselt number may be defined. For a cylinder, however, the total amount of heat is infinite (since the length in cylindrical geometry is presumed to be infinite), and there is not a tendency to approach an asymptotic temperature. (A similar statement can be made for plane geometry.) Verification of the above statements is left as a problem at the end of the chapter.

7-6.   BANKS OF TUBES

Heat exchangers frequently involve flow across a large set of tubes. If the set is large enough, and if the tubes are arranged in a repeating pattern, then a flow pattern will become established. Thus, we obtain correlations for large sets of tubes and apply corrections for small sets.

Two types of tube arrangement will be considered. Figure 7-6-1 shows a rectangular array (or in-line arrangement), while Fig. 7-6-2 shows a triangular array (or staggered arrangement). The triangular array makes it possible to put more tubes into a given volume.

Heat transfer again be expressed in the form

$$\text{Nu} = C\,\text{Re}^n\text{Pr}^{1/3} \tag{7-49}$$

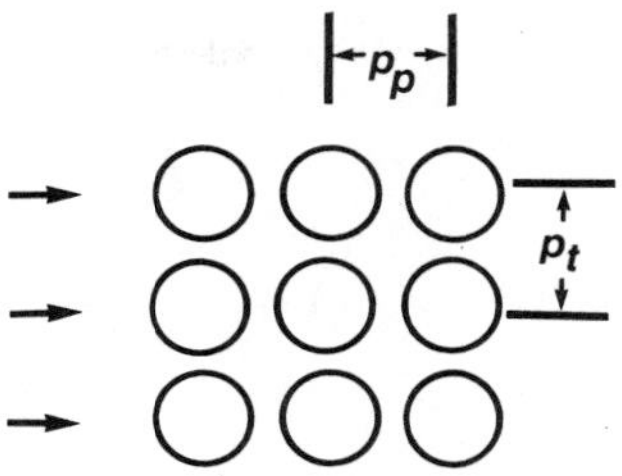

FIGURE 7-6-1. Flow directed at a rectangular (in-line) array of tubes.

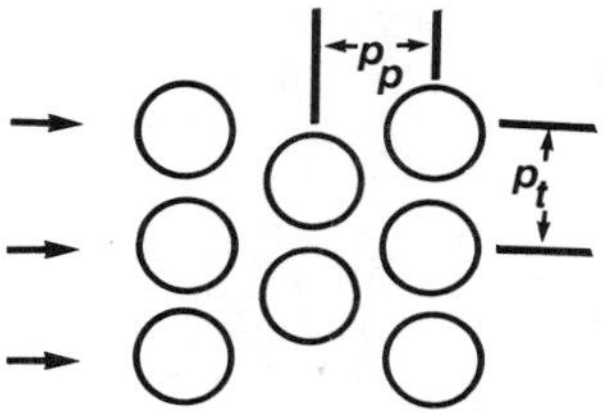

FIGURE 7-6-2. Flow directed at a triangular (staggered) array of tubes.

Values of C and n for rectangular and triangular arrays at least 10 rows deep are given in Tables 7-6-1 and 7-6-2. The original work by Grimison (1937) was for gases and did not include Prandtl number dependence. The values in Tables 7-6-1 and 7-6-2, as used in texts by Holman (1981) and by Kreith and Black, (1980), modify the original coefficients provided by Grimison so as to provide agreement with the original work for gases and to provide Prandtl number correction for liquids. Corrections when there are fewer rows are given by Kays and Lo (1952, 1954) in Table 7-6-3. A convenient analytical expression to use is, for M rows,

$$\mathrm{Nu}(M) = \mathrm{Nu}(10)\left(\frac{M}{10}\right)^{0.12} \tag{7-50}$$

TABLE 7-6-1
Correlation Parameters for Rectangular Tube Arrays

	p_t/D							
	1.25		1.50		2.00		3.00	
p_p/D	C	n	C	n	C	n	C	n
1.25	0.386	0.592	0.305	0.608	0.111	0.704	0.0703	0.752
1.50	0.407	0.586	0.278	0.620	0.112	0.702	0.0753	0.744
2.00	0.464	0.570	0.332	0.602	0.254	0.632	0.220	0.648
3.00	0.322	0.601	0.396	0.584	0.415	0.581	0.317	0.608

TABLE 7-6-2

Correlation Parameters for Triangular Tube Arrays

P_p/D	P_t/D							
	1.25		1.50		2.00		3.00	
	C	n	C	n	C	n	C	n
0.6							0.236	0.636
0.9					0.495	0.571	0.445	0.581
1.0			0.552	0.558				
1.125					0.531	0.565	0.575	0.560
1.25	0.575	0.556	0.561	0.554	0.576	0.556	0.579	0.562
1.50	0.501	0.565	0.511	0.562	0.502	0.568	0.542	0.568
2.00	0.448	0.572	0.462	0.568	0.535	0.556	0.498	0.570
3.00	0.344	0.592	0.395	0.580	0.488	0.562	0.467	0.574

TABLE 7-6-3

Correction Factors for Fewer Than 10 Rows

Number of rows	Rectangular	Triangular	Approximate formula
1	0.64	0.68	0.76
2	0.80	0.75	0.82
3	0.87	0.83	0.87
4	0.90	0.89	0.90
5	0.92	0.92	0.92
6	0.94	0.95	0.94
7	0.96	0.97	0.96
8	0.98	0.98	0.97
9	0.99	0.99	0.99
10	1.00	1.00	1.00

This expression corresponds well to the correction factor data, especially for rectangular arrays, except for the case of one row.

The Reynolds number is evaluated, for use in these tables, using tube diameter as the length dimension and the maximum velocity (associated with the minimum flow area) as the velocity term. We note that, by continuity, the mass flow in kg/sec is constant and is

$$\dot{m} = \rho u A \tag{7-51}$$

The maximum velocity in the direction of flow occurs when

$$\rho u_{max} = \frac{\dot{m}}{A_{min}} = G_{max} \tag{7-52}$$

If the total area of the heat exchanger is A and there are N transverse rows of diameter D and length L, then

$$\rho u_{max} = \frac{\dot{m}}{A - NDL} \qquad (7\text{-}53)$$

and the Reynolds number is

$$\text{Re} = \frac{\rho u_{max} D}{\mu} = \frac{\dot{m} D}{(A - NDL)\mu} \qquad (7\text{-}54)$$

If we take the total area A as N times the transverse pitch p_t times length, then

$$\text{Re} = \frac{\dot{m} D}{NL(p_t - D)\mu} \qquad (7\text{-}55)$$

Overall fits to the data by Zukauskas (1970) have yielded formula

$$\text{Nu} = C \, \text{Re}^n \text{Pr}_\infty^{0.36} \left(\frac{\text{Pr}_\infty}{\text{Pr}_w} \right)^{1/4} \qquad (7\text{-}56)$$

where, for a triangular array and $10^2 < \text{Re} < 2 \times 10^5$,

$$n = 0.6$$

$$C = \begin{cases} 0.35 \left(\dfrac{p_t}{p_p} \right)^{0.2} & \dfrac{p_t}{p_p} > 2 \qquad (7\text{-}57) \\[2em] 0.4 & \dfrac{p_t}{p_p} > 2 \qquad (7\text{-}58) \end{cases}$$

In the same range of Reynolds numbers, for a rectangular array

$$n = 0.63 \qquad (7\text{-}59)$$

$$C = 0.27 \qquad (7\text{-}60)$$

For Reynolds numbers greater than 2×10^5, n is 0.84 for either array, while C is 0.021 for rectangular and 0.022 for triangular arrays. Thus, at high Reynolds numbers, both arrays give similar heat transfer.

The ratio of Nusselt numbers for triangular (assuming $p_t/p_p \geq 2$) and rectangular arrays is plotted in Fig. 7-6-3. At low Reynolds numbers, the triangular array provides substantially greater (up to about 30%) heat transfer, with the benefit declining with increasing Reynolds number. For the triangular array, flow tends to be directed so as to hit tubes head on. For the rectangular array, there is a tendency for flow to bypass tubes behind the leading tube. As the Reynolds number increases, turbulence and greater mixing lead to reduc-

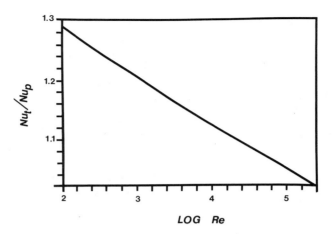

FIGURE 7-6-3. Ratio of Nusselt numbers in triangular and rectangular banks of tubes.

tion in the bypassing tendency, and the relative advantage of the triangular array declines.

We also tend to be concerned with pressure drop. The pressure drop over a bank of tubes with N rows is given by Jakob (1938) as

$$\Delta p = \frac{2f'G_{max}^2 N}{\rho} \left(\frac{\mu_w}{\mu_b} \right)^{0.14} \tag{7-61}$$

where the friction factor f' for a triangular array is

$$f' = \left\{ 0.25 + \frac{0.118}{[(p_t - D)/D]^{1.08}} \right\} Re^{-0.16} \tag{7-62}$$

and for a rectangular array is

$$f' = \left\{ 0.044 + \frac{0.08 p_p/D}{[(p_t - D)/D]^{(0.43 + 1.13D/p_p)}} \right\} Re^{-0.15} \tag{7-63}$$

Note that the pressure drops in these two configurations can differ considerably. The triangular array, which can be used to pack tubes more tightly together, can also give rise to greater form drag.

The case of a bank of tubes illustrates the range of design questions that can be associated with heat transfer. The engineer is interested in increasing heat transfer, but doing so will also increase friction and required pumping power. The engineer may also be interested in keeping down the space required for the equipment, but this too can have pressure drop implications. Design decisions will be determined by relative priorities assigned to the various considerations.

EXAMPLE 7-7. Air 300°K and atmospheric pressure flows at 10 m/sec across a rectangular bank of tubes five rows deep with ten transverse rows with outside surface temperature of 400°K. The tubes are 5 cm in diameter, 1 m long, and are on a pitch of 8 cm in the transverse direction and in the parallel direction. Find the (a) heat transfer coefficient and (b) pressure drop.

SOLUTION. (a) For the heat transfer coefficient, let us use the overall correlation

$$\mathrm{Nu} = C \, \mathrm{Re}^n \mathrm{Pr}_\infty^{0.36} \left(\frac{\mathrm{Pr}_\infty}{\mathrm{Pr}_w} \right)^{1/4} K_M$$

where K_M will correct for M being fewer than ten rows. The Reynolds number is

$$\mathrm{Re} = \frac{\dot{m} D}{NL(p_t - D)\mu} = \frac{\rho u N L p_t D}{NL(p_t - D)\mu} = \frac{uD}{\nu} \frac{p_t}{p_t - D}$$

Evaluating the Reynolds number yields

$$\mathrm{Re} = \frac{(10)(0.05)}{1.59 \times 10^{-5}} \left(\frac{0.08}{0.08 - 0.05} \right) = 8.39 \times 10^4$$

For this Reynolds number, we have

$$\mathrm{Nu} = 0.27(8.39 \times 10^4)^{0.63}(0.7)^{0.36}(0.99)(0.92) = 274$$

The heat transfer coefficient is then given by

$$h = \mathrm{Nu} \frac{k}{D} = 274 \left(\frac{0.03}{0.05} \right) = 164 \ \mathrm{W/m^2 \cdot {}^\circ K}$$

(b) The pressure drop is given by

$$\Delta p = \frac{2f'G_{max} N}{\rho} \left(\frac{\mu_n}{\mu_s} \right)^{0.14}$$

$$f' = 0.044 + \frac{0.08 p_p/D}{[(p_t - D)/D](0.43 + 1.13D/p_p)} \mathrm{Re}^{-0.15}$$

194

Chapter Seven

The coefficient f' may be evaluated as

$$f' = 0.44 + \frac{(0.08)(0.08/0.05)}{[(6.08 - 0.05)/0.05][0.43 + 1.151(0.05/0.08)]}(8.39 \times 10^4)^{-0.5}$$

$$= 0.042$$

$$G_{max} = \rho u_{max} = \frac{\rho u_{max} D}{\nu} = \frac{\nu}{D} \text{Re} = 26.7 \text{ kg/m}^2 \cdot \text{sec}$$

$$\Delta p = \frac{2(0.042)(26.7)^2(10)}{1}(1.02) = 611 \text{ N/m}^2$$

This problem can be done in a spreadsheet with an IF test to see if the Reynolds number is less than 2×10^5. It is easier to set up a spreadsheet for the approach taken in this example than for the approach of using Table 7-6-1.

COMPUTER PROJECT 7-3.   Extend your computer program for evaluating heat transfer coefficients to include flow over bluff bodies and over banks of tubes.

COMPUTER PROJECT 7-4.   Extend your computer program for evaluating heat transfer coefficients to include options for evaluating friction coefficients and pressure drops.

7-7. LIQUID METALS

As noted in Chapter 6, liquid metals have very low Prandtl numbers and thus tend to have somewhat different characteristics from gases and other liquids. The analysis of a laminar flow over a flat plate leads us to expect that heat transfer may be correlated effectively using the Peclet number (the product of the Reynolds and Prandtl numbers) and heat transfer correlations applicable to liquid metals have been given in earlier sections of this chapter.

A rough argument, in which even in turbulent flow heat transfer is dominated by conduction, indicates that we should expect a Nusselt number on the order of 8. This argument says that we may repeat the analysis used in Section 7-2 in laminar flow except that for turbulent flow, we assume the velocity distribution is uniform. This type of flow is called slug flow. We thus see

$$\frac{1}{r}\frac{\partial}{\partial r}\left(kr\frac{\partial T}{\partial r}\right) = \rho c u \frac{\partial T}{\partial x} \tag{7-64}$$

with the right-hand side of the equation a constant. This is the same as the equation in Chapter 3 for a cylinder with a uniform heat source, so we may write directly

$$T_w - T(r) = \frac{\rho c u\, \partial T/\partial x}{4k}(R^2 - r^2) \tag{7-65}$$

The bulk temperature may be obtained from

$$T_w - T_b = \frac{\int_0^R 2\pi r\, dr\, \rho u c \left[T_w - T(r)\right]}{\int_0^L 2\pi r\, dr\, \rho u c} \qquad (7\text{-}66)$$

We then define

$$h = \frac{q}{A}(T_w - T_b) = \frac{k(\partial T/\partial r)(R)}{T_w - T_b} \qquad (7\text{-}67)$$

We may evaluate

$$k\frac{\partial T}{\partial r} = k\frac{\rho c u\, \partial T/\partial x}{4k}2R \qquad (7\text{-}68)$$

We then obtain

$$\text{Nu} = \frac{hD}{k} = 8 \qquad (7\text{-}69)$$

In actual turbulent flow, we expect fluid motion to contribute in addition to conduction, and the correlation of Eq. (7-41) yields appropriate values. The pure conduction analysis is instructive even for fluids other than liquid metals because it provides a basis for understanding the degree to which flow considerations enhance heat transfer relative to pure conduction. Knowledge of the pure conduction Nusselt number can also be helpful for identifying errors. If you were to calculate the Nusselt number for turbulent flow in a circular pipe to be 0.7, you would be wise to check your calculations.

EXAMPLE 7-8. Liquid sodium (Pr = 0.005, $\nu = 3.27 \times 10^{-7}$, $k = 72$) flows at 3 m/sec through a channel 1 cm in diameter in which there is a constant heat flux. Evaluate the heat transfer coefficient.

SOLUTION. Equation (7-41) provides, for constant heat flux in turbulent flow,

$$\text{Nu} = 4.82 + 0.0185\, \text{Pe}^{0.827}$$

The Reynolds number is

$$\text{Re} = \frac{uD}{\nu} = \frac{(3)(0.01)}{3.27 \times 10^{-7}} = 9.17 \times 10^4 > 2300$$

so the flow is indeed turbulent. The Peclet number is

$$\text{Pe} = \text{Re}\,\text{Pr} = (9.17 \times 10^4)(0.005) = 459$$

The Nusselt number is

$$Nu = 4.82 + 0.0185(459)^{0.827} = 7.76$$

This is very close to the value of 8 predicted by the slug flow model of this section. The heat transfer coefficient is then

$$h = Nu\frac{k}{D} = 7.76\left(\frac{72}{0.01}\right) = 5.59 \times 10^4$$

PROBLEMS

7-1. Consider the two parallel flat plates of Fig. 7-3-1. If the separation between the plates is D and the width W of the channel is very large, show that the condition that the boundary layers grow to comparable size is expressed by a condition on the parameter $Re_D D/L$.

7-2. Repeat Example 7-1 if the entire tube is the 10 cm long heated section.

7-3. Prepare a spreadsheet for Example 7-1.

7-4. Suppose in Example 7-1 that there is a desire to reduce the wall temperature from 125°C to some lower value. Tabulate or plot the variation of the heat transfer coefficient and the heat flow as the wall temperature is reduced to 75°C in steps of 10°C.

7-5. Repeat Problem 7-2 if the fluid is engine oil instead of air.

7-6. Fluid flows in a pipe whose wall temperature is 350°K. The average bulk temperature is 300°K. Evaluate the correction term associated with viscosity in Eq. (7-31) for water, engine oil, and ethylene glycol.

7-7. Suppose in Example 7-2 that you were given that the air inlet temperature is 500°K and that the fluid speed is 2 m/sec, but were not given the outlet temperature of the air. How much heat is transferred to the air?

7-8. Suppose in Example 7-3 that you were asked to find what roughness should have been placed in the pipe to achieve a heat transfer coefficient one-third greater than that which would be achieved in a smooth pipe.

7-9. What is the energy lost due to friction in the tube of Example 7-3?

7-10. Suppose the fluid speed in Problem 7-9 is doubled. By what ratios would the energy transferred and the energy lost due to friction change?

7-11. In Example 7-3, the Nusselt number was taken to be 698. If the uncertainty in the heat transfer correlation is taken to be 15%, what flow rate would be needed to assure the same heat transfer?

7-12. You wish to consider several design variations relative to Problem 7-7. Examine the implications for heat transfer and frictional loss. Each variation leads to a doubling of flow rate.
a. You double the fluid speed.
b. You double the air pressure.
c. You increase the tube diameter so as to double the flow cross-sectional area.

7-13. Suppose in Example 7-2 that you were given that the average air speed is 2 m/sec, but were not given the length of the tube. What tube length is required?

7-14. Suppose in Example 7-2 that you were given that the average air speed is 2 m/sec, but were not given the tube diameter. What tube diameter is needed?

7-15. One kg/sec of water with an average temperature of 60°C flows through a tube 2 cm in diameter and 10 m long. A constant heat flux is imposed on the surface of the tube.

Experimental testing indicates that the heat transfer coefficient is $20{,}000 \text{ W}/\text{m}^2 \cdot {}^\circ\text{K}$. What is the average roughness of the tube?

7-16. Instead of the air of Examples 7-5 and 7-6, other fluids at the same temperature are used, with fluid speeds modified so as to yield the same Reynolds number. Tabulate or plot the Nusselt number as a function of the Prandtl number. Include, in the range of your table, water, engine oil, and mercury.

7-17. In Example 7-5, what is the drag force per meter of pipe length?

7-18. Extend the spreadsheet for Example 7-5 to provide for the calculation of drag force.

7-19. A spherical surface is at a constant temperature T_i. This surface is surrounded by a stagnant conducting medium in the shape of a spherical shell whose outer surface temperature is T_o.
   a. Find the heat transfer between the inner and outer surfaces.
   b. Find the limiting value of the heat transfer as the outer radius of the shell goes to zero.
   c. Define a heat transfer coefficient that yields the same heat flow as in part b.
   d. If the diameter of the inner sphere is the length dimension, show that the Nusselt number corresponding to the heat transfer coefficient in part c is 2.

7-20. Using the approach of Problem 7-19, show that the Nusselt number for slab and plane geometries would be zero if there were an infinite stagnant surrounding medium.

7-21. For the conditions of Example 7-5, vary the air speed between 1 and 20 m/sec. Tabulate or plot the heat transfer, the drag force, and the ratio of the two.

7-22. For the conditions of Example 7-5, vary the pipe diameter between 5 and 20 cm. Tabulate or plot the heat transfer, the drag force, and the ratio of the two.

7-23. Show that according to the Dittus–Boelter heat transfer correlation [Eq. (7-28)], heat transfer in turbulent flow in a pipe varies as $h \sim \rho^{4/5} k^{3/5} C_p^{2/5} \mu^{-2/5}$.

7-24. Using the result of Problem 7-23, define a new figure-of-merit property conveying tendency to transfer heat in turbulent flow in a pipe to add to the database of Appendix A. Rank order the fluids according to that property at $300{}^\circ\text{K}$. Do not include liquid metals.

7-25. Based on the result of Problem 7-23, how would you expect heat transfer to be affected by
   a. increasing pressure for a gas;
   b. increasing temperature for a gas; and
   c. increasing temperature for a liquid.

7-26. Tabulate or plot the properties μ, C_p, k, ν, Pr, and α for air as a function of temperature. Note which properties vary strongly with temperature (and in which direction). Note which properties vary weakly.

7-27. Repeat Problem 7-26 for water.

7-28. Based on Problems 7-26 and 7-27 compare how properties vary for air and water. Consult data for other liquids and gases, and note whether the trends for air and water apply to gases and liquids in general.

7-29. Consider the figure of merit for forced convection heat transfer developed in Problems 7-23 and 7-24. Tabulate or plot this figure of merit as a function of temperature for air. Does the "quality" of air as a heat transfer fluid improve with temperature.

7-30. Repeat Problem 7-29 water.

7-31. To the extent that Eq. (7-26) applies as a friction factor correlation, show that a parameter that describes the tendency of a fluid to lose energy by friction is $\mu^{1/5} \rho^{4/5}$.

7-32. Based on the parameter in Problem 7-31, rank order fluids according to susceptibility to friction losses.

7-33. It is of interest to know the degree to which increasing heat transfer also increases friction. Show that a parameter that can be used to characterize the ratio of friction to heat transfer for a particular fluid for turbulent flow in a pipe is $\text{Pr}/C_p^{5/3}$.

7-34. Rank order fluids (excluding liquid metals) according to the parameter of Problem 7-33 at 300°K.

7-35. For air, tabulate or plot the parameter of Problem 7-33 as a function of temperature. Does the friction to heat transfer ratio increase or decrease with temperature? Do you expect a similar trend for other gases?

7-36. For water, tabulate or plot the parameter of Problem 7-33 as a function of temperature. Do you expect a similar trend for other fluids?

REFERENCES

1. V. S. Arpaci and P. S. Larsen, *Convection Heat Transfer*, Prentice-Hall, Englewood Cliffs, NJ, 1984.
2. A. Bejan, *Convection Heat Transfer*, Wiley, New York, 1984.
3. L. C. Barmeister, *Convection Heat Transfer*, Wiley, New York, 1983.
4. T. Cebeci and P. Bradshaw, *Physical and Computational Aspects of Convective Heat Transfer*, Springer, New York, 1984.
5. S. W. Churchill and M. Bernstein, "A Correlating Equation for Forced Convection from Gases and Liquids to a Circular Cylinder in Crossflow," *J. Jeat Transfer*, **99**, 300 (1977).
6. A. P. Colburn, "A Method of Correlating Forced Convection Heat Transfer Data and a Comparison with Fluid Friction," *Trans. AIChE*, **29**, 174 (1933).
7. F. W. D. Dittus and L. M. K. Boelter, *Univ. California (Berkeley) Publ. Eng.*, **2**, 443 (1930).
8. W. H. Giedt, "Investigation of Variation of Point Unit Heat Transfer Coefficient Around a Cylinder Normal to an Air Stream," *ASME Trans.*, **71**, 375 (1949).
9. E. C. Grimison, "Correlation and Utilization of New Data on Flow Resistance and Heat Transfer for Cross-Flow of Gases over Tube Banks," *Trans. ASME*, **59**, 583 (1937).
10. S. E. Haaland, "Simple and Explicit Formulas for the Friction Factor in Turbulent Pipe Flow," *J. Fluids Eng.*, **105**, 89 (1983).
11. H. Hausen, "Dartellung des Warmeuberganges in Rohren durch verallgemeinerte potenz besiehungen," *VDIZ*, **4**, 91 (1943).
12. J. P. Holman, *Heat Transfer*, 5th ed., McGraw-Hill, New York, 1981.
13. M. Jakob "Heat Transfer and Flow Resistance in Cross-Flow of Gases over Tube Banks," *Trans. ASME*, **60**, 384 (1938).
14. M. Jakob, *Heat & Transfer*, Vol. 1, Wiley, New York, 1949.
15. W. M. Kays and M. E. Crawford, *Convective Heat and Mass Transfer*, McGraw-Hill, New York, 1980.
16. W. M. Kays and R. K. Lo, "Basic Heat Transfer and Friction Data for Gas Flow Normal to Banks of Staggered Tubes: Use of a Transient Technique," Stanford U. Tech Rept. 15, 1952; see also *Trans. ASME*, **76**, 387 (1954).
17. F. Kreith and W. Z. Black, *Basic Heat Transfer*, Harper & Row, New York, 1980.
18. S. Nakai and T. Okazaki, "Heat Transfer from a Horizontal Circular Wire at Small Reynolds and Grashof Numbers—I. Pure Convection," *Int. J. Heat Mass Transfer*, **18**, 387 (1975).
19. R. H. Norris, "Some Simple Approximate Heat Transfer Correlations for Turbulent Flow in Ducts with Rough Surfaces," in A. E. Bergles and R. L. Webb (eds), *Augmentation of Convective Heat and Mass Transfer*, ASME Symposium, 1970.
20. W. Nusselt, "Warmeaustausch zwicher Wand and Wasser in Rohr," *Forsch. Geb. Ingenieuswes*, **2**, 309 (1931).
21. B. W. Petukhov, Heat Transfer and Friction in Turbulent Pipe Flow with Variable Physical Properties, in J. P. Harlett and T. F. Irvine (eds), *Advances in Heat Transfer*, Academic Press, New York, 1970.
22. W. Rohsenow and J. Hartnett, *Handbook of Heat Transfer*, McGraw-Hill, New York, 1973.
23. H. Schlichting, *Boundary Layer Theory*, 7th ed., McGraw-Hill, New York, 1979.

24. R. A. Seban and T. T. Shimazaki, "Heat Transfer to a Fluid Flowing Turbulently in a Smooth Pipe with Walls at Constant Temperature," *Trans. ASME*, **73**, 803 (1951).

25. R. K. Shah and S. L. London, *Laminar Flow Forced Convection in Ducts*, Academic Press, New York, 1979.

26. E. N. Sieder and C. E. Tate, "Heat Transfer and Pressure Drop of Liquids in Tubes," *Ind. Eng. Chem.*, **28**, 1429 (1936).

27. E. Skupinshi, J. Hortel, and L. Vautrey, "Determination des coefficients de convection d'un alliage sodium–potassium dans un tube circulaire," *Int. J. Heat Mass Transfer*, **8**, 937 (1965).

28. S. Whitaker, "Forced Convection Heat—Transfer Correlations for Flow in Pipes, Past Flow in Packed Beds and Tube Bundles," *AIChE J.*, **18**, 361 (1972).

29. F. White, *Heat Transfer*, Addison-Wesley, Reading, MA, 1983.

30. A. Zukauskas, Heat Transfer from Tubes in Crossflow, in J. P. Hartlett and T. F. Irvine, Jr. (eds.), *Advances in Heat Transfer*, Academic Press, New York, 1970.

Chapter Eight

NATURAL CONVECTION

8-1. INTRODUCTION

In modeling forced convection, we introduced one extreme view of the relationship between fluid flow and heat; namely, we assumed that the fluid flow distribution was independent of the heat flow. When analyzing boundary layers, we found the velocity profile before we introduced any information about heat. With pure natural convection, we introduce an opposite extreme; namely, we assume that we will not have any flow except for that caused by heating.

Heating stimulates flow because fluid properties vary with temperature. When a fluid is heated, its density usually decreases (i.e., it gets lighter), so that it tends to rise. Thus, the force of gravity becomes a significant factor. When the fluid rises, it leads to a situation where fluid is in motion in the vicinity of the hot surface. The degree of motion influences the process of heat transfer.

In this chapter, we are concerned with natural convection due to gravity. We recognize, however, that analogous natural convection can exist if there is an existing force field that depends on density. For example, rotating systems impose centrifugal forces that depend on density. Centrifugal force is the analogue of gravity in such a situation.

The manner in which natural convection occurs depends on the geometry. Next to a vertical hot wall, fluid will expand and rise along the wall. This is analogous to the boundary layer behavior encountered with forced convection.

Because of previous consideration of boundary layers, this case is considered first. The types of dimensionless parameter that characterize natural convection problems are determined. Above horizontal surfaces, a heated fluid rises directly. However, the types of dimensionless parameter derived from the boundary layer problem are also useful in dealing with other natural convection problems.

8-2. GRAVITY AND BOUNDARY LAYER THEORY

Consider a vertical flat plate immersed in a fluid, as shown in Fig. 8-2-1. The surface of the plate is maintained at a temperature T_w greater than the asymptotic fluid temperature T_∞. The fluid near the wall is heated and, since it becomes less dense, tends to rise. Asymptotically, the fluid is stagnant at a uniform temperature. Therefore, fluid flow occurs primarily in the vicinity of the plate and can be modeled as a boundary layer phenomenon.

The momentum equation for the situation may be written

$$\frac{\partial}{\partial z}(\rho u^2) + \frac{\partial}{\partial y}(\rho uv) = \frac{\partial p}{\partial z} - \rho g + \mu \frac{\partial^2 u}{\partial y^2} \tag{8-1}$$

where u is the component of velocity in the vertical (z) direction. The right-hand side includes the pertinent forces, which for this case are pressure gradient, gravity, and viscosity. In the stagnant case, there is no flow, and pressure gradient balances gravity (e.g., air pressure decreases with altitude in the atmosphere). We assume that the pressure gradient is determined by the asymptotic conditions: that is

$$\frac{\partial p}{\partial z} = -\rho_\infty g \tag{8-2}$$

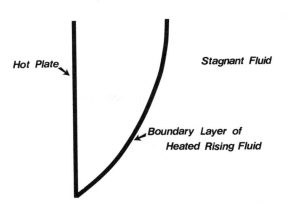

FIGURE 8-2-1. Heated vertical plate.

In other words, we assume that the existence of heat transfer does not affect the pressure gradient.

With the above assumptions, the momentum equation becomes

$$\frac{\partial}{\partial z}(\rho u^2) + \frac{\partial}{\partial y}(\rho u v) = g(\rho_\infty - \rho) + \mu \frac{\partial^2 u}{\partial y^2} \tag{8-3}$$

The difference in density may be related to a fluid property called the coefficient of expansion β

$$\beta = \frac{1}{V} \frac{\partial V}{\partial T} \tag{8-4}$$

Since volume and density are related by

$$V = \frac{M}{\rho} \tag{8-5}$$

where M is mass, we may write

$$\beta = \rho \frac{\partial}{\partial T}\left(\frac{1}{\rho}\right) = -\frac{1}{\rho} \frac{\partial \rho}{\partial T} \tag{8-6}$$

For an ideal gas

$$\rho = \frac{p}{RT} \tag{8-7}$$

$$\beta = -\frac{1}{\rho} \frac{\partial \rho}{\partial T} = \frac{1}{T} \tag{8-8}$$

For fluids in general, we approximate

$$\beta = -\frac{1}{\rho} \frac{\partial \rho}{\partial T} = \frac{\rho_\infty - \rho}{\rho(T - T_\infty)} \tag{8-9}$$

Substituting for $\rho_\infty - \rho$ in the momentum equation yields

$$\frac{\partial}{\partial z}(\rho u^2) + \frac{\partial}{\partial y}(\rho u v) = g\rho\beta(T - T_\infty) + \mu \frac{\partial^2 u}{\partial y^2} \tag{8-10}$$

The momentum equation is now expressed such that the difference between boundary layer and asymptotic temperature is the driving force producing fluid motion.

We note that since temperature drives fluid motion, we cannot deal separately with momentum and energy as we did with forced convection.

Thus, we must consider the energy equation

$$\frac{\partial}{\partial z}(uT) + \frac{\partial}{\partial y}(vT) = \frac{k}{\rho c}\frac{\partial^2 T}{\partial y^2} \tag{8-11}$$

simultaneously with the momentum equation.

When we dealt with forced convection, we assumed that boundary layer thickness could be different for velocity and temperature. This was because the velocity boundary layer would exist for a flowing fluid even if there were no heat transfer. In the natural convection case, there is no partial independence of phenomena. Where the temperature has not yet reached its asymptotic value, that is, within the thermal boundary layer, heated fluid will have a density lower than the asymptotic value and will tend to rise. Therefore, for the natural convection case, it is reasonable to assume that the boundary layer thicknesses are the same for temperature and velocity.

8-3. ANALYSIS FOR THE VERTICAL FLAT PLATE

We note that the boundary layer for a vertical plate should have the following characteristics

$$u(0) = u(\delta) = 0 \tag{8-12}$$

$$\frac{\partial u}{\partial y}(\delta) = 0 \tag{8-13}$$

$$T(0) = T_w \tag{8-14}$$

$$T(\delta) = T_\infty \tag{8-15}$$

$$\frac{\partial T}{\partial y}(\delta) = 0 \tag{8-16}$$

The above conditions are the same as those for forced convection except that velocity is zero at the outside of the boundary layer since the asymptotic velocity condition is stagnation. Thus, whereas the velocity profile in the forced convection boundary layer was monotonically increasing from zero at the wall to a maximum at the outside, the velocity profile in the natural convection boundary layer will rise from zero at the wall to some maximum value and then decline back to zero, as shown in Fig. 8-3-1. Consistent with these conditions, we postulate profiles in polynomial form:

$$u = u_0 \frac{y}{\delta}\left(1 - \frac{y}{\delta}\right)^2 \tag{8-17}$$

$$\frac{T - T_\infty}{T_w - T_\infty} = \left(1 - \frac{y}{\delta}\right)^2 \tag{8-18}$$

The quantity u_0 is as yet undetermined.

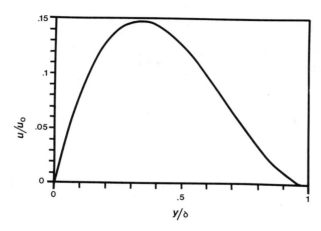

FIGURE 8-3-1

Let us now integrate the momentum equation over the boundary layer:

$$\int_0^\delta dy \frac{\partial}{\partial z}(\rho u^2) + \rho u v |_0^\delta = \int_0^\delta dy \, g\rho\beta(T - T_\infty) + \mu \frac{\partial u}{\partial y}\bigg|_0^\delta \qquad (8\text{-}19)$$

Since $u = 0$ at $y = 0$ and $y = \delta$, the second term on the left-hand side vanishes. For the same reason, application of the Leibnitz rule to interchange integration and differentiation permits moving the derivative outside the integral in the first term. On the right-hand side, $\partial u/\partial y = 0$ at $y = \delta$. The equation becomes

$$\frac{d}{dz}\int_0^\delta dy \, \rho u^2 = \int_0^\delta dy \, g\rho\beta(T - T_\infty) - \mu \frac{\partial u}{\partial y}(y = 0) \qquad (8\text{-}20)$$

The last term on the right-hand side is the viscous stress at the wall.
Integrating the energy equation yields

$$\int_0^\delta dy \frac{\partial}{\partial z}(uT) + vT|_0^\delta = \alpha \frac{\partial T}{\partial y}\bigg|_0^\delta \qquad (8\text{-}21)$$

Since speeds vanish at both sides of the boundary layer and $\partial T/\partial y = 0$ at $y = \delta$, and noting that

$$\frac{\partial}{\partial z}(uT) = \frac{\partial}{\partial z}[u(T - T_\infty)] \qquad (8\text{-}22)$$

we obtain

$$\frac{d}{dz}\int_0^\delta dy \, u(T - T_\infty) = -\alpha \frac{\partial T}{\partial y}(y = 0) \qquad (8\text{-}23)$$

Substituting the polynomials for speed and temperature into the integrated momentum and energy equations, and evaluating the integrals assuming constant properties, yields the equations

$$\frac{1}{105}\frac{d}{dz}(u_0^2\delta) = \tfrac{1}{3}g\beta(T_w - T_\infty)\delta - \nu\frac{u_0}{\delta} \tag{8-24}$$

$$\frac{d}{dz}(u_0\delta) = \frac{60\alpha}{\delta} \tag{8-25}$$

These two equations are simultaneous equations for δ and u_0. We may assume solutions of the form

$$\delta(z) = c_1 z^n \tag{8-26}$$

$$u_0(z) = c_2 z^m \tag{8-27}$$

This assumption, when substituted into the equations, yields $c_1, c_2, n,$ and m such that

$$\delta = 3.93\left(0.952 + \frac{\nu}{\alpha}\right)^{1/4}\left[\frac{g\beta(T_w - T_\infty)}{\nu^2}\right]^{-1/4}\left(\frac{\nu}{\alpha}\right)^{-1/2}z^{1/4} \tag{8-28}$$

$$u_0 = 5.17\nu\left(0.952 + \frac{\nu}{\alpha}\right)^{-1/2}\left[\frac{g\beta(T_w - T_\infty)}{\nu^2}\right]^{1/2}z^{1/2} \tag{8-29}$$

Defining a new parameter that we call the Grashof number Gr by

$$Gr = \frac{g\beta(T_w - T_\infty)z^3}{\nu^2} \tag{8-30}$$

we obtain

$$\frac{\delta}{z} = 3.93\,\text{Pr}^{-1/2}(0.952 + \text{Pr})^{1/4}\text{Gr}^{-1/4} \tag{8-31}$$

Noting that

$$h = \frac{-k(\partial T/\partial y)(y=0)}{T_w - T_\infty} = \frac{2k}{\delta} \tag{8-32}$$

we may evaluate the Nusselt number as

$$\text{Nu} = 0.508\,\text{Pr}^{1/2}(0.952 + \text{Pr})^{-1/4}\text{Gr}^{1/4} \tag{8-33}$$

As a rough approximation, assume that the Prandtl number is near 1 (as for gases) so that we may approximate

$$0.952 + \text{Pr} \approx 2\,\text{Pr} \tag{8-34}$$

Then we obtain a Nusselt number proportional to $(\text{Gr Pr})^{1/4}$. It is common practice to correlate heat transfer for natural convection in terms of

$$\text{Nu} = C(\text{Gr Pr})^{n} \tag{8-35}$$

although we should expect some separate sensitivity to Prandtl number. A formula [based on more elaborate analysis than is Eq. (8-33)] incorporating Prandtl number dependence has been given by LeFevre (1956)

$$\text{Nu} = \frac{0.75\,\text{Pr}^{1/4}}{(2.435 - 4.884\sqrt{\text{Pr}} + 4.953\,\text{Pr})^{1/4}}(\text{Gr Pr})^{1/4} \tag{8-36}$$

which implies a Prandtl number dependence for the coefficient C; that is,

$$C = 0.75\left(\frac{\text{Pr}}{2.435 + 4.88\sqrt{\text{Pr}} + 4.953\,\text{Pr}}\right)^{1/4} \tag{8-37}$$

This value of C is for the local value of the heat transfer coefficient. The value of C that would be associated with the average over the plate would be somewhat different.

The above analysis thus far has been for laminar behavior in the boundary layer. Let us obtain a Reynolds-like number by taking the characteristic speed and using it in

$$\frac{u_0 z}{\nu} = 5.17(0.952 + \text{Pr})^{1/2}\left[\frac{g\beta(T_w - T_\infty)z^3}{\nu^2}\right]^{1/2}$$

$$= 5.17(0.952 + \text{Pr})^{-1/2}\text{Gr}^{1/2} \tag{8-38}$$

The square root of the Grashof number provides the type of information carried by the Reynolds number. Thus, we expect flow to be laminar when the Grashof number is low and turbulent when the Grashof number is high. This is indeed the case.

If we consider the ratio of kinetic energy and viscous force, we obtain, since δ is the characteristic length dimension for viscous stress,

$$\frac{u_0 \delta}{\nu} = \frac{u_0 z}{\nu}\frac{\delta}{z} \tag{8-39}$$

which, from Eqs. (8-31) and (8-38), is proportional to the one-fourth power of the Grashof number. We thus expect a fairly high Grashof number to be required for transition to turbulent flow. This is the case, with values for transition being in the neighborhood of 4×10^8.

It may be observed that

$$h \sim \left(\frac{1}{z}\right)^{1/4} \tag{8-40}$$

Thus, as with forced convection over a flat plate, heat transfer is greatest at the leading edge, but the rate of decline is slower. As in the forced convection case, there is a decline in the rate of conduction into the fluid as the boundary layer gets thicker. However, with natural convection, there is a compensating increase in the flow area as the boundary layer gets thicker. It may be easily shown that the average heat transfer coefficient on a vertical plate in laminar flow is 4/3 the heat transfer coefficient at the end.

The Grashof number, as defined, includes a single wall temperature. If, instead of a constant wall temperature case, we were to encounter a uniform heat flux case, we might wish a somewhat modified dimensionless parameter. Noting that, on a dimensional basis, heat flux and temperature difference may be related by

$$\Delta T = \frac{q''H}{k} \tag{8-41}$$

where H is some length dimension, we may define a modified Grashof number Gr^* by using z as this length dimension, which yields

$$\mathrm{Gr}^* = \frac{g\beta q_w'' z^4}{k\nu^2} \tag{8-42}$$

EXAMPLE 8-1.   Find the boundary layer thickness, the heat transfer coefficient, and the characteristic speed u_0 at the center of a vertical 20 c plate maintained at a uniform temperature of $325°\,\mathrm{K}$, with surrounding air at $275°\,\mathrm{K}$.

SOLUTION.   Let properties be evaluated at the film temperature, that is, the average of the wall and asymptotic temperatures. The film temperature is

$$T_f = \tfrac{1}{2}(T_w + T_\infty) = \tfrac{1}{2}(325 + 275) = 300°\,\mathrm{K}$$

The Grashof number may be evaluated as

$$\mathrm{Gr} = \frac{g\beta(T_w - T_\infty)z^3}{\nu^2} = \frac{g(T_w - T_\infty)z^3}{\nu^2 T_f}$$

where we have taken air to be an ideal gas and used Eq. (8-8). Evaluating the Grashof number yields

$$\mathrm{Gr} = \frac{(9.8)(50)(0.1)^3}{(1.566 \times 10^{-5})^2(300)} = 6.66 \times 10^6$$

The Prandtl number is 0.711, so

$$0.952 + \mathrm{Pr} = 1.663$$

From Eq. (8-31)

$$\frac{\delta}{z} = 3.39(0.711)^{-1/2}(1.663)^{1/4}(6.66 \times 10^6)^{-1/4} = 0.104$$

$$\delta = (0.104)(0.1) = 0.0104 \text{ m} = 1.04 \text{ cm}$$

Note that the boundary layer, while thin relative to the overall plate dimension, is thicker than boundary layers we encountered in dealing with forced convection.
The heat transfer coefficient is

$$h = \frac{2k}{\delta} = \frac{2(0.026)}{0.01041} = 5 \text{ w/m}^2 \cdot {}^\circ \text{K}$$

The characteristic speed from Eq. (8-38) is

$$u_0 = \frac{1.566 \times 10^{-5}}{0.1} 5.17(1.663)^{-1/2}(6.66 \times 10^6)^{1/2} = 1.62 \text{ m/sec}$$

$$= 162 \text{ cm/sec}$$

The average speed can be shown (a problem at the end of this chapter) to be one-twelfth the characteristic speed, or 4 cm/sec. We therefore expect that it would take about 1 sec for air to flow by the plate.

8-4. VERTICAL AND HORIZONTAL SURFACES

As noted previously, we expect the Nusselt number for free convection from a vertical plate to depend on the Grashof number and the Prandtl number. It has been common to represent the average Nusselt number for the surface as

$$\text{Nu} = C(\text{Gr Pr})^n \qquad (8\text{-}43)$$

The product Gr Pr is called the Rayleigh number Ra. The height of the surface is the length dimension used. Table 8-4-1 shows values of C and n gathered by Holman (1979). A more general correlation for laminar flow (Ra $< 10^9$) is given by Churchill and Chu (1975a, b)

$$\overline{\text{Nu}} = 0.68 + \frac{0.670 \, \text{Ra}^{1/4}}{\left[1 + (0.492/\text{Pr})^{9/16}\right]^{4/9}} \qquad (8\text{-}44)$$

This equation, in addition to providing Prandtl number dependence, provides a formula for use when the Rayleigh number is less than 10^4, and C, n pairs are not available from Table 8-4-1.
Note that for high Ra (turbulent flow), the value of n is $\frac{1}{3}$. The length dimension L in the Nusselt number and the $(L^3)^{1/3}$ from the Grashof number then cancel, and the heat transfer coefficient is independent of the height of

TABLE 8-4-1
Values of C and n for Natural Convection

	Ra range	C	n
Vertical plates	10^4–10^9	0.50	$\frac{1}{4}$
	10^9–10^{13}	0.10	$\frac{1}{3}$
Horizontal cylinders	10^4–10^9	0.53	$\frac{1}{4}$
	10^9–10^{12}	0.13	$\frac{1}{3}$
Horizontal plates			
Hot surface up	2×10^4–8×10^6	0.54	$\frac{1}{4}$
or cold surface down	8×10^6–10^{11}	0.15	$\frac{1}{3}$
Hot surface down or cold surface up	10^5–10^{11}	0.58	$\frac{1}{5}$

the plate. This is an empirical result, and other research has led to somewhat different C and n values. There is an implication that the heat transfer coefficient is insensitive to plate height. A more general correlation for turbulent flow provided by Churchill and Chu is

$$\overline{\mathrm{Nu}} = \left\{ 0.825 + \frac{0.387\,\mathrm{Ra}^{1/6}}{\left[1 + (0.492/\mathrm{Pr})^{9/16}\right]^{8/27}} \right\}^2 \tag{8-45}$$

This correlation actually was developed to cover both laminar and turbulent flow. For large Rayleigh numbers this correlation tends toward the $\mathrm{Ra}^{1/3}$ form with

$$C = \frac{0.15}{\left[1 + (0.492/\mathrm{Pr})^{9/16}\right]^{16/27}} \tag{8-46}$$

These expressions apply for liquid metals, that is, for very low Prandtl numbers. We may observe that for low Prandtl numbers, Eq. (8-44) tends toward dependence on $\mathrm{Gr}^{1/4}\mathrm{Pr}^{1/2}$. (The algebra is left as an exercise.) Since we found that $\mathrm{Gr}^{1/2}$ is analogous to the Reynolds number, $\mathrm{Gr}^{1/4}\mathrm{Pr}^{1/2}$ is analogous to $\mathrm{Pe}^{1/2}$ (recall that $\mathrm{Pe} = \mathrm{Re\,Pr}$), the type of dependence encountered for forced convection laminar flow over a flat plate.

It has been found that the coefficients of the modified Grashof number of the constant heat flux case differ from the coefficients of the Grashof number of the constant temperature case. These differences may be understood as follows. The modified Grashof number may be expressed in the form

$$\mathrm{Gr}^* = \frac{g\rho q_w'' z^4}{k\nu^2} = \frac{g\beta\,\Delta T z^3}{\nu^2}\frac{q_w'' z^4}{\Delta T k} = \frac{g\beta\,\Delta T z^3}{\nu^2}\frac{hz}{k} \tag{8-47}$$

Thus, we may write, at any elevation z,

$$Gr^* = Gr\,Nu \qquad (8\text{-}48)$$

If, at that elevation, we were to find

$$Nu \sim Gr^{*m} = (Gr\,Nu)^m \qquad (8\text{-}49)$$

we would expect to find

$$Nu \sim Gr^{m/(1-m)} = Gr^n \qquad (8\text{-}50)$$

The local heat transfer coefficients have been correlated by Vliet (1969) and by Vliet and Lin (1969) for low and high values of Ra* (the Rayleigh number using Gr* instead of Gr) according to

$$Nu = 0.60(Gr^*Pr)^{1/5} \qquad (8\text{-}51)$$

$$Nu = 0.17(Gr^*Pr)^{1/4} \qquad (8\text{-}52)$$

with the two correlations yielding the same value at Ra* $\approx 10^{11}$. Thus, the value of $m = \frac{1}{5}$ corresponds to a value of $n = m/(1 - m) = \frac{1}{4}$, which indeed is the value provided for the isothermal wall. Similarly, in turbulent flow, $m = \frac{1}{4}$ corresponds to $n = \frac{1}{3}$. Thus, for the constant heat flux case in addition to the isothermal surface case, the correlations imply a lack of sensitivity of the heat transfer coefficient to plate height.

A recommended alternative is that correlations developed for isothermal conditions be applied for constant heat flux conditions using the ΔT at the center to evaluate the Grashof number. This implies an iterative process, since the ΔT is not known in advance. One may estimate the heat transfer coefficient, evaluate ΔT, evaluate the heat transfer coefficient, and continue to consistency. Note that in the turbulent flow region, the heat transfer coefficient is insensitive to position, so that the constant heat flux case is also isothermal. This is not true in the laminar region.

Suppose, on the basis of an assumed average heat transfer coefficient, we obtain an average temperature difference $\overline{\Delta T}$ by

$$\overline{\Delta T} = \frac{q_w}{\bar{h}} \qquad (8\text{-}53)$$

For laminar conditions, we may evaluate (since $h\,\Delta T$ is constant)

$$\Delta T(z) = Cq_w z^{1/4} \qquad (8\text{-}54)$$

where C is a constant. Evaluating the average temperature difference as

$$\overline{\Delta T} = \frac{1}{L}\int_0^L Cq_w z^{1/4}\,dz = Cq_w\left(\tfrac{4}{5}L^{1/4}\right) \qquad (8\text{-}55)$$

Since $\overline{\Delta T}$ has been evaluated in terms of $\overline{h}$, this equation can be used to get C and substitute into Eq. (8-54) to yield

$$\Delta T\left(\frac{L}{2}\right) = Cq_w\left(\frac{L}{2}\right)^{1/4} = \frac{5}{(4)(2^{1/4})}\overline{\Delta T} = 1.05\,\overline{\Delta T} \qquad (8\text{-}56)$$

This difference is small and may not be meaningful given the accuracy of correlations.

A particular procedure that may be used with constant heat flux would be the following:

1. Use a local correlation in terms of Ra* to estimate h at the center.
2. Evaluate ΔT at the center from the local h.
3. Use this ΔT to evaluate average h from the general correlation.
4. Iterate to consistency between average h and average ΔT.

The above procedure uses the cruder correlation (that does not have Prandtl number dependence) to get a good start on iterating to consistency with the sophisticated correlation.

EXAMPLE 8-2.   Evaluate the average heat transfer coefficient for the conditions of Example 8-1 using Eq. (8-43).

SOLUTION.   For this case, the length dimension to use is the full height of the plate, so

$$Gr = \frac{(9.8)(50)(0.2)^3}{(1.56 \times 10^{-5})^2(300)} = 5.33 \times 10^7$$

With $Pr = 0.711$

$$Ra = Gr\,Pr = 3.79 \times 10^7$$

For this case, the appropriate values from Table 8-4-1 yield

$$Nu = \frac{hL}{k} = 0.59\,Ra^{1/4} = 46.3$$

$$\overline{h} = Nu\frac{k}{L} = (46.3)\left(\frac{0.026}{0.2}\right) = 6.0$$

To compare this with the procedures of Example 8-1, we would find

$$h(L) = h(z)\left(\frac{z}{L}\right)^{1/4} = 5\left(\frac{0.1}{0.2}\right)^{1/4} = 4.2$$

$$\overline{h} = \tfrac{4}{3}h(L) = 5.6$$

The results are not identical, but they are within the accuracy expected of correlations and approximate solutions.

EXAMPLE 8-2 (Spreadsheet).   Because of the large number of cases to which the form of Eq. (8-41) has been applied, it is convenient to have this equation developed in a spreadsheet as follows:

	A	B
1	GRAVITY	9.8
2	BETA	1.0 / 300.0
3	DELTA – TEMP	50.0
4	LENGTH	.2
5	NU	1.566E – 5
6	LENGTH CUBE	+B4 ∧ 3
7	NU SQUARE	+B5 ∧ 2
8	GRASHOF	+B1*B2*B3*B6 / B7
9	PRANDTL	.711
10	RAYLEIGH	+B8*B9
11	C – CONSTANT	.59
12	N – CONSTANT	.25
13	NUSSELT	+B11*(B10 ∧ B12)
14	CONDUCTIVITY	.026
15	H	+B13*B14 / B4

With different numbers entered, this spreadsheet can be used for a variety of natural convection problems. In the above spreadsheet, the values of .59 and .25 are entered when the value of the Rayleigh number is displayed in B10. It is, of course, possible to use checking logic (e.g., @ IF statements) to make a general evaluation that determines problem type, selects proper constants, and so on. As will be discussed later in this section, there is a more general form of correlation that can be used with spreadsheets.

EXAMPLE 8-3.   Based on the discussion of Section 8-3, develop a table of values of C versus Prandtl number for laminar conditions at a vertical plate. Compare with the value of C in Table 8-4-1. Include in the table values of Pr corresponding to air and water at 300° K.

SOLUTION.   We found that for the local heat transfer coefficient

$$C_{\text{local}} = 0.75 \left(\frac{\text{Pr}}{2.435 + 4.884\sqrt{\text{Pr}} + 4.953\,\text{Pr}} \right)^{1/4}$$

Since h goes as $z^{-1/4}$, for the average heat transfer coefficient, it follows that (proof is left as an exercise)

$$C = \tfrac{4}{3} C_{\text{local}} = \left(\frac{\text{Pr}}{2.435 + 4.88\sqrt{\text{Pr}} + 4.953\,\text{Pr}} \right)^{1/4}$$

Evaluating this formula yields the following:

Identification	Prandtl	C
200°C Sodium	0.0072	0.224
300°K Air	0.708	0.516
	1.0	0.535
	2.0	0.569
300°K Freon	3.5	0.592
300°K Water	5.85	0.609
	1000	0.668

The results are plotted in Fig. 8-4-1. There does appear to be significant Prandtl number dependence, and the coefficient of 0.59 in Table 8-4-1 seems more characteristic of liquids (other than liquid metals) than of gases. Thus, there is incentive to use the more general correlations of Eqs. (8-44) and (8-45).

EXAMPLE 8-3 (Spreadsheet). This problem, involving repetitive evaluation of a specified formula, can be treated by the spreadsheet below, with Column B replicated for each value of Prandtl number desired for the table. This spreadsheet can be merged with that of Example 8-2 to evaluate $C\,Ra^{1/4}$.

	A	B
1	NATURAL	CONVECTION
2	C FROM	FORMULA
3	LAMINAR	FLOW
4	PRANDTL	
5	ROOT	+B4 ∧ .5
6	DENOMINATOR	2.435 + (4.884*B5) + 4.953*B4
7	RATIO	+B4 / B6
8	C	+B7 ∧ .25

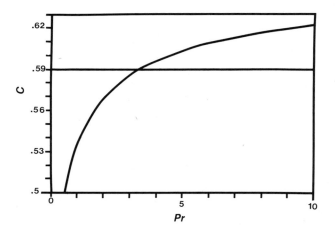

FIGURE 8-4-1. Dependence of C for laminar conditions at a vertical plate on the Prandtl number.

In the above example (and in an analogous case for turbulent flow in a problem at the end of the chapter), it was observed that the Prandtl number dependence can be significant, although except for typical gases and liquids (except for liquid metals) the simpler correlation is in reasonable agreement with more elaborate formulas. In general, it would appear to be appropriate to include Prandtl number dependence. However, it is not "wrong" to use the simpler correlation. The practicing engineer, on the basis of experience and judgment, is responsible for choosing the formulas applicable to the situation at hand based on factors such as accuracy required and correspondence of the situation of the problem to the idealized situation under which the formula was developed.

For vertical cylinders, it is generally acceptable to use the correlations for vertical planes, provided the diameter of the cylinder is large enough so that curvature is not a major factor. For laminar flow, we may express this provision by requiring that

$$\frac{D}{\delta} \gg 1 \tag{8-57}$$

that is the boundary layer is much thinner than the diameter. This may be written using our formula for δ/L,

$$\frac{D \, \mathrm{Gr}^{1/4}}{L(3.93 \, \mathrm{Pr}^{1/2})(0.952 + \mathrm{Pr})^{1/4}} \gg 1 \tag{8-58}$$

In practice, it has been found that

$$\frac{D}{L} \geq \frac{35}{\mathrm{Gr}^{1/4}} \tag{8-59}$$

is a suitable criterion. Again, note that the significant length dimension for evaluating the Nusselt number and Grashof number for a vertical cylinder is the length (not diameter) of the cylinder.

From Eqs. (8-58) and (8-59), it may be expected that $\mathrm{Gr}^{1/4}L$ would constitute a useful dimensionless parameter by which to characterize deviation of cylinder heat transfer from plane heat transfer. A formula provided by Minkowycz and Sparrow (1974) is

$$\overline{\mathrm{Nu}}_{\mathrm{cyl}} = \overline{\mathrm{Nu}}_{\mathrm{plate}}\left[1 + 1.43\left(\frac{L}{D \, \mathrm{Gr}^{1/4}}\right)^{0.9}\right] \tag{8-60}$$

Thus, the curvature of cylindrical geometry, providing additional volume for fluid moving away from the surface, leads to a higher heat transfer coefficient.

It may be noted that the formula implies large Nusselt number and heat transfer coefficient as the diameter gets very small, all other things being equal. However, since the total heat transferred depends on the product of heat transfer coefficient and area, the formula will predict zero natural convection heat transfer from a zero diameter cylinder.

For horizontal surfaces, the situation is somewhat different. Heated lighter fluid may rise directly upward from the surface, as illustrated in Fig. 8-4-2, rather than travel in a narrow layer along the surface. We again assume that heat transfer may be correlated in terms of the Rayleigh number, and coefficients are given in Table 8-4-1.

One ambiguity present in dealing with horizontal surfaces is that there may not be a single clearly defined length dimension. Observations to date have led to recommendations that, for a square surface, the characteristic dimension should be the length of one side and, for a circular disk, the dimension should be 0.9D. These are both consistent with treating the length dimension as the square root of the area. (Note that $\sqrt{\pi/4} \approx 0.9$). However, for a non square rectangle, a literature recommendation is to use the average of length and width. For other horizontal surfaces, it has been recommended (see Lloyd and Moran, 1974; Goldstein, Sparrow and Jones, 1973) that the length dimension to be taken as the ratio of area to perimeter

NON-SQUARE rectangles
$$L = \frac{A}{P} \tag{8-61}$$

We note, however, that this relationship would not reproduce the recommended values for squares and cylinders. One reason for the differences in recommended values may be the insensitivity of the heat transfer coefficient to the length dimension, especially for high Rayleigh numbers. Some in the literature have recommended the single length formula of Eq. (8-61) for all surfaces. The use of alternate length formulas yields different Rayleigh numbers and thus could yield different assessments as to which correlation coefficients C and n to apply, for example, if the Rayleigh number were near 8×10^6. However, the correlations in Table 8-4-1 yield similar Nusselt numbers near the transition Rayleigh number (some in the literature have used different transition Rayleigh numbers).

Note that for horizontal surfaces, we have a different situation prevailing depending on whether the surface facing up (or down) is heated or cooled. A hot surface facing down will cause the fluid near the surface to expand and thus will stimulate some fluid motion as shown in Fig. 8-4-3. However, gravity does not promote flow as when the plate faced up. Therefore, the heat transfer coefficient will be lower as per Table 8-4-1.

FIGURE 8-4-2. Free convection flow at a hot plate facing up.

Square
L = length of one side
circular disk
$L = 0.9 D$

A horizontal cylinder is a more complicated case since it has heat removal in all directions. The pertinent length dimension is taken as the diameter, and results are given in Table 8-4-1. It is reasonable, for length much longer than diameter, to expect flow to be primarily around the cylinder, as illustrated in Fig. 8-4-4, rather than along its length.

For the horizontal cylinder, Prandtl number dependence also applies. A general correlation given by Churchill and Chu is (1975a, b)

$$Nu = \left\{ 0.60 + \frac{0.387 \, Ra^{1/6}}{\left[1 + (0.559/Pr)^{9/16} \right]^{8/27}} \right\}^2 \qquad \text{(8-62)}$$

which is similar in form to Eq. (8-45) presented earlier for a vertical plate. A correlation restricted to laminar flow ($Ra < 10^9$) by Churchill and Chu is

$$Nu = 0.36 + \frac{0.518 \, Ra^{1/4}}{\left[1 + (0.559/Pr)^{9/16} \right]^{4/9}} \qquad \text{(8-63)}$$

As with the vertical plate, these correlations apply for liquid metals, with a trend toward dependence on a Peclet-like number $Gr^{1/2}Pr$ at low Prandtl numbers.

While some of these correlations appear complicated, they are simple to treat on a personal computer in a program or in a spreadsheet. Note that

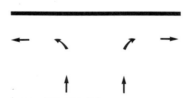

FIGURE 8-4-3. Free convection flow at a hot plate facing down.

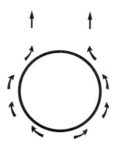

FIGURE 8-4-4. Free convection flow at a hot horizontal cylinder.

correlations covering a variety of conditions can be put in the form

$$Nu = \left\{ A_1 + \frac{A_2 Ra^{n_1}}{\left[1 + (A_3/Pr)^{n_2}\right]^{n_3}} \right\}^{n_4} \tag{8-64}$$

Thus, a program or spreadsheet prepared to treat any of the correlation in this general form can be modified easily to treat any of the others by changing the values of the constants used. The constants are summarized in Table 8-4-2. When using a spreadsheet or program, the effort of using Table 8-4-2 is not much greater than the effort of using Table 8-4-1, and Table 8-4-2 with more general correlations should be preferable. Note that if we set A_1 and A_3 to zero and n_4 to 1, we recover the form used in Table 8-4-1. Thus, the form of equation used in Table 8-4-1 may be considered a special case of the form used in Table 8-4-2.

EXAMPLE 8-4.   A long horizontal pipe 2.5 cm in diameter with a surface temperature of $330°K$ passes through a cold air space at a temperature of $270°K$. How much heat is lost per unit length of pipe?

SOLUTION.   The film temperature is $\frac{1}{2}(270 + 330) = 300°K$. The Grashof number is

$$Gr = \frac{g\beta(T_w - T_\infty)D^3}{\nu^2} = \frac{g(T_w - T_\infty)D^3}{\nu^2 T_f}$$

$$= \frac{(9.8)(60)(0.025)^3}{(16.84 \times 10^{-6})^2(300)} = 1.08 \times 10^5$$

$$Pr = 0.708$$

$$Ra\, Gr\, Pr = 7.65 \times 10^4$$

TABLE 8-4-2
Coefficients for General Natural Convection Formula

	A_1	A_2	A_3	n_1	n_2	n_3	n_4
Vertical plate							
Laminar	0.68	0.670	0.492	$\frac{1}{4}$	$\frac{9}{16}$	$\frac{4}{9}$	1
Turbulent	0.825	0.387	0.492	$\frac{1}{6}$	$\frac{9}{16}$	$\frac{8}{27}$	2
Horizontal cylinder							
Laminar	0.36	0.518	0.559	$\frac{1}{4}$	$\frac{9}{16}$	$\frac{4}{9}$	1
Turbulent	0.60	0.387	0.559	$\frac{1}{6}$	$\frac{9}{16}$	$\frac{8}{27}$	2

With this value of the Rayleigh number, we use

$$Nu = 0.36 + \frac{0.518 \, Ra^{1/4}}{\left[1 + (0.559/Pr)^{9/16}\right]^{4/9}} = 6.87$$

This leads to a heat transfer coefficient of

$$h = \frac{Nu \, k}{D} = \frac{(6.87)(0.0262)}{0.025} = 7.14$$

The heat lost per meter length is

$$q = \frac{hA \, \Delta T}{L} = h\pi D \Delta T = (7.14)\,\pi(0.025)(60) = 33.65 \text{ W/m}$$

Note that the simpler correlation

$$Nu = 0.53 \, Ra^{1/4} = 8.81$$

implies significantly higher heat transfer.

EXAMPLE 8-4 (Spreadsheet). We may use the spreadsheet developed for Example 8-2 as a starting point, since Rows 1–10 evaluate the Rayleigh number. From there we proceed to evaluate

$$Nu = \left\{ A_1 + \frac{A_2 \, Ra^{n_1}}{\left[1 + (A_3/Pr)^{n_2}\right]^{n_3}} \right\}^{n_4}$$

with the following addition

	C	D
1	A1	.36
2	A2	.518
3	N1	.25
4	NUMERATOR	+D3*(B10 ∧ D3)
5	A3	.559
6	N2	9 / 16
7	PRANDTL TERM	+(D5 / D9) ∧ D6
8	N3	4 / 9
9	DENOMINATOR	(1 + D7) ∧ D8
10	N4	1
11	NUSSELT	(D1 + (D4 / D9)) ∧ D10
12	H	+D11*B14 / B4

This spreadsheet is in a general form so that it may be applied to turbulent flow and to vertical plates and cylinders by changing the constants.

8-5. INCLINED SURFACES

As discussed in previous sections, horizontal and vertical surfaces provide somewhat different heat transfer mechanisms. Near a vertical surface, flow tends to be confined to a thin strip near the surface, while near a horizontal surface, buoyancy can lead to direct upward flow from the surface. Near an inclined surface then, we may expect the situation to be somewhat complicated.

For an inclined plate with upward facing heated face, it has been found, by Fuji and Imura (1972), for small Rayleigh numbers, that

$$\mathrm{Nu} = 0.56(\mathrm{Gr}\,\mathrm{Pr}\cos\theta)^{1/4} \qquad 15° < \theta < 75° \qquad (8\text{-}65)$$

where θ is the angle made with the vertical direction. This formula, in effect, says that under appropriate circumstances the vertical direction formula applies if we use the vertical component of acceleration of gravity.

If the Grashof number increases past a critical value Gr_o, we obtain, for the range $10^5 < \mathrm{Gr}\,\mathrm{Pr}\cos\theta < 10^{11}$, from the work of Fuji and Imura

$$\mathrm{Nu} = 0.14\left[(\mathrm{Gr}\,\mathrm{Pr})^{1/3} - (\mathrm{Gr}_c\mathrm{Pr})^{1/3}\right] + 0.56(\mathrm{Gr}\,\mathrm{Pr}\cos\theta)^{1/4} \qquad (8\text{-}66)$$

The critical Grashof number is a strong function of an angle as shown in Table 8-5-1. For $\theta = 15°$, Gr_c is 5×10^9. By the time θ reaches 75°, Gr_c is only 10^6. Thus, the more nearly vertical the plate, the easier it is to sustain a mode of heat transfer characteristic of vertical plates.

For a heated plate facing downward, we do not face a natural tendency for buoyancy to move fluid away from the plate. It has been found by Fuji and Imura that

$$\mathrm{Nu} = 0.58(\mathrm{Gr}\,\mathrm{Pr}\cos\theta)^{1/5} \qquad \theta < 88° \qquad (8\text{-}67)$$

Thus, until the plate becomes virtually horizontal, the mode of heat transfer is determined by the vertical plate aspect of the situation.

TABLE 8-5-1
Critical Grashof Number versus Tilt Angle

Tilt angle θ (degrees)	Gr_c
15	5×10^9
30	2×10^9
60	10^8
75	10^6

8-6. ENCLOSED SPACES

Natural convection can take place within enclosed spaces (cavities) having surfaces at different temperatures. In this section, we discuss the standard geometries illustrated in Fig. 8-6-1.

For the case of a horizontal rectangular cavity with the hot surface at the floor, the fluid dynamics regime and the heat transfer coefficient depend on the

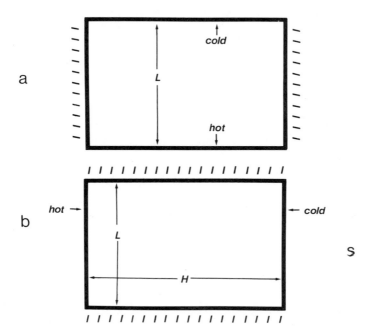

FIGURE 8-6-1. Standard geometrics for heated enclosures (a) heated floor, cooled ceiling, insulated walls (b) insulated floor and ceiling, heated and cooled walls.

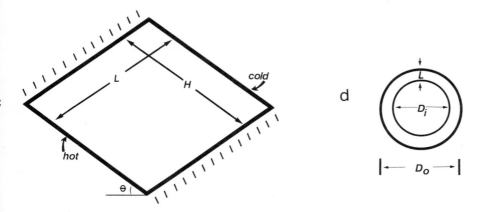

FIGURE 8-6-1. Standard geometrics for heated enclosures (c) tilted enclosure (d) concentric cylinders or spheres.

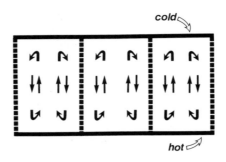

FIGURE 8-6-2.

Rayleigh number. There is a natural tendency for lighter, hotter fluid to rise, and for heavier, colder fluid to sink. There is also a tendency for viscosity to restrain motion. At sufficiently low Rayleigh numbers (below 1700), the driving force is not sufficient to overcome the restraining force, and a stable or stagnant condition results. Heat transfer is by conduction, and the Nusselt number is 1.

When the Rayleigh number exceeds 1700, moderate flow begins to take place in a pattern of internal circulation loops. A cross-section of the enclosure may have a pattern like that shown in Fig. 8-6-2. The particular nature of the circulation depends on several factors.

As the Rayleigh number continues to increase, the flow becomes turbulent, destroying the simple circulating loops. At high Rayleigh numbers, Globe and Dropkin (1959) recommend

$$\text{Nu} = 0.069\,\text{Ra}^{1/3}\text{Pr}^{0.074} \qquad 3 \times 10^5 < \text{Ra} < 7 \times 10^9 \qquad \textbf{(8-68)}$$

At lower Rayleigh numbers, the Nusselt number may be estimated from Eq. (8-70) for tilted enclosures with a zero tilt angle. For more detailed work, correlations in the form of Eq. (8-43) holding over several narrow ranges of Rayleigh number, with separate coefficients for gases and liquids, have been listed in the text of Osizik (1985).

If the heated surface of the rectangular cavity is the ceiling, we do not have a natural driving force for convection, and the Nusselt number is 1; that is, heat transfer is by conduction.

Note the distinction between the enclosed and open spaces. We did observe that in an open space there is convection heat transfer from a hot surface facing down. This is because a heated fluid will increase in pressure and will expand into adjacent lower-pressure space. In an enclosed cavity, the space is fixed. There is no adjacent space. A steady state will develop in which the cavity pressure is compatible with the amount of heating. Heat transfer in this stagnant fluid will be by conduction.

If the heated and cooled surfaces are vertical and the insulated surfaces are horizontal, we also get convective heat transfer. A circuit will be established where cool fluid at the vertical hot wall rises as it is heated and warm fluid at the vertical cold wall falls as it is cooled. These driving forces will tend to move the cold fluid along the floor to the hot wall and the hot fluid along the ceiling to the cold wall. Heat transfer correlations are given by Catton (1978) as

$$\text{Nu} = A\left(\frac{\text{Pr}}{0.2 + \text{Pr}}\text{Ra}_L\right)^{n_1}\left(\frac{H}{L}\right)^{n_2} \tag{8-69}$$

where H is the horizontal dimension (separation between walls) and L is the vertical dimension. For moderate wall separation $(1 < H/L < 2)$, $A = 0.18$, $n = 0.28$, and $n_2 = 0$. These values apply for most fluids $(10^{-3} < \text{Pr} < 10^5)$ and for $\text{Ra}_L \text{Pr}/(0.2 + \text{Pr}) > 1000$. For larger wall separation $(2 < H/L < 10)$, $A = 0.22$, $n = 0.29$, and $n_2 = -\frac{1}{4}$. These values hold for $\text{Pr} < 10^5$ and for $\text{Ra}_L < 10^{10}$.

It is also possible for the cavity to be tilted. Inclined rectangular spaces are of interest in dealing with solar collectors, for example. If the heated surface is inclined at angle θ to the horizontal as shown in Fig. 8-6-1C, then convection will not take place unless Ra exceeds $1700/\cos\theta$, for $\theta < 80°$ and $H/L > 10$. The heat transfer coefficient may be obtained from the correlation of Hollands et al. (1976):

$$\text{Nu}_L = 1 + 1.44\left(1 - \frac{1708}{\text{Ra}_L\cos\theta}\right)\left\{1 - \frac{1708[\sin(1.8\theta)]^{1.6}}{\text{Ra}_L\cos\theta}\right\}$$

$$+ \left[\left(\frac{Ra\cos\theta}{5830}\right)^{1/3} - 1\right] \tag{8-70}$$

The above equation holds if each of the factors on the right-hand side is positive. If any are negative, Nu is set equal to 1. (Convection can make heat transfer more effective than conduction alone but should not make it less effective.)

It was mentioned earlier that Eq. (8-70) with a zero tilt angle may be used when Eq. (8-68) does not apply. Actually, the ranges of applicability overlap, with Eq. (8-70) tending to $Ra^{1/3}$ behavior at high Rayleigh numbers. Thus, Eq. (8-70) can serve as a general purpose correlation. Usually, when correlations have been developed for narrow ranges, they tend to be more reliable than correlations covering broad ranges and multiple effects.

Convective heat transfer between concentric cylinders and between concentric spheres tends to be expressed in terms of the factor by which one must multiply the thermal conductivity to obtain the true heat transfer. Expressions for this ratio of effective-to-actual conductivity are given by Raithby and Hollands (1975) in the form

$$\frac{k_{\text{eff}}}{k} = A\left(\frac{\text{Pr}}{0.861 + \text{Pr}}\right)^{1/4}\text{Ra}^{1/4}\zeta^{1/4} \tag{8-71}$$

where A is 0.386 for cylinders and 0.74 for spheres. The length dimension in evaluating the Rayleigh number is the difference between outer and inner radii of the cavity. The dimensional corrections ζ are, for the cylinder,

$$\zeta_{\text{cyl}} = \frac{\left[\ln(D_o/D_i)\right]^4}{(L/D_o)^3\left[1 + (D_o/D_i)^{3/5}\right]^5}$$ (8-72)

and, for the sphere,

$$\zeta_{\text{sph}} = \frac{L}{D_i}\left(\frac{D_o}{D_i}\right)^3 \frac{1}{\left[1 + (D_o/D_i)^{7/5}\right]^5}$$ (8-73)

The effective conductivity would be used with a heat flow equation based on conduction; that is, for heat flow per unit length of cylinder

$$q' = \frac{2\pi k_{\text{eff}}}{\ln D_o/D_i}(T_i - T_o)$$ (8-74)

and for the sphere,

$$q = \frac{\pi D_i D_o}{L}k_{\text{eff}}(T_i - T_o)$$ (8-75)

EXAMPLE 8-5. A solar collector is to be placed at a 45° angle to the horizontal as shown in Fig. 8-6-1c. The collector has a hot (black) surface at 340°K. Across an air cavity 5 cm thick ($L = 0.05$ m) is a glass face through which sunlight enters. On a particular cold and windy day, the glass is assumed to be at 260°K. The collector height H is 2 m, and its width is 1 m. How much heat is transferred by convection across the gap?

SOLUTION. We apply with $H/D = 2/0.05 = 40$,

$$\text{Nu} = 1 + 1.44\left(1 - \frac{1708}{\text{Ra}\cos\theta}\right)\left\{1 - \frac{1708[\sin(1.8\theta)]^{1.6}}{\text{Ra}\cos\theta}\right\} + \left[\frac{\text{Ra}\cos\theta^{1/3}}{5830} - 1\right]$$

Air properties are evaluated at $\frac{1}{2}(340 + 260)$, or 800°K,

$$\text{Ra} = \frac{q\beta(T_1 - T_2)L^3}{\nu^2}\text{Pr} = \frac{(9.8)(1/300)(340 - 260)(0.05)^3}{(1.57 \times 10^{-5})^2}(0.712)$$

$$\text{Ra} = 9.44 \times 10^5$$

$$\text{Ra at 45°} = 6.77 \times 10^5$$

$$\text{Na} = 1 + 1.433 + 3.855 = 6.288$$

The heat transfer coefficient is

$$h = \frac{\text{Nu}\, k}{L} = \frac{(6.288)(0.026)}{0.05} = 3.28$$

$$q = hA\,\Delta T = 3.28(2)(1)(80) = 525\ \text{W}$$

This loss of 263 W/m^2 is a significant fraction of potential incident sunlight (see Chapter 10). It implies that design consideration should be given to inhibiting convective loss for collectors to be used in cold weather conditions. More will be said about this problem in Chapter 12, where we deal with multinode heat transfer problems.

It should be noted that when the fluid in the cavity is a gas, radiation heat transfer also may be important and may take place simultaneously.

8-7. FREE AND FORCED CONVECTION

We speak of forced convection when fluid motion is induced by an external agent (e.g., a pump). We speak of free (natural) convection when fluid motion arises as part of the heat transfer process. It is possible that both types of motion could take place simultaneously. Buoyancy forces may add to the fluid motion imposed externally.

The ratio of kinetic energy (or of inertial force) to viscous force for forced convection was found to be characterized by the Reynolds number. Similarly, the ratio of induced kinetic energy (or of buoyancy force) to viscous force for natural convection was found to be the square root of the Grashof number. Consequently, we may expect natural and forced convections to be of comparable importance when

$$\text{Gr}^{1/2} \approx \text{Re} \qquad\qquad (8\text{-}76)$$

If the Reynolds number is much larger than $\text{Gr}^{1/2}$, then natural convection should be unimportant. If the Reynolds number is much smaller than $\text{Gr}^{1/2}$, then the imposed flow is of little consequence.

One effect of combining free and forced convections is that the transition from laminar to turbulent conditions will be affected. In mixed convection, the transition to turbulent flow takes place at a lower Reynolds number than for pure forced convection and at a lower Grashof number than for pure free convection.

It also should be recognized that when both natural and free convections exist, the buoyance effects may reinforce the forced flow (aiding flow) or may impede the forced flow (opposing flow). For example, if flow is forced upward along a vertical plate, buoyancy will tend to produce natural convection flow upward, thus reinforcing the forced convection. It also is possible for the flows to be normal to one another (neither aiding nor opposing). This would be the case if flow were forced horizontally along a vertical plate.

In aiding flow, the effects of free and forced convections are reinforcing but are not simply additive. An approximate rule for interpolation for aiding flow by a vertical plate or a circular cylinder is

$$\text{Nu} = \left(\text{Nu}_{fc}^3 + \text{Nu}_{nc}^3\right)^{1/3} \tag{8-77}$$

where the Nu_{fc} and Nu_{nc} are Nusselt numbers calculated for forced and natural convections as if the other effect did not exist.

For flow in horizontal tubes, we may estimate, according to Brown and Gauvin (1965),

$$\text{Nu} = 1.75\left[\text{Gz} + 0.012\left(\text{Gz}\,\text{Gr}^{1/3}\right)^{4/3}\right]^{1/3}\left(\frac{\mu}{\mu_w}\right)^{0.14} \tag{8-78}$$

for laminar flow and

$$\text{Nu} = 4.69\,\text{Re}^{0.27}\text{Pr}^{0.21}\text{Gr}^{0.07}\left(\frac{D}{L}\right)^{0.36} \tag{8-79}$$

for turbulent flow. The transition Reynolds number may be evaluated as

$$\text{Re}_{\text{tr}} = \begin{cases} 2000 & \text{Ra}\,D/L < 2 \times 10^4 \\ 800 & \text{Ra}\,D/L > 2 \times 10^4 \end{cases} \tag{8-80}$$

Note that at large Rayleigh numbers, the transition to turbulence occurs at a low Reynolds number.

In keeping with the general guideline that in convection problems concern must be given to determining the type of problem encountered, it is good practice, when dealing with forced convection problems, to check to see if mixed convection effects apply. This is especially true if the forced convection flow rate is low, or if the characteristic length dimension is large (the ratio $\text{GR}^{1/2}/\text{Re}$ increases with the square root of that length dimension).

PROBLEMS

8-1. Derive expressions for the average and maximum speeds in the laminar boundary layer at a vertical flat plate relative to the characteristic speed U_0. At what location (y/δ) does the maximum occur?

8-2. Repeat Example 8-1 using water instead of air. For the same temperature difference and plate dimension, note the differences in Grashof number, boundary layer thickness, heat transfer coefficient, and characteristic speed.

8-3. Prepare a spreadsheet to calculate C for turbulent conditions from Eq. (8-46). Use this spreadsheet to make a table of C versus Prandtl number at 300°K for a broad range of Prandtl number from liquid metals to gases to highly viscous liquids. At what Prandtl number (characteristic of what fluid) does your table agree with the value of C given in Table 8-4-1. By how much do the values in your table depart from the values in Table 8-4-1?

8-4. Show that when the Prandtl number gets very small (as with liquid metals), Eq. (8-44) tends toward dependence on $Gr^{1/4}Pr^{1/2}$, that is, to the square root of the natural convection equivalents of the Peclet number.

8-5. Compare the predictions of Eqs. (8-44) and (8-45) for $Ra < 10^9$ for a Prandtl number of 0.7. The spreadsheet of Example 8-4 may be convenient for this comparison.

8-6. The vertical plate of Example 8-1 may be cooled by natural convection with gases whose properties are given in Appendix A. Which gas will yield the best heat transfer?

8-7. It has been suggested that natural convection heat loss from a person be treated as natural convection from a vertical cylinder. Assess whether this suggestion is reasonable.

8-8. It is suggested that the heat transfer for the plate of Example 8-1 be improved by increasing the air pressure. Prepare a plot of the heat transfer coefficient as a function of pressure between 1 and 30 atm.

8-9. A cylindrical cask 1 m in diameter and 3 m long is to have a surface temperature of 325°K with ambient air at 275°K. Compare the heat transfer if the cask if positioned horizontally or vertically.

8-10. Heat sources behind a plate 20 cm high are such that there will be a uniform heat flux of 300 W/m^2 at the surface of the plate. Estimate the average surface temperature of the plate if ambient air is at 20°C.

8-11. An ice skating rink 30 m long and 10 m wide has its ice at -5°C. The ambient air is at a temperature of 5°C. Estimate the cooling requirement for the rink.

8-12. A swimming pool is heated for night swimming. The pool is 20 m by 10 m. The water is kept at 23°C. The air temperature is 18°C. The circulating pool water is well mixed. Neglecting evaporative losses (i.e., mass transfer, as in Chapter 6), at what rate is heat lost from the pool?

8-13. For the case of Problem 8-12, estimate the heat loss by evaporation and compare it with the heat loss by natural convection.

8-14. A pipe 4 cm in diameter contains steam at 100°C. The pipe passes vertically through a large room 2.5 m high. The pipe wall is of a good conducting metal, so the pipe temperature is essentially at 100°C. Estimate the heat transfer to the room.

8-15. For safety reasons, it is desired to lower the temperature with which people may come in contact, relative to the condition of Problem 8-14. A 1 cm layer of insulating material with $k = 0.1$ W/m² · °K is wrapped around the pipe. Find the surface temperature and the heat transfer to the room.

8-16. Show that for common fluids (Pr not too different from 1) a figure of merit (FOM) for natural convection heat transfer at a vertical plate with laminar flow may be defined as

$$\text{FOM} = \frac{g\beta\rho^2 C_p k^3}{\mu}$$

8-17. Using the figure of merit of Problem 8-16, compare the relative suitability of water and air for natural convection using properties at 300°K.

8-18. Tabulate or plot the figure of merit problem 8-16 for air as a function of temperature at atmospheric pressure. Note whether or not air improves with temperature as a free convection fluid.

8-19. Examine the data for gases other than air. Should the trend for the figure of merit with increasing temperature be similar for other gases?

8-20. Tabulate or plot the figure of merit problem 8-16 for water as a function of temperature. Note whether or not water improves with temperature as a free convection liquid.

8-21. Develop a figure of merit to use for all fluids (any Prandtl number) for applicability to liquid metals and high-viscosity fluids at low Rayleigh numbers (laminar flow) over vertical plates.

8-22. Use the figure of merit developed in Problem 8-21 to compare the relative merits of air, water, and liquid sodium as natural convection fluids at 400°K.

8-23. Ethylene glycol (sometimes called antifreeze) can be a competitor to water for certain applications. Compare the relative merits (according to the figure of merit of Problem 8-21) of ethylene glycol and water as a function of temperature for natural convection.

8-24. Consider the figure of merit used in Problem 8-16. How would this figure of merit change when the Rayleigh number increases to the point that flow becomes turbulent?

8-25. If solar collectors are embedded in roofing, then the angle of inclination of the collector will be determined by the roof design. Repeat Example 8-5 for angles relative to the horizontal of 30° and 60°.

8-26. It is suggested that one way of modifying heat transfer in the solar collector of Example 8-5 is by changing the thickness of the air gap. What would the heat transfer be if the air gap thickness were doubled to 10 cm?

8-27. On a summer day, the glass temperature of the solar collector in Example 8-5 is 300°K. Repeat Example 8-5 for this condition.

8-28. For a calm day (no wind), estimate the heat transferred from a roof inclined at 45° (actually two separate pieces, each inclined at 45°). Each roof segment is a rectangle 15 m long (the length of the house) by 6 m. The outside air temperature is 250°K and the roof surface temperature is 275°K.

8-29. A house has an unheated enclosed porch. One wall, the house wall, is at 275°K. The outer wall on a cold winter day is at 250°K. The floor and ceiling of the porch are very well insulated. The walls are separated by 5 m, and the height of the porch is 3 m. How much heat is lost per unit length of house?

8-30. A cylindrical tank as an outer surface maintained at 400°K. The tank diameter is 1 m. The tank is contained within a second cylinder whose temperature is 300°K. The second tank has an inner diameter of 2 m. How much heat is transferred from the inner cylinder to the outer one?

8-31. What would the diameter of the second tank in Problem 8-30 have to be for the effective thermal conductivity to be double the actual thermal conductivity?

REFERENCES

1. V. S. Arpaci and P. S. Larsen, *Convection Heat Transfer*, Prentice-Hall, Englewood Cliffs, NJ, 1984.
2. A. Bejan, *Convection Heat Transfer*, Wiley, New York, 1984.
3. C. K. Brown and W. H. Gauvin, "Combined Free and Forced Convection," *Can. J. Chem. Eng.*, **43**, 306 (Part I) and 313 (Part II) (1965).
4. L. C. Burmeister, *Convective Heat Transfer*, Wiley, New York, 1983.
5. I. Catton, "Natural Convection in Enclosures," in *Proceedings of the 6th International Heat Transfer Conference*, 1978, Hemisphere, Washington, D.C.
6. T. Cebeci and P. Bradshaw, *Physical and Computational Aspects of Convective Heat Transfer*, Springer, New York, 1984.
7. S. W. Churchill and H. H. S. Chu, "Correlating Equations for Laminar and Turbulent Free Convection from a Vertical Plate" *Int. J. Heat Mass Transfer*, **18**, 1323 (1975a).
8. S. W. Churchill and H. H. S. Chu, "Correlating Equations for Laminar and Turbulent True Convection from Horizontal Cylinder," *Int. J. Heat Mass Transfer*, **18**, 1049 (1975b).
9. T. Fuji and H. Imura, "Natural Convection Heat Transfer from a Plate with Arbitrary Inclination," *Int. J. Heat Mass Transfer*, **15**, 755 (1972).
10. S. Globe and D. Dropkin, "Natural Convection Heat Transfer in Liquids Confined Between Two Horizontal Plates," *J. Heat Transfer*, **81**, 24 (1959).

11. R. J. Goldstein, E. M. Sparrow, and D. C. Jones, "Natural Convection Mass Transfer Adjacent to Horizontal Plates," *Int. J. Heat Mass Transfer*, **16**, 1025 (1973).

12. K. G. T. Hollands, S. E. Unny, G. D. Raithby, and L. Konicek, "Free Convective Heat Transfer Across Inclined Air Layers," *J. Heat Transfer*, **98**, 189 (1976).

13. J. P. Holman, *Heat Transfer*, 5th ed., McGraw-Hill, New York, 1979.

14. Y. Jaluria, *Natural Convection Heat and Mass Transfer*, Pergamon, Elmsford, NY, 1980.

15. W. M. Kays and M. E. Crawford, *Convective Heat and Mass Transfer*, McGraw-Hill, New York, 1980.

16. E. J. LeFevre, "Laminar Free Convection from a Vertical Plane Surface," in *Proceedings of the 9th International Congress of Applied Mechanics*, 1956, Vol. 4, p. 168.

17. J. R. Lloyd and W. R. Moran, "Natural Convection Adjacent to Horizontal Surface of Various Planforms," ASME Paper 74-WA/HT-66, 1974.

18. W. J. Minkowycz and E. M. Sparrow, "Local Nonsimilar Solutions for Natural Convection on a Vertical Cylinder," *J. Heat Transfer*, **96**, 178 (1974).

19. M. N. Osizik, *Heat Transfer, A Basic Approach*, McGraw-Hill, New York, 1985.

20. G. D. Raithby and K. G. T. Hollands, "A Great Method of Obtaining Approximate Solutions to Laminar and Turbulent Free Convection Problems," in T. F. Irvine and J. P Hartnett (eds.), *Advances in Heat Transfer*, Vol. 11, Academic Press, New York, 1975.

21. G. C. Vliet, "Natural Convection Local Heat Transfer on Constant Heat Flux Inclined Surfaces," *J. Heat Transfer*, **91**, 511 (1969).

22. G. C. Vliet and C. K. Lin, "An Experimental Study of Turbulent Natural Convection Boundary Layers," *J. Heat Transfer*, **91**, 517 (1969).

Chapter Nine

CONVECTION WITH PHASE CHANGES

9-1. INTRODUCTION

Thus far, in studying convection, we have considered homogeneous fluids whose characteristics remain the same throughout the process under consideration. In this chapter, we are concerned with fluids that change as a result of heat transfer. Upon receiving heat, a liquid may boil, that is, change in phase from liquid to vapor. Since liquid and vapor have substantially different densities, we expect gravity to be a pertinent parameter. In addition, since boiling involves formation of bubbles, we expect surface tension to be a significant parameter.

Boiling and condensation heat transfer are important processes. From thermodynamics, we know that high efficiency is associated with adding heat to a working fluid at a constant high temperature and rejecting waste heat at a constant low temperature. As a practical matter, heat exchange at constant temperature implies heat exchange with a change in phase. Thus, power plants usually involve heat addition via boiling and heat rejection via condensation.

9-2. CONDENSATION ON A VERTICAL SURFACE

When vapor comes in contact with a cold surface such that the vapor is cooled below the saturation temperature corresponding to its pressure (e.g., at atmo-

231

spheric pressure, the saturation temperature for water is 100°C), the vapor condenses on the surface. If the surface is clean and uniform, then a film of liquid forms from the condensation and tends to flow downward due to gravity. Thus, the film is thicker at the bottom of the surface than at the top, as illustrated in Fig. 9-2-1.

If the surface is nonuniform or coated with a substance that inhibits wetting, condensation occurs as individual drops. These drops tend to grow and to flow downward. This is referred to as dropwise, as opposed to filmwise, condensation.

Once condensation begins and a film forms, further condensation must result from vapor in contact with the film of "condensate." This condensate acts as a thermal resistance that inhibits further condensation. A steady state is achieved when the rate of condensation is balanced by the rate of downward flow.

For simplicity, we consider a case where the vapor is at the saturation temperature, and the surface is at a lower temperature. Thus, transfer of heat leads to condensation. (More generally, if the vapor is above the saturation temperature, it loses heat by a lowering of temperature before condensing.) In addition, we neglect shear stress between liquid and vapor, therby permitting the film to flow downward without dragging vapor along with it. Finally, we assume that transfer across the thin liquid film is by conduction.

We assume that within the fluid there must be a balance between net shear stress and net gravitational force so that

$$\mu_l \frac{\partial^2 u}{\partial y^2} = -g(\rho_l - \rho_v) \tag{9-1}$$

At the wall, as usual, we require that

$$u(0) = 0 \tag{9-2}$$

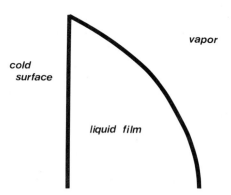

FIGURE 9-2-1. Condensation on a cold vertical surface.

At the outside of the film, we require that

$$\frac{\partial u}{\partial y}(\delta) = 0 \tag{9-3}$$

With these two boundary conditions, we may use a polynomial profile

$$u(y) = \frac{g(\rho_l - \rho_v)\delta^2}{\mu_l} \frac{y}{\delta}\left[1 - \frac{1}{2}\left(\frac{y}{\delta}\right)\right] \tag{9-4}$$

The mass flow rate per unit depth of surface is obtained by integrating over the film:

$$\dot{m}(x) = \rho_l \int_0^{\delta(x)} u(y)\, dy = \frac{g\rho_l(\rho_l - \rho_v)\delta^3}{3\mu_l} \tag{9-5}$$

As noted earlier, in steady state the rate at which film flows is balanced with the rate at which vapor condenses. Thus, if an increment $d\dot{m}(x)$ occurs in the flow rate in the interval dx, then the heat transferred per unit depth of surface by condensation must be

$$dq = h_{fg}\, d\dot{m} \tag{9-6}$$

This heat must be conducted across the film to the cold surface

$$dq = k_l \frac{dx(T_{\text{sat}} - T_s)}{\delta} \tag{9-7}$$

The flow rate may then be expressed as

$$\frac{d\dot{m}}{dx} = \frac{k_l(T_{\text{sat}} - T_s)}{h_{fg}\delta} \tag{9-8}$$

Differentiating our other expression for flow rate

$$\frac{d\dot{m}}{dx} = \frac{g\rho_l(\rho_l - \rho_v)\delta^2}{\mu_l}\frac{d\delta}{dx} \tag{9-9}$$

Combining these last two equations provides an equation for film thickness,

$$\delta^3\frac{d\delta}{dx} = \frac{k_l\mu_l(T_{\text{sat}} - T_s)}{g\rho_l(\rho_l - \rho_v)h_{fg}} \tag{9-10}$$

This may be integrated to yield

$$\delta(x) = \left[\frac{4k_l\mu_l(T_{\text{sat}} - T_s)x}{g\rho_l(\rho_l - \rho_v)h_{fg}}\right]^{1/4} \tag{9-11}$$

Thus, the film thickness increases slowly with distance down the surface.

If we wish to assign an overall heat transfer coefficient to the process, we can define

$$\frac{dq}{dx} = h(T_{\text{sat}} - T_s) = \frac{k_l}{\delta}(T_{\text{sat}} - T_s) \qquad (9\text{-}12)$$

and the heat transfer coefficient is

$$h = \left[\frac{g\rho_l(\rho_l - \rho_v)k_l^3 h_{fg}}{4\mu_l(T_{\text{sat}} - T_s)x}\right]^{1/4} \qquad (9\text{-}13)$$

The average heat transfer coefficient for a surface of height L is

$$\bar{h} = \frac{1}{L}\int_0^L h(x)\, dx = \tfrac{4}{3}h(l) \qquad (9\text{-}14)$$

It may be observed that the transfer coefficient increases as T_{sat} and T_s become close, tending to infinity as T_s approaches T_{sat}. However, heat flow depends on the product of the coefficient and the temperature difference and thus is proportional to the temperature difference to the three-fourths power. Thus, heat flow declines as the temperature difference gets smaller.

The analysis we performed, due originally to Nusselt (1916), involved several important assumptions. These lead to some inaccuracy in the formula that results. However, the general functional form of the formula for average heat transfer coefficient is in reasonable agreement with experimental results. Empirically, McAdams (1954) has found that reasonable predictions are obtained using a modified coefficient involving a 20% increase; that is,

$$\bar{h} = 1.13\left[\frac{g\rho_l(\rho_l - \rho_v)k_l^3 h_{fg}}{\mu_l(T_{\text{sat}} - T_s)L}\right]^{1/4} \qquad (9\text{-}15)$$

This formula may also be used for nonplane surfaces when the dimensions characterizing curvature are much larger than the film thickness (e.g., for a cylinder of radius R, when $R \gg \delta$). This is analogous to application of vertical surface formulas in regard to natural convection. The formula may also be applied to slightly inclined surfaces, as with natural convection by letting g go to $g\cos\theta$, where θ is the angle with the vertical direction.

While the above formula is a conventional means of expressing the heat transfer coefficient for condensation, an alternate representation is possible in terms of dimensionless groups. As noted already, there is similarity between condensation and natural convection. Let us define, analogous to the coefficient of expansion, a modified expansion coefficient β' by

$$\beta' = \frac{\rho_l - \rho_v}{\rho_l(T_{\text{sat}} - T_s)}$$

Let us regroup parameters as follows:

$$\text{Nu} = \frac{\bar{h}L}{k} = \bar{h}\left(\frac{L^4}{k^4}\right)^{1/4}$$

$$= 1.13 \left\{ \left[\frac{g\rho_l^2\beta'(T_{sat} - T_s)L^3}{\mu_l^2}\right]\left[\frac{c\mu_l}{k_l}\right]\left[\frac{h_{fg}}{C(T_{sat} - T_s)}\right]\right\}^{1/4}$$

The first grouping on the right-hand side has the form of a generalized Grashof number and is called the Galileo number Ga. The second grouping may be recognized to be the Prandtl number. The product of the first two groupings thus may be viewed as a generalized Rayleigh number. The third grouping on the right-hand side relates latent heat to specific heat and temperature. This type of grouping (often applied for boiling where T_s exceeds T_{sat}) is the reciprocal of the Jakob number Ja. This last grouping introduces information about the energy in the phase change process not contained in the buoyancy-type Grashof number.

When cast in this alternate form of dimensionless groups, condensation heat transfer may be viewed as a natural convection process with an additional feature of latent heat. The density differential between phases replaces the density differential at different temperatures. When the substitution of density changes is made, the same dimensionless groupings describe the phenomena. An additional dimensionless grouping then is required to represent the influence of latent heat.

When the flow rate in the film becomes high, flow in the film can become turbulent. Transition to turbulence is again governed by the Reynolds number. It is conventional when dealing with condensation to define the Reynolds number in terms of the mean speed u_m as

$$\text{Re} = \frac{\rho u_m(4\delta)}{\mu_l} \tag{9-16}$$

where 4δ is the hydraulic diameter of the film flow area. Flow is turbulent when the Reynolds number exceeds 1800. Noting that the mass flow rate per unit depth is

$$\dot{m} = \rho u_m\delta = \frac{g\rho_l(\rho_l - \rho_v)\delta^3}{3\mu_l} \tag{9-17}$$

and that

$$\delta^3 = \left[\frac{4k_l\mu_l(T_{sat} - T_s)x}{g\rho_l(\rho_l - \rho_v)h_{fg}}\right]^{3/4} \tag{9-18}$$

the Reynolds number may be expressed as

$$\text{Re} = \frac{4}{3}\mu_l\left[4k_l\frac{(T_{sat} - T_s)x}{h_{fg}}\right]^{3/4}\left[\frac{g\rho_l(\rho_l - \rho_v)}{\mu_l}\right]^{1/4} \tag{9-19}$$

This last expression in terms of x yields the location at which transition occurs. This expression also shows the dependence of the Reynolds number on temperature difference and other quantities. Unlike the situation for natural convection where a new parameter, the Grashof number, was defined for condensation, it is common practice to continue to use the Reynolds number. For turbulent flow, it has been found empirically by Kirkbride (1934) that

$$\bar{h} = 0.0077 \left[\frac{g\rho_l(\rho_l - \rho_v)k_l^3}{\mu_l^2} \right]^{1/3} \mathrm{Re}^{0.4} \tag{9-20}$$

An alternate way of evaluating the Reynolds number is to note that the average heat transfer above a certain location determines the mass flow through that location, that is,

$$\bar{h}L(T_{\mathrm{sat}} - T_s) = \dot{m}h_{fg} = \rho u_m \delta h_{fg} = \frac{\mathrm{Re}\,\mu}{4} h_{fg} \tag{9-21}$$

Thus, the Reynolds number is

$$\mathrm{Re} = \frac{4\bar{h}L(T_{\mathrm{sat}} - T_s)}{\mu_l h_{fg}} \tag{9-22}$$

The heat transfer coefficient of Eq. (9-15) is 20% higher than would be obtained without empirical adjustment. The Reynolds number of Eq. (9-20) was obtained without adjustment. Thus, this evaluation of the Reynolds number should be multiplied by 1.2 to test whether flow is laminar or turbulent.

If flow is laminar, then Eq. (9-15) can be applied directly. If flow is turbulent, then the Reynolds number obtained with the adjusted laminar formula, while providing the information that flow is turbulent, is not applicable in Eq. (9-21) for the heat transfer coefficient. From Eq. (9-23), it may be observed that the Reynolds number depends on the heat transfer coefficient. While Eq. (9-21) is the common form in which heat transfer coefficient is expressed, it may be more convenient in performing problems to solve for $\bar{h}$ explicitly as

$$\bar{h} = \left\{ 0.0077 \left[\frac{g\rho_l(\rho_l - \rho_v)k_l^3}{\mu_l^2} \right]^{1/3} \left[\frac{4L(T_{\mathrm{sat}} - T_s)}{\mu_l h_{fg}} \right]^{0.4} \right\}^{5/3} \tag{9-21a}$$

EXAMPLE 9-1. Saturated steam at 400°K is in contact with a vertical wall at 398°K. The wall is 1 m high. Determine (a) the heat transfer per meter of wall depth and (b) the amount of condensate per meter of wall depth.

SOLUTION. As with other types of convection problem, we must determine which heat transfer correlation to apply. We evaluate the Reynolds number

$$Re = \frac{4}{3\mu_l}\left[\frac{4k_l(T_{sat} - T_s)L}{h_{fg}}\right]^{3/4}\left[\frac{\rho_l(\rho_l - \rho_v)}{\mu_l}\right]^{1/4} \quad (1.2)$$

Properties will be evaluated at 400°K. (In general, liquid properties should be evaluated at the film temperature, that is, at the average of T_{sat} and T_s. In this case, the temperature difference is small, so we pick T_s.) This yields

$$Re = \frac{4}{3(2.19 \times 10^{-4})}\left[\frac{4(0.685)(2)(1)}{2.18 \times 10^6}\right]^{3/4}\left[\frac{(9.8)(937)(937 - 0.55)}{2.19 \times 10^{-4}}\right]^{1/4} \quad (1.2)$$

Note that the vapor density is very small compared to the liquid density.

$$Re = \frac{4}{3(2.19 \times 10^{-4})}(6.31 \times 10^{-5})(4.45 \times 10^2) = 205$$

This value is less than 1800, so we use the laminar formula

$$\bar{h} = 1.13\left[\frac{g\rho_l(\rho_l - \rho_v)k_l^3 h_{fg}}{\mu_l(T_{sat} - T_s)L}\right]^{1/4}$$

which can be evaluated as

$$\bar{h} = 1.13\left[\frac{(9.8)(937)(937.55)(0.685)^3(2.183 \times 10^6)}{(2.19 \times 10^{-4})(2)(1)}\right]^{1/4} = 1.22 \times 10^4$$

The heat transfer coefficient is large compared to many heat transfer coefficients encountered in this text. Phase changes tend to be associated with large heat transfer coefficients. The heat transfer per unit length is

$$q' = hL(T_{sat} - T_s) = (1.22 \times 10^4)(1)(2) = 2.44 \times 10^4 \text{ W/m}$$

The mass flow of condensate is then given by

$$\dot{m} = \frac{q'}{h_{fg}} = \frac{2.44 \times 10^4}{2.183 \times 10^6} = 0.011 \text{ kg/sec}$$

The above solution procedure is logical and straightforward in the usual sense for convection problems; that is, evaluate the Reynolds number, test for turbulent versus laminar flow, and select the appropriate heat transfer correlation. When working with spreadsheets, however, an alternate approach is more efficient.

EXAMPLE 9-1 (Spreadsheet). In approaching spreadsheet examples with choices in earlier chapters, both evaluation procedures are provided in the spreadsheet, and a choice between two evaluated heat transfer coefficients is made on the basis of the appropriate selection criterion. For this problem then, let us proceed by evaluating the heat transfer and mass flow rate on the basis of laminar flow. Let us then evaluate the Reynolds number by

$$\mathrm{Re} = \frac{4\bar{h}L(T_{\mathrm{sat}} - T_s)}{\mu_l h_{fg}} = \frac{4}{\mu_l}\dot{m} = \frac{4(0.11)}{2.19 \times 10^{-4}} = 201$$

This value differs slightly from the original Reynolds evaluation because of rounding of intermediate calculations. On the basis of this value, we choose to accept the results obtained. If this value were to exceed 1800, then we would proceed to evaluate the turbulent formula.

This problem is a good one to put in spreadsheet form because the heat transfer correlation for laminar flow at a horizontal tube is identical except for a constant with the formula used here. Thus, the effort involved in setting up the spreadsheet is applicable to a broader range of problems.

In setting up the spreadsheet, we will, in evaluating terms for the laminar formula, anticipate groupings of parameters that will be used in the turbulent formula. The spreadsheet follows.

Note that in the turbulent calculations, TERM 1 is evaluated by dividing TERM A of the laminar calculation by viscosity. Note also that TERM 2 is obtained by dividing the laminar Reynolds number by the laminar heat transfer coefficient.

When the turbulent flow formula is evaluated for low Reynolds numbers (as in this example), it can predict lower heat transfer than the laminar formula. This is because the turbulent correlation is not applicable at low Reynolds numbers. For Reynolds numbers over 1800, the turbulent formula predicts greater heat transfer than the laminar formula, as is to be expected.

	A	B	C	D
1	CONDENSATION	VERTICAL	WALL	
2	LAMINAR	FORMULAS		
3	G	9.8	TURBULENT	CALCULATION
4	DENSITY LIQ	937	TERM 1	+B9 / B8
5	DENSITY VAP	.55	CUBE ROOT	+D4 ∧ (1 / 3)
6	KLIQUID	.685	TERM 2	+B20 / B16
7	KCUBE	+B6 ∧ 3	POWER	+D6 ∧ .4
8	VISCOSITY	2.19 E−4	PRODUCT	.0077*D5*D7
9	TERM A	+B3*B4*(B4−B5)*B7 / B8	H	+D8 ∧ (5 / 3)
10	HFG	2.183 E6	RATIO	+D9 / B16
11	TSAT	400	HEATFLOW	+D10*B18
12	TS	398	MASSFLOW	+D10*B19
13	L	1	REYNOLDS	+D10*B20
14	TERM B	+B9*B10 / ((B11−B12)*B13)		
15	ROOT	+B14 ∧ .25	SELECTION	
16	H	1.13*B15	STANDARD	1800 > B20
17	LAMINAR	VALUES	TEST	@ IF (D16, 1, 2)
18	HEATFLOW	+B16*B13*(B11−B12)	HEATFLOW	@ CHOOSE (D17, B18, D11)
19	MASSFLOW	+B18 / B10	MASSFLOW	@ CHOOSE (D17, B19, D12)
20	REYNOLDS	4*B19 / B8	REYNOLDS	@ CHOOSE (D17, B20, D13)

If a horizontal tube (that may contain a cold fluid) is used to condense vapor, the vapor forms a film about the tube and tends to flow down where it can be collected, as illustrated in Fig. 9-3-1. The film tends to be thicker at the bottom of the tube than at the top but tends to be uniform along the length of the tube (if the tube surface temperature is constant). Thus, we expect a horizontal tube to be more effective than a vertical tube as a condensing surface. Using tube diameter as a characteristic length dimension, it has been found by Dhir and Lienhard (1971) that

$$\bar{h} = 0.729 \left[\frac{g\rho_l(\rho_l - \rho_v)k_l^3 h_{fg}}{\mu_l(T_{sat} - T_s)D} \right]^{1/4} \tag{9-23}$$

The coefficient 0.729 is lower than the coefficient 1.13, but the diameter of the tube will usually be much smaller than its length.

If there is a bank of tubes one above the other (Fig. 9-3-2), the film flowing off the top tube flows onto the tube below, increasing the film

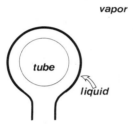

FIGURE 9-3-1. Condensation on a horizontal tube.

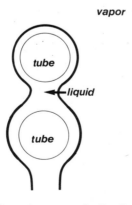

FIGURE 9-3-2. Condensation on vertically aligned horizontal tubes.

thickness on that tube. For N rows of tubes, the length dimension D is replaced by ND, and the average heat transfer per tube is reduced.

The similarity of the equation for condensation of horizontal cylinders to that for condensation at a vertical wall makes the heat transfer coefficient calculation in Columns A and B of the spreadsheet of Example 9-1 directly applicable. The length dimension in Row 13 is to be interpreted as calling for ND. Entry B16 should be edited to change 1.13 to 0.729. To get heat flow in Row 18, the length dimension should be circumference rather than diameter, so entry B18 should be edited by multiplying by π. The mass flow calculation (Row 19) is then applicable. The Reynolds number calculation of Row 20 is not needed.

It is also possible to have vapor condensing inside a tube. For low Reynolds numbers (below 35,000, with properties evaluated for vapor at inlet), it has been found that

$$\bar{h} = 0.555 \left[\frac{g\rho_l(\rho_l - \rho_v)k_l^3 h_{fg}'}{\mu_l(T_{\text{sat}} - T_s)D} \right]^{1/4} \tag{9-24}$$

$$h_{fg}' = h_{fg} + 0.375\, C_{p,l}(T_{\text{sat}} - T_s) \tag{9-25}$$

Adaptation of the spreadsheet for flow over horizontal cylinders (adapted from that for vertical walls as discussed above) for this case can be accomplished by changing 0.728 to 0.555 in B13 and inserting rows to introduce specific heat and calculate a modified h_{fg}. At higher Reynolds numbers, the two-phase flow regime changes. A correlation that has been suggested is

$$\text{Nu} = 0.026\, \text{Pr}_l^{1/3} \text{Re}_m^{0.8} \tag{9-26}$$

where the Reynolds number is defined for the two-phase mixture

$$\text{Re}_m = \frac{D}{\mu_l} \left[\rho_l u_l + \rho_v u_v \left(\frac{\rho_l}{\rho_v} \right)^{1/2} \right] \tag{9-27}$$

where the mass velocities (ρu) are calculated as if each occupied the entire flow area.

In general, we expect a desire for condensation to take place on the outsides of tubes, since the vapor to be condensed tends to require significant volume. Thus, where the cold fluid that causes the condensing is a liquid, as in a power plant condenser using river water to condense steam, the cold fluid tends to be inside tubes and the hot condensing fluid tends to be outside. On the other hand, when the cold fluid is a gas, as with condensers in refrigeration and air conditioning systems, then the volume requirements of the noncondensing fluid can be greater than those of the condensing fluid, and the condensing fluid is placed inside tubes. There may also be other considerations, regarding handling of the specific fluids.

EXAMPLE 9-2.   Pressurized ammonia with a saturation temperature of 30°C is to be condensed on a 2.5 cm diameter horizontal piping having a surface temperature of 29°C.

a. How much heat is transferred per length of piping?
b. If there are five rows of piping vertically aligned, what is the average heat transfer per length of piping per pipe?

SOLUTION.   This problem may be treated with

$$h = 0.729 \left[\frac{-g\rho_l(\rho_l - \rho_v)k_l^3 h_{fg}}{\mu_l ND(T_{sat} - T_s)} \right]^{1/4}$$

For (a), $N = 1$ and (neglecting ρ_v)

$$h = 0.729 \left[\frac{g(596)^2(0.507)^3(1.15 \times 10^6)}{(2.08 \times 10^{-4})(0.025)(30 - 29)} \right]^{1/4} = 12,900$$

$$q' = \pi Dh(T_{sat} - T_s) = 1010 \text{ W/m}$$

With five pipes vertically aligned, $N = 5$ and

$$h = 8630$$

The average heat transfer for the five pipes is

$$q' = 678 \text{ W/m}$$

As discussed earlier in the section, this type of example is easy to treat by adapting the spreadsheet provided for Example 9-1. The adaptation is left as an exercise.

9-4. BOILING PROCESS

When the temperature of a surface exceeds the boiling point of a liquid with which it is in contact, the liquid at the surface may be heated to the boiling point and evaporate. Evaporation involves formation of a bubble which then migrates from the surface, as illustrated in Fig. 9-4-1. There the surface is horizontal and bubbles rise under the influence of gravity. The situation of a stagnant body of liquid boiling is called pool boiling.

If the liquid average temperature is less than the saturation temperature, the bubble transfers heat to the colder liquid and condenses. This is called subcooled boiling. The reader undoubtedly has observed, when boiling water at home, that initially bubbles form at the bottom of the pot, but do not make it to the top, while eventually bubbles do come all the way to the top. This is because water near the bottom is heated to saturation before water in the rest of the pot. Ultimately, all the water is heated to saturation, and bubbles rise through the liquid. This is called saturated pool boiling. (Refinements of this discussion will be considered later in regard to surface tension.)

A driving force in boiling heat transfer is the excess temperature

$$T_e = T_s - T_{sat} \tag{9-28}$$

One might at first, think that the greater this driving force, the greater the tendency for heat to flow. The situation, however, is not so simple.

When a hot wire is placed in a pool of water, heat flux varies with excess temperature as shown in Fig. 9-4-2. When the excess temperature is very small, slightly superheated liquid rises by natural convection to the surface, where it tends to evaporate. As the excess temperature is increased, this natural convection is enhanced, as is to be expected, since natural convection depends

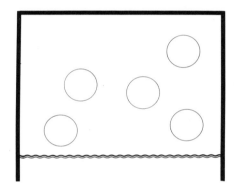

FIGURE 9-4-1.   Boiling bubbles rising through liquid.

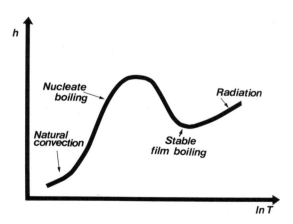

FIGURE 9-4-2.   A typical boiling curve.

on the difference between wall temperature and fluid temperature. At some point, the heat flux becomes large enough that liquid acquires the latent heat of vaporization before it can move away by convection.

When bubbles form, the tendency for convection increases, since the bubbles of vapor are much lighter than equal volumes of liquid at the same temperature. Initially, these bubbles rise into colder fluid and condense. As the excess temperature increases, the bubbles become sufficiently superheated that they rise all the way to the surface. Under these conditions, heat transfer increases rapidly with excess temperature.

The initial boiling consists of bubbles forming at discrete isolated sites. This process is called nucleate boiling. As the excess temperature increases further, the bubble formation and growth leads to a combining of bubbles and the formation of a vapor film on the surface. This leads to a reduction in heat transfer, and we observe a peak in Fig. 9-4-2 followed by a decline.

Recall that even in convection heat transfer, we depend on conduction in the fluid to get heat from the wall into the fluid. The thermal conductivity of water vapor, for example, is an order of magnitude smaller than the thermal conductivity of liquid water. We thus have reduced capability of getting heat into the boiling liquid. Heat must be conducted across the film to get to the liquid to cause more boiling.

At first there will be a mixture of nucleate and film boiling. As the excess temperature increases, a region of stable film boiling is obtained. As excess temperature continues to increase, heat flux again begins to increase because radiation heat transfer becomes significant. The surface is hot enough for radiation to be important, and the vapor is transparent enough for the radiation to cross into the liquid. Heat transfer continues to increase with increasing excess temperature.

The phenomena just discussed are very important for the designer to consider in regard to boiling heat transfer. The regime of nucleate boiling provides high values of heat transfer coefficients. On the other hand, the peak of the curve which signals the departure from nucleate boiling is a serious concern. If too much heat is being generated so that surface temperature rises past the peak of the curve, then surface temperature will continue to increase toward the point where radiation will take over. Melting of material and damage to equipment may occur before this point is reached, since thousands of degrees may be required before it becomes possible to transfer the same amount of heat.

This potential for damage must be considered by designers of equipment in which heat is generated. Indeed, the situation is illustrative of trade-offs that must be made in design. On the one hand, a designer would like to be as efficient as possible in the use of resources and therefore would like to achieve as close to the theoretical maximum heat transfer as is possible. On the other hand, a designer would like to avoid the prospect of damage by having a margin of safety, thereby accepting a lower heat transfer rate.

In our discussion above, we have stated that when boiling occurs, bubbles form. Sizes of bubbles are determined by surface tension, that is, attractive forces between molecules of liquid. A spherical bubble has a radius such that the energy required for a pressure difference to change the volume is balanced by the energy required to change the area; that is,

$$(p_v - p_l)\,dV = \sigma\,dA \tag{9-29}$$

For a sphere, inserting expressions for dV and dA yields

$$(p_v - p_l)4\pi r^2\,dr = \sigma 8\pi r\,dr \tag{9-30}$$

which may be simplified to

$$p_v - p_l = \frac{2\sigma}{r} \tag{9-31}$$

Let us consider some implications. For a bubble to exist, the vapor pressure must exceed the liquid pressure. Indeed, this pressure differential is needed to overcome the surface tension that tends to bring a liquid together. The temperature of the vapor should be at or above saturation temperature corresponding to the vapor pressure. Thus, when boiling begins, we do not get simply vapor at the saturation temperature corresponding to the liquid pressure, but rather we get vapor at a slightly higher pressure and temperature.

At the outset of boiling then, the vapor is hotter than neighboring liquid. Heat is transferred from the bubble to the liquid, thereby causing some of the vapor to condense. Since vapor takes more volume than liquid, the remaining vapor can expand, with a lowering of pressure. The resulting pressure differential no longer balances the surface tension, so the bubble shrinks. This process leads to the collapse of the bubble.

For bulk boiling to occur, a situation of heat loss by the bubble must be avoided. The bubble does not lose heat to the liquid if the liquid is at or above the vapor temperature. Such a situation occurs if the liquid becomes superheated.

In a slow, equilibrium process, we expect all liquid to be converted to vapor before temperature continues to rise. In a rapid, nonequilibrium situation, however, this need not be the case. Heat is conducted from the wall into the fluid. Some of the heat goes into the formation of bubbles. However, because a significant amount of heat (the latent heat) is required to vaporize liquid, some liquid may simply increase in temperature, that is, become superheated. Conduction and convection in liquid can increase the amount of superheated liquid. If the liquid is superheated sufficiently, the liquid temperature may exceed the vapor temperature. This leads to heat transfer from liquid to bubble and to bubble growth.

9-5. BOILING HEAT TRANSFER—POOL BOILING

From the discussion of Section 9-4, we expect boiling heat transfer to depend on the density difference between liquid and vapor, surface tension, and other properties. It has been found by Rohsenow (1952) that nucleate pool boiling can be correlated in the form

$$q'' = \mu_l h_{fg} \left[\frac{g(\rho_l - \rho_v)}{g_c \sigma} \right]^{1/2} \left[\frac{c_{pl} T_e}{C_{sf} h_{fg} \mathrm{Pr}^n} \right]^3 \tag{9-32}$$

where C_{sf} and n depend on liquid and surface. For water, $n = 1$, while for other liquids, $n = 1.7$. Specific values for C_{sf} are given in Table 9-5-1.

We may combine some of the parameters into new dimensionless groups that illustrate the physical processes involved. First, define the Jakob number by

$$\mathrm{Ja} = \frac{c_{pl} T_e}{h_{fg}} \tag{9-33}$$

The Jakob number is a measure of the relative ease by which heat can be accounted for through temperature increase as opposed to vaporization.

Let us next introduce a characteristic dimension L to the equation to obtain

$$q'' L = \mu_l h_{fg} \left[\frac{g(\rho_l - \rho_v) L^2}{g_c \sigma} \right]^{1/2} \left[\frac{\mathrm{Ja}}{C_{sf} \mathrm{Pr}^n} \right]^3 \tag{9-34}$$

The Bond number Bo is defined as

$$\mathrm{Bo} = \frac{g(\rho_l - \rho_v) L^2}{g_c \sigma} \tag{9-35}$$

TABLE 9-5-1
The C_{sf} Coefficient for Pool Boiling

Liquid–Surface	C_{sf}
Water–copper	0.0130
Water–platinum	0.0130
Water–chemically etched stainless steel	0.0133
Water–ground and polished stainless steel	0.0080
Water–brass	0.0060
Water–nickel	0.0060
Benzene–chromium	0.0100
Ethanol–chromium	0.0027
n-Pentane–chromium	0.0150
n-Pentane–energy polished copper	0.0154
n-Pentane–lapped copper	0.0049
Carbon tetrachloride–copper	0.0130
Carbon tetrachloride–energy polished copper	0.0070
Isopropanol–copper	0.0025
n-Butanol–copper	0.0030

This parameter is a measure of the relative influence of buoyancy (gravity and density differential) and surface tension.

If we note that heat flux is given by

$$q'' = hT_e \tag{9-36}$$

we may regroup the heat flux equation in the form

$$\frac{hL}{k_l} = \mathrm{Nu} = \frac{c_{pl}\mu_l}{k_l} \frac{h_{fg}}{c_{pl}T_e} \mathrm{Bo}^{1/2} \left[\frac{\mathrm{Ja}}{C_{sf}\mathrm{Pr}^n}\right]^3 \tag{9-37}$$

Thus, an alternate way of expressing the correlation is

$$\mathrm{Nu} = \frac{1}{C_{sf}^3} \mathrm{Bo}^{1/2} \frac{\mathrm{Ja}^2}{\mathrm{Pr}^{3n-1}} \tag{9-38}$$

While this expression is compact in terms of dimensionless groups, the boiling literature frequently expresses heat flux in terms of Eq. (9-32).

Let us see how the compact form in terms of dimensionless groups illustrates the physical phenomena. First, we note that heat transfer increases when the Bond number increases. The Bond number increases when buoyancy increases either through acceleration of gravity (i.e., gravity force is greater in a deep mine and weaker on a mountain top) or through liquid–vapor density differential. It is reasonable to expect boiling to be aided by buoyancy. The Bond number decreases when surface tension increases. This is reasonable, since surface tension holds a liquid together and thereby retards vapor. Consequences in terms of vapor pressure and temperature were discussed in Section 9-4.

Second, we note that heat transfer increases when the Jakob number increases. In the nucleate boiling regime, we expect heat transfer to increase with excess temperature, which is in the numerator of the Jakob number. The Jakob number decreases when latent heat of vaporization increases, so heat transfer decreases as latent heat increases. This is reasonable, since the larger the latent heat, the smaller the amount of liquid that can be boiled with a given amount of heat.

Heat transfer decreases with Prandtl number of the liquid. This is reasonable since liquid conductivity is important for getting heat into the liquid over the fraction of the surface not covered by vapor, and conductivity is in the denominator of the Prandtl number. Similarly, it is expected that bubbles would move more freely in a liquid of low viscosity than in one of high viscosity.

The pool boiling formula holds for nucleate boiling, that is, before the peak of Fig. 9-4-2. The peak heat flux has been correlated in the form

$$q''_{max} = 0.15\rho_v^{1/2}h_{fg}\left[g(\rho_l - \rho_v)\sigma\right]^{1/4} \tag{9-39}$$

This formula, with a slightly different coefficient, has been predicted theoretically by Kutateladze (1951) and by Zuber (1958).

The formula is interesting in that it does not depend directly on conductivity, viscosity, or specific heat. In addition, we note that maximum flux increases with latent heat and surface tension. Thus, certain factors that affect heat transfer adversely enhance the maximum allowable heat transfer. A large latent heat and a large surface tension make it more difficult to achieve a situation where no liquid is left in contact with the surface.

The nature of the maximum heat flux formula may be explained in terms of the competing effects. As film boiling starts to become significant, a potentially unstable situation is created with heavy liquid positioned over light vapor. Heavy liquid will fall into the vapor film region. This collapse of the unstable film is called a Taylor instability.

There is a characteristic dimension to this instability, that is, associated with sizes of regions through which liquid falls and vapor rises. Since there are competing forces of gravity and surface tension, a characteristic length may be defined by

$$L = C_1 \left[\frac{\sigma g_c}{g(\rho_l - \rho_v)} \right]^{1/2}$$

where C_1 is some undetermined constant. As usual, we set $g_c = 1$.

When the vapor rises and removes heat, it does so at some characteristic speed. A dimensionless speed involving surface tension is

$$u = C_2 \left(\frac{\sigma}{\rho_v L} \right)^{1/2}$$

Using the characteristic length above,

$$u = C_3 \frac{\sigma^{1/4}}{g^{1/2} \rho_v^{1/2}} (\rho_l - \rho_v)^{1/4}$$

The heat flux is in the form

$$q'' = C_4 \rho_v u h_{fg}$$

When these equations are combined, the form of the maximum heat flux correlation is obtained with an undetermined constant.

EXAMPLE 9-3. Water undergoes nucleate boiling over a copper surface at 120°C. Air over the pool is at atmospheric pressure. Evaluate the Jakob, Bond, and Nusselt numbers for the pool having a characteristic dimension of 10 cm. Evaluate the heat flux.

SOLUTION. The Jakob number is

$$Ja = \frac{C_{pl}T_e}{h_{fg}} = \frac{(4210)(120 - 100)}{2.26 \times 10^6} = 0.037$$

The Bond number is

$$Bo = \frac{g(\rho_l - \rho_v)L^2}{g_e\sigma} = \frac{(9.8)(958)(0.1)^2}{(1)(58.8 \times 10^{-3})} = 1600$$

The Nusselt number is

$$Nu = \frac{1}{C_{sf}^3} Bo^{1/2} \frac{Ja^2}{Pr^{3n-1}} = \frac{1}{(0.013)^3}(1600)^{1/2}\frac{(0.037)^2}{(1.76)^2} = 8050$$

The heat transfer coefficient is

$$h = Nu\frac{k}{L} = 8050\left(\frac{0.682}{0.1}\right) = 5.49 \times 10^4$$

Again, with a change of phase we obtain a high heat transfer coefficient. The heat flux is

$$q'' = hT_e = (5.49 \times 10^4)(120 - 20) = 1.10 \times 10^6 \text{ W/m}^2$$

Recall that the length dimension does not affect the calculation of h and q'', although it enters into the Bond and Nusselt numbers.

EXAMPLE 9-4. A 1 cm diameter electric heater is to be used to boil water at atmospheric pressure. The heater material has $C_{sf} = 0.013$.

a. What is the maximum heat per length of heater that can be provided at moderate temperature?
b. If the heat to be delivered by the heater is 10% of the maximum, what will be the surface temperature of the heater?

SOLUTION. (a) The maximum heat flux is obtained from

$$q''_{max} = 0.15\rho_v^{1/2}h_{fg}[g(\rho_l - \rho_v)\sigma]^{1/4}$$

$$q''_{max} = 0.15(0.597)^{1/2}(2.26 \times 10^6)[9.8(958 - 0.6)(58.8 \times 10^{-3})]^{1/4}$$

$$q''_{max} = 1.27 \times 10^6 \text{ W/m}^2$$

The maximum heat per unit length is

$$q''_{max} = q''_{max}\pi D = (1.27 \times 10^6)(\pi)(0.01) = 4 \times 10^4 \text{ W/m} = 400 \text{ W/cm}$$

(b) Set $q'' = 0.1 q''_{max} = 1.27 \times 10^5$ W/m². Rearrange the formula

$$q'' = \mu_l h_{fg} \left[\frac{g(\rho_l \rho_v)}{g_c \sigma} \right]^{1/2} \left[\frac{C_{pl} T_e}{C_{sf} h_{fg} \text{Pr}^n} \right]^3$$

to solve for T_e; that is,

$$T_e = \frac{C_{sf} h_{fg} \text{Pr}^*}{C_{pl}} \left(\frac{q''}{\mu_l h_{fg}} \right)^{1/3} \left[\frac{g(\rho_l - \rho_u)}{g_c \sigma} \right]^{-1/6}$$

$$T_e = \frac{(0.013)(2.26 \times 10^6)(1.72)}{4219} \left[\frac{1.27 \times 10^5}{(2.78 \times 10^{-4})(2.26 \times 10^6)} \right]^{1/3} \left[\frac{(9.8)(957)}{0.0588} \right]^{-1/6}$$

$$= (11.98)(5.87)(0.136) = 9.55°\text{C}$$

$$T_s = T_{sat} + T_e = 109.55°\text{C}$$

A more complete analysis of this problem would involve checking into electrical characteristics such as electric current required.

9-6. FORCED CONVECTION BOILING

The discussion so far has been in regard to boiling in a pool, that is, a natural convection situation. Boiling is frequently associated with forced convection heat transfer. For flow over a hot surface, it has been recommended by Rohsenow and Bergles (1973) that the total heat flux be obtained from

$$q'' = \left(q''^2_{Fc} + q''^2_{Pb} - q''^2_i \right)^{1/2} \qquad (9\text{-}40)$$

In this expression q''_{Fc} is the heat flux that would be calculated on the basis of forced convection alone if boiling were not taking place, q''_{pb} is the pool boiling that would be predicted for that surface and liquid, and q''_i is the heat flux from natural convection at the inception of boiling on the pool boiling curve. The relationship of these quantities is illustrated in Fig. 9-6-1. This approach can be used for situations with moderate vapor production. Note that if the flowing liquid has not yet reached saturation, the forced convection heat flux should be based on $T_s - T_b$, where T_b is the bulk temperature.

The formula would be applied in the following way. Suppose, for example, that liquid flows across a heated cylinder. To evaluate q''_{Fc}, an appropriate single-phase heat transfer correlation (based on the Reynolds number and the Prandtl number for the liquid) should be selected from Chapter 7 and

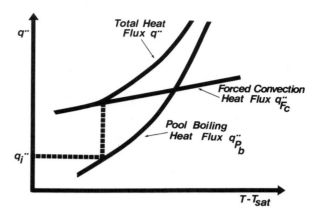

FIGURE 9-6-1. Relation of quantities for forced convection boiling heat flux.

evaluated. To evaluate q_{Pb}'', the pool boiling correlation of this chapter should be evaluated. To evaluate q_i'', the inception point for boiling should be known. For water at atmospheric pressure, this point is at about 5°C of excess temperature. An applicable natural convection correlation based on the Rayleigh and Prandtl numbers of the liquid is then selected and evaluated. Since q_{Pb}'' increases rapidly with T_e (varying as T_e^3), the q_i'' correction may turn out to be small.

The boiling heat transfer formula, involving forced convection, pool boiling, and natural convection formulas, can require considerable hand calculations for evaluation. However, if the various formulas have been provided for in individual spreadsheets, it is a simple matter to combine these individual spreadsheets into a single spreadsheet. This is another example of how spreadsheets can be used to build on work done previously. Preparation of such a combined spreadsheet, building on the spreadsheets of Chapters 7 and 8, is left as an exercise.

Forced convection flow over hot surfaces is characterized by a boiling curve similar to the pool boiling curve of Fig. 9-4-2. Thus, there is concern about the maximum heat flux. For cross-flow over a cylinder, Lienhard and Eichhorn ((1979) obtained

$$\frac{q_{max}}{\rho_v h_{fg} u_\infty} = \frac{1}{\pi}\left[1 + \left(\frac{4}{We}\right)^{1/3}\right] \qquad \frac{q_{max}}{\rho_v h_{fg} u_\infty} > \frac{0.275}{\pi}\left(\frac{\rho_l}{\rho_v}\right)^{1/2} + 1 \quad \textbf{(9-41)}$$

$$\frac{q_{max}}{\rho_v h_{fg} u_\infty} = \frac{(\rho_l/\rho_v)^{1/2}}{\pi}\left[\frac{(\rho_l/\rho_v)^{1/4}}{169} + \frac{1}{19.2\,We^{1/3}}\right]$$

$$\textbf{(9-42)}$$

$$\frac{q_{max}}{\rho_v h_{fg} u_\infty} < \frac{0.275}{\pi}\left(\frac{\rho_l}{\rho_v}\right)^{1/2} + 1$$

These formulas involve the Weber number

$$\text{We} = \frac{\rho_v u_\infty^2 D}{\sigma} \tag{9-43}$$

where the diameter D is the significant length dimension. This number relates inertial force (kinetic energy) to surface tension.

The first formula is referred to as the low-velocity formula. It predicts that when u_∞ goes to zero, q_{max} goes to zero. Actually, there will be a finite limiting value for q_{max} associated with stagnant pool boiling. Thus, evaluating q_{max}, one should check whether the low-velocity or high-velocity formula applies. If the low-velocity formula applies, one should check to see if the pool boiling limit applies.

For forced convection flow inside tubes of ducts, the situation can be complicated because various flow regimes are possible. When subcooled liquid enters a tube, heat is transferred to it according to the single-phase heat transfer relationship until boiling occurs. Nucleate boiling will then occur and there will be a mixture of single-phase convection and nucleate boiling. A situation of annular flow may then arise, with a film of liquid at the tube wall and vapor at the center. Heat is conducted through the liquid film, and liquid evaporates into the vapor section in what has been called forced convection vaporization.

A correlation developed by Chen (1966) that is applicable to both fully developed nucleate boiling and to forced convection vaporization is of the form

$$h = h_1 + h_2 \tag{9-44}$$

The term h_1 is a modified single-phase convection coefficient given by

$$\text{Nu}_1 = \frac{h_l D}{k_l} = 0.023 \, \text{Re}^{0.8} \text{Pr}^{0.4} F \tag{9-45}$$

where the Reynolds number used is a liquid-phase Reynolds number

$$\text{Re} = \frac{G(1 - x)D}{\mu_l} \tag{9-46}$$

where G is mass flux (kg/m^2 · sec) and x is quality (ratio of vapor mass flux to total mass flux). The parameter F is a ratio to provide an effective two-phase Reynolds number; that is,

$$F = \left(\frac{\text{Re}_{2p}}{\text{Re}} \right)^{0.8} \tag{9-47}$$

This empirical quantity depends on a quantity called the Martinelli parameter

X_{tt} (a measure of the ratio of liquid and vapor pressure drops) which is given by

$$X_{tt} = \left(\frac{1-x}{x}\right)^{0.9}\left(\frac{\rho_v}{\rho_l}\right)^{0.5}\left(\frac{\mu_v}{\mu_l}\right)^{0.1} \tag{9-48}$$

The F parameter may be approximated within the quoted uncertainty as

$$F = 2.608\left(\frac{1}{X_{tt}}\right)^{0.7434} \qquad 0.7 < \frac{1}{X_{tt}} < 100$$

$$F = 2.2709\left(\frac{1}{X_{tt}}\right)^{0.3562} \qquad 0.1 < \frac{1}{X_{tt}} < 0.7 \tag{9-49}$$

This formula approximation is useful if evaluation is to be performed on a computer and is plotted in Fig. 9-6-2.

The second part of the heat transfer coefficient is a nucleate boiling coefficient adapted from an equation of Foster and Zuber (1955),

$$h_2 = 0.00122\frac{k_l^{0.79}C_{pl}^{0.45}\rho_l^{0.49}}{\sigma^{0.5}\mu_l^{0.29}h_{fg}^{0.24}\rho_v^{0.24}}T_e^{0.24}\Delta p^{0.75}\,S \tag{9-50}$$

where the pressure difference Δp is

$$\Delta p = p_{\text{sat}}(T_s) - p_{\text{sat}}(T_{\text{sat}}) \tag{9-51}$$

The empirical factor S may be approximated within the quoted uncertainty as

$$S = \begin{cases} 3.6278 - 0.64557\log_{10}\text{Re}_{TP} & 2\times10^4 < \text{Re}_{TP} < 3\times10^5 \\ 0.1 & \text{Re}_{TP} > 10^5 \end{cases} \tag{9-52}$$

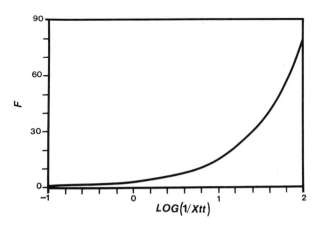

FIGURE 9-6-2. Variation of the S-Parameter with two-phase Reynolds number.

The variation of S is shown in Fig. 9-6-3. The data for S became sparse at Re_{TP} beyond 2×10^5. In principle, it is expected that S would tend to 0 at very high flow rates and would tend to 1 at very low flow rates.

The above form is a common way of encountering h_2 in the literature. It may be observed, with a little algebra, that the parameters can be reorganized into dimensionless groups. It is left as an exercise to show that

$$\text{Nu}_2 = \frac{h_2 D}{k_l} = 0.00122 \, \text{Pr}_l^{0.21} \text{Ja}^{0.24} \left(\frac{\rho_l}{\rho_v} \right)^{0.24} \left(\frac{\Delta p \, D}{\sigma} \right)^{0.5} \left(\frac{\rho_l \Delta p \, D^2}{\mu_l^2} \right)^{0.25} S \quad (9\text{-}53)$$

If we recall that pressure is related to kinetic energy via the Euler number

$$\text{Eu} = \frac{\Delta p}{\frac{1}{2} \rho_l u^2} \quad (9\text{-}54)$$

then, for fluid speed u

$$\frac{\rho_l \Delta p D^2}{\mu_l^2} = \frac{1}{2} \frac{\Delta p}{\frac{1}{2} \rho_l u^2} \left(\frac{\rho u D}{\mu_l} \right)^2 = \frac{1}{2} \text{Eu} \, \text{Re}_{\text{eff}}^2 \quad (9\text{-}55)$$

In other words, we have Euler and Reynolds-like numbers. Similarly,

$$\frac{\Delta p \, D}{\sigma} = \frac{1}{2} \frac{\Delta p}{\frac{1}{2} \rho_l u^2} \frac{\rho_l u^2 D}{\sigma} = \frac{1}{2} \text{Eu} \, \text{We} \quad (9\text{-}56)$$

The dimensionless right-hand side includes the Weber number, which, as noted earlier, relates inertial force to surface tension.

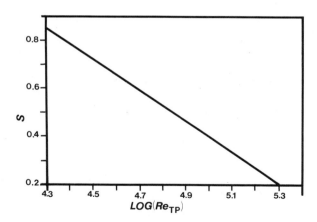

FIGURE 9-6-3. Variation of F-Parameter with Martinelli parameter.

For forced convection boiling in tubes or channels, concern again must be given to maximum heating. However, the situation is more complicated than the maximum heating situations discussed earlier. In particular, the earlier situations involved single, well-defined limits on local heat flux. In a long channel, however, various additional considerations come into play. As liquid proceeds along the channel, boiling takes place. The quality varies along the channel, for example, so the boiling fluid is not the same at every location. It is also possible for heating along the channel to be nonuniform. Limiting values on heating have been understood in terms of a critical quality achieved within a critical boiling length. The quality achieved in a given length of channel is determined by the total heating in the channel. The shorter the boiling length to achieve a given quality, the higher the total heating. The critical quality concept says that for a given boiling length, there is a maximum allowable quality. Experimental results support this hypothesis, with critical quality data being insensitive to heat flux profile (except for extreme profiles). Indeed, the critical condition can exist at a location in the channel where the heat flux is below the maximum local value. The critical quality for a particular boiling length depends on geometry (channel hydraulic diameter), pressure, and flow rate.

Design practice can be complicated, and the interested reader is referred to the references at the end of the chapter for further detail. A brief qualitative discussion is provided here. A boiling channel clearly should not lead to critical quality being attained at any location within the channel. The engineer must determine that an adequate margin exists for operational purposes. It is logical then to ask by what ratio the total power into the channel would have to be increased (assuming the same pressure, flow rate, and inlet conditions) for a critical condition to be achieved. This ratio is called the critical power ratio. Design objectives or safety standards (e.g., in regulating boiling water nuclear reactors) may call for a specified minimum critical power ratio.

The discussion and results of this section illustrate the complexity of boiling heat transfer. References for further study into this broad and important topic are provided at the end of this chapter. Boiling heat transfer, while complicated, is of great importance. For one thing, devices from power plants to refrigerators involve boiling as a significant component of the thermodynamic cycle. For another, as the examples of this chapter illustrate, boiling leads to high heat transfer rates, and boiling therefore can be useful in enhancing heat removal.

EXAMPLE 9-5.   The water of Example 9-4 is made to flow over the heater at 1 m/sec. What is the maximum heat flux, and how does it compare to the value of Example 9-4?

SOLUTION.   First, let us evaluate the parameter that determines whether the low-velocity or high-velocity formula applies:

$$\frac{0.275}{\pi}\left(\frac{958}{0.597}\right)^{1/2} + 1 = 4.51$$

Next, let us evaluate the low-velocity formula

$$\frac{q''_{max}}{\rho_v h_{fg} u_\infty} = \frac{1}{\pi}\left[1 + \left(\frac{4}{We}\right)^{1/3}\right]$$

$$We = \frac{\rho_v u_\infty^2 D}{\sigma} = \frac{(0.597)(1)^2(0.01)}{0.0588} = 0.102$$

$$\frac{q''_{max}}{\rho_v h_{fg} u_\infty} = 1.40 < 4.51$$

The low-velocity formula does not apply and we apply the high-velocity formula

$$\frac{q''_{max}}{\rho_v h_{fg} u_\infty} = \frac{(\rho_l/\rho_g)^{1/2}}{\pi}\left[\frac{(\rho_l/\rho_g)^{1/4}}{169} + \frac{1}{19.2\,We^{1/3}}\right]$$

$$= \frac{40}{\pi}\left(\frac{6.33}{169} + 0.111\right) = 1.90$$

$$q''_{max} = (1.90)(0.597)(2.26 \times 10^6)(1) = 2.56 \times 10^6 \ W/m^2$$

This is roughly double the pool boiling maximum found in Example 9-4. Note that the maximum heat flux in forced convection depends on the diameter (through the Weber number) whereas the pool boiling limit did not.

EXAMPLE 9-6.   Saturated liquid at 100°C flowing at 0.3 m/sec enters a 2 cm diameter channel whose walls are at 120°C. Find the heat transfer coefficient at locations where quality is 0.05, 0.1, 0.2, and 0.3.

SOLUTION.   The heat transfer coefficient is made up of two parts. First, attention is given to the modified single-phase convection coefficient obtained from

$$Nu_1 = 0.023 \ Re^{0.8} Pr^{0.4} F$$

The Reynolds number is

$$Re = \frac{G(1-x)D}{u_l} = \frac{\rho u(1-x)D}{\mu_l} = \frac{(958)(0.3)(1-x)(0.02)}{2.78 \times 10^{-4}}$$

$$Re = 2.07 \times 10^4 (1-x)$$

so we obtain

x	0.05	0.1	0.2	0.3
Re	1.96×10^4	1.86×10^4	1.65×10^4	1.45×10^4

To evaluate F, attention is first directed to the Martinelli parameter

$$X_{tt} = \left(\frac{1-x}{x}\right)^{0.9}\left(\frac{\rho_v}{\rho_l}\right)^{0.5}\left(\frac{\mu_v}{\mu_l}\right)^{0.1} = \left(\frac{1-x}{x}\right)^{0.9}\left(\frac{0.597}{958}\right)^{0.5}\left(\frac{2.78 \times 10^{-4}}{1.27 \times 10^{-5}}\right)^{0.1}$$

$$X_{tt} = 0.0338\left(\frac{1-x}{x}\right)^{0.9}$$

$$\frac{1}{X_{tt}} = 29.6\left(\frac{x}{1-x}\right)^{0.9}$$

which yields, using Eq. (9-49) for F,

x	0.05	0.1	0.2	0.3
F	4.51	7.44	12.8	18.41

We can now evaluate

$$h_1 = \frac{k_l}{D}\text{Nu}_1 = \frac{0.681}{0.02}(0.023)(1.72)^{0.4}\text{Re}^{0.8}F = 0.973\,\text{Re}^{0.8}F$$

which can be evaluated at the indicated qualities to yield

x	0.05	0.1	0.2	0.3
h_1	1.19×10^4	1.89×10^4	2.95×10^4	3.82×10^4

The modified single-phase convection coefficient is large and increases quality. The second part of the heat transfer coefficient is

$$\text{Nu}_2 = 0.00122\,\text{Pr}_l^{0.21}\,\text{Ja}^{0.24}\left(\frac{\rho_l}{\rho_v}\right)^{0.24}\left(\frac{\Delta p\,D}{\sigma}\right)^{0.5}\left(\frac{\rho_l\,\Delta p\,D}{\mu_l^2}\right)^{0.25}S$$

The only quantity that varies with quality is S. The other groupings are

$$\text{Pr}_l^{0.21} = (1.72)^{0.21} = 1.12$$

$$\text{Ja}^{0.24} = \left(\frac{C_{pl}T_e}{h_{fg}}\right)^{0.24} = \left[\frac{(4219)(20)}{2.26 \times 10^6}\right]^{0.24} = (0.0373)^{0.24} = 0.454$$

$$\left(\frac{\rho_l}{\rho_v}\right)^{0.24} = \left(\frac{958}{0.597}\right)^{0.24} = 5.88$$

The pressure difference is

$$\Delta p = p_{\text{sat}}(120°C) - p_{\text{sat}}(100°C)$$

$$= (1.9854 - 1.0733)10^5 = 9.72 \times 10^4 \,\text{N/m}^2$$

$$\left(\frac{\Delta p \, D}{\sigma}\right)^{0.5} = \left[\frac{(9.72 \times 10^4)(0.02)}{0.0588}\right]^{0.5} = (3.307 \times 10^4)^{1/2} = 182$$

$$\left(\frac{\rho_l \, \Delta p \, D^2}{\mu_l^2}\right)^{0.25} = \left[\frac{(958)(9.72 \times 10^4)(0.02)^2}{(2.78 \times 10^{-4})^2}\right]^{0.25} = 833$$

$$h_2 = \frac{0.681}{0.02}(0.00122)(1.12)(0.454)(5.88)(182)(833)\,S$$

$$= 1.88 \times 10^4 \, S$$

The two-phase Reynolds number is

$$\text{Re}_{TP} = \text{Re} \, F^{1.25}$$

which yields

x	0.05	0.1	0.2	0.3
Re_{TP}	1.28×10^5	2.29×10^5	3.99×10^5	5.53×10^5

From Eq. (9-52)

$$S = \begin{cases} 3.6278 - 0.64557 \log_{10}\text{Re}_{TP} & 2 \times 10^4 < \text{Re}_{TP} < 3 \times 10^5 \\ 0.1 & \text{Re}_{TP} > 3 \times 10^5 \end{cases}$$

we obtain

x	0.05	0.1	0.2	0.3
S	0.33	0.17	0.10	0.10

We can then obtain

x	0.05	0.1	0.2	0.3
h_2	6.2×10^3	3.2×10^3	1.9×10^3	1.9×10^3

This heat transfer coefficient declines with quality. The overall heat transfer coefficient is

x	0.05	0.1	0.2	0.3
h	1.81×10^4	2.21×10^4	3.14×10^4	4.01×10^4

In this example, the modified single-phase coefficient is more influential than the boiling coefficient, and the overall heat transfer coefficient increases with quality. At higher saturation temperatures and pressures, the properties of water are such that the

Martinelli parameters for a given quality are lower and F multiples are lower, and the S multiples are higher. As a result, the h_1 contributions are lower and the h_2 contributions are higher. A problem at the end of the chapter will call for investigating these relative influences at different saturation temperatures.

Because of the large amount of calculation in this problem and the repetitive aspect of the calculations at different qualities, spreadsheet application to this example is useful. A spreadsheet for this example follows evaluated for a quality of 0.05. The other values of this example can be obtained by changing the entry in B7.

	A	B	C	D
1	FORCED	CONVECTION	BOILING	
2	DENSITY LIQUID	958	PR POWER	+B23 ∧ .21
3	SPEED	.3	CP	4219
4	G	+B2*B3	TEXCESS	20
5	DIAMETER	.02	HFG	2.26 E6
6	VISCOSITY L	2.78 E−4	JAKOB	+D3*D4 / B5
7	QUALITY	.05	JAKOB POWER	+D6 ∧ .24
8	``1−QUALITY´´	1−B7	DENSITY TERM	(B2 / B10) ∧ .24
9	REYNOLDS L	+B4*B8*B5 / B6	PSAT(TS)	1.9854 E5
10	DENSITY VAPOR	.597	PSAT(T)	1.0133 E5
11	RATIO DENSITY	(B10 / B2) ∧ .5	SURF TENSE	.0588
12	VISCOSITY V	1.27 E−5	RATIO	(D9−D10)*B5 /
13	RATIO VISCOSITY	(B6 / B12) ∧ .1	ROOT	@ SQRT (D12)
14	RATIO QUALITY	(B8 / B7) ∧ .9	RATIO	(D9−D10)*B2* [(B5 / B6) ∧ 2
15	MARTINELLI	+B14*B11*B13	ROOT	+D14 ∧ .25
16	INVERSE	1 / B15	REYNOLDS TP	+B9*(B21 ∧ 1.2.
17	F	2.608*(B16 ∧ .7434)	S	3.6278 − .64557 @ * LOG (D16
18	F	2.2709*(B16 ∧ .3562)	S	.1
19	STANDARD	B16>.7	STANDARD	D16<3E5
20	TEST	@ IF (B19, 1, 2)	TEST	@ IF (D19, 1,
21	F CHOICE	@ CHOOSE (B20, B17, B18)	S CHOICE	@ CHOOSE (D20, D18)
22	K LIQUID	.681	H2	.00122*B22*D7 *D8*D13*D15 *D21 / B5
23	PRANDTL LIQ	1.72	H	+B26 + D22
24	PR POWER	+B23 ∧ .4		
25	RE POWER	+B9 ∧ .8		
26	H1	.023*B22*B21* B24*B25 / B5		

PROBLEMS

9-1. Repeat Example 9-1 for walls 0.5 m and 2 m high.

9-2. Repeat Example 9-1 for a wall temperature of 375°K.

9-3. For the conditions of Example 9-1, evaluate the film thickness at the middle and the bottom of the plate.

9-4. Suppose Example 9-1 involved pressurized ammonia with a saturation temperature of 30°C and a wall temperature of 29°C. Find the heat transfer and amount of condensate per meter of wall depth.

9-5. You wish to provide a vertical wall whose heat transfer coefficient will be the same as that for the set of horizontal tubes of Example 9-2. How high should this wall be?

9-6. If the tubes of Example 9-2 are 1 m long, how wide should the wall of Problem 9-5 be for the total heat transfer to be the same?

9-7. Saturated steam at 400°K is to be condensed on a horizontal pipe 10 cm in diameter having a surface temperature of 398°K. Find the heat transfer coefficient.

9-8. A condenser is to use horizontal tubes 2.5 cm in diameter to condense steam at 100°C. What would be the heat transfer coefficient if the tube temperature were 20°C?

9-9. Because of variations in outside temperature during the year, the tube temperature varies from 5 to 25°C. What is the range of heat transfer coefficients?

9-10. The condenser of Problems 9-8 and 9-9 is to condense 150 kg/sec of steam. How many tubes will be required if the tubes are to be 2 m long?

9-11. To save space, the condenser of Problem 9-10 is to have 10 vertical rows of tubing. How many tubes will be required?

9-12. Repeat Example 9-4 for a heater 2 cm in diameter. Note which quantities are affected by a change in diameter and which are not.

9-13. The surface tension of water in contact with vapor at the saturation temperature has been expressed in the form

$$\sigma = 0.2358 \left(1 - \frac{T_{\text{sat}}}{T_c} \right)^{1.256} \left[1 - 0.625 \left(1 - \frac{T_{\text{sat}}}{T_c} \right) \right] \text{N/m}$$

where T_c is the critical temperature 647.2°K above which there is no phase change from liquid to vapor. Tabulate or plot the variation of the surface tension between 100°C and the critical temperature.

9-14. The Bond number is the ratio of the squares of a characteristic system length and an effective length dimension L_{eff} determined by surface tension and gravity

$$L_{\text{eff}} = \frac{\sigma}{g(\rho_l - \rho_v)}$$

Tabulate or plot this effective length dimension for water from 100 to 600°C.

9-15. The Jakob number relates heat capacity of liquid to latent heat of vaporization. For a constant excess temperature of 20°C, tabulate or plot the Jakob number for water from 100 to 600°C.

9-16. A speed parameter characterizing bubble motion was expressed in the form

$$u = \frac{\sigma^{1/4}}{g^{1/4}\rho_v^{1/2}} (\rho_l - \rho_v)^{1/4}$$

Evaluate this speed parameter for water at 100°C.

9-17. Consider the effective length dimension of Problem 9-14. Evaluate this dimension at atmospheric pressure for water, ethylene glycol (a common antifreeze), and freon-12 (a common refrigerant).

9-18. Consider the Jakob number. Evaluate it at atmospheric pressure and an excess temperature of 20°C for water, ethylene glycol, and freon-12.

9-19. The configuration of Example 9-4 (involving a 1 cm diameter heater, an excess temperature of 9.55°C, and atmospheric pressure) is to be used in a space station on the moon, where the acceleration of gravity is one-sixth that on earth. The air pressure within the space station is maintained equal to earth's atmospheric pressure.
  a. What is the heat flux at the heater?
  b. What percentage is this of the maximum heat flux?

9-20. For the conditions of Example 9-5, what is the actual heat flux? Note the relative magnitudes of the individual heat flux components.

9-21. Repeat Example 9-5 for fluid speeds of 0.5 and 2 m/sec.

9-22. Repeat Problem 9-20 for fluid speeds of 0.5 and 2 m/sec.

9-23. Repeat Problem 9-20 for excess temperatures of 5, 15. and 20°C.

9-24. Repeat Example 9-6 with saturated liquid at 300°C and the channel wall at 320°C.

9-25. Repeat Problem 9-24 with a tube diameter of 4 cm.

9-26. Repeat Problem 9-24 with a flow speed of 1 m/sec.

9-27. Show that Eq. (9-50) can be reorganized into dimensionless groups as in Eq. (9-53).

9-28. Express Eq. (9-20) in terms of dimensionless groups.

9-29. Express Eq. (9-23) in terms of dimensionless groups.

9-30. Verify that Eq. (9-21a) follows from Eq. (9-21).

REFERENCES

1. J. C. Chen, "Correlation for Boiling Heat Transfer to Saturated Liquids in Convective Flow," *Int. Eng. Chem. Process Des. Dev.*, **5**, 322 (1966).
2. V. K. Dhir and J. H. Lienhard, "Laminar Film Condensation on Plane and axisymmetric Bodies in Non-Uniform Gravity," *J. Heat Transfer*, **93**, 97 (1971).
3. H. K. Forster and N. Zuber, "Dynamics and Vapor Bubbles and Boiling Heat Transfer," *AIChE J.*, **1**, 531 (1955).
4. C. G. Kirkbride, "Heat Transfer by Condensing Vapors," *Trans. AIChE*, **30**, 170 (1934).
5. S. S. Kutateladze, "A Hydrodynamic Theory of Changes in Boiling Process under Free Convection," *Iz. Akad. Nauk SSSR Otd. Tehk. Nauk*, **4**, 524 (1951).
6. J. H. Lienhard and R. Eichhorn, "Peak Boiling Heat Flux on Cylinders in a Cross Flow," *Int. J. Heat Mass Transfer*, **23**, 774 (1979).
7. W. H. McAdams, *Heat Transmission*, 3rd ed., McGraw-Hill, New York, 1954.
8. W. Nusselt, "Die Oberflachenkondensation des Wasserdampfes," *Z. Ver. Dent. Ing.*, **60**, 541 (1916).
9. W. M. Rohsenow, "A Method of Correlating Heat Transfer Data for Surface Boiling of Liquids," *Trans. ASME*, **74**, 969 (1952).
10. W. M. Rohsenow and A. E. Bergles, in W. Rohsenow and J. Hartnett (eds.), *Handbook of Heat Transfer*, McGraw-Hill, New York, 1973.
11. N. Zuber, "On the Stability of Boiling Heat Transfer," *J. Heat Transfer*, **80**, 711 (1958).

Chapter Ten

RADIATION

10-1. INTRODUCTION

All bodies emit and receive electromagnetic radiation. Electromagnetic radiation can have a variety of characteristics, depending on the frequency and wavelength of the radiation, as illustrated in Fig. 10-1-1. At very short wavelengths (high frequencies), we have the gamma radiation of nuclear physics. That type of radiation can be harmful to people directly exposed, but it also has useful industrial and medical applications. At higher wavelengths we speak of X-rays, which are used extensively to take "pictures" of the internal structures of people (e.g., chest and dental X-rays) and industrial equipment. These too can be harmful in excessive doses. At higher wavelengths we encounter visible light, the form of electromagnetic radiation of greatest familiarity to most people. At still higher wavelengths we encounter the radiation used to transmit radio broadcasts.

Thermal radiation tends to include a range of wavelengths from higher than those of visible light (the infrared region) to lower than those of visible light (the ultraviolet region). This is not to say that other electromagnetic radiation does not contain energy. What it does say is that there is a distribution of frequencies of radiation normally given off in the natural radiation of bodies, and this distribution tends to peak in the range cited. Thus, radiation heat transfer as a subject tends to identify this range as thermal radiation. Other electromagnetic radiation, such as gamma radiation,

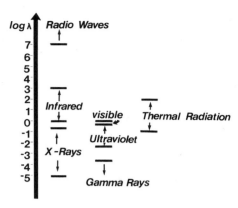

FIGURE 10-1-1. Types of electromagnetic including thermal radiation. Wavelength is expressed in micrometers.

can contain considerable energy and can provide substantial heating to a material through which it passes.

10-2. BLACK BODY RADIATION

A black body is an ideal emitter and absorber of radiation. All thermal radiation incident on the body is absorbed. Since black as a color actually denotes the absence of color (just as white denotes a combination of all colors), the term black body is used to denote a situation in which no radiation is reflected.

A black body will emit energy in a distribution of wavelengths, such that the energy per unit wavelength is

$$\frac{dE_b}{d\lambda} = \frac{2\pi hc^2\lambda^{-5}}{e^{\frac{hc}{\lambda kT}} - 1} \tag{10-1}$$

where h is Planck's constant with value

$$h = 6.625 \times 10^{-34} \text{ J} \cdot \text{sec} \tag{10-2}$$

(not to be confused with the h used for convective heat transfer), k is Boltzmann's constant

$$k = 1.38066 \times 10^{-23} \text{ J/°C} \tag{10-3}$$

(not to be confused with the k used for thermal conductivity), and c is the speed of light in vacuum

$$c = 3 \times 10^8 \text{ m/sec} \tag{10-4}$$

(not to be confused with using c for specific heat). The total energy emitted by the black body is

$$E_b = \int_0^\infty d\lambda \frac{dE_b}{d\lambda} = \sigma T^4 \qquad (10\text{-}5)$$

$$\sigma = 5.669 \times 10^{-8} \text{ W/m}^2 \cdot {}^\circ\text{K}^4 \qquad (10\text{-}6)$$

For most of our applications in this chapter, we will be concerned with total energy. However, when we consider certain specific topics such as transmission, we will be concerned with phenomena specifically sensitive to wavelength dependence.

EXAMPLE 10-1. Evaluate the energy density per unit wavelength at 0, 1, 1.5, 2, 2.5, 3, 4, and 5 μm (10^{-6} m) for temperatures of 1400 and 2000°K.

SOLUTION. This problem is a good candidate for spreadsheet solution since it involves repetitive application of a single formula—Eq. (10-1)—with slight variation. Note first that

$$2\pi hc^2 = 2\pi(6.625 \times 10^{-34})(3 \times 10^8) = 3.746 \times 10^{-16} \text{ W/m}^2$$

To work with λ in micrometers (μm), we may use instead

$$2\pi hc^2 = 3.746 \times 10^8 \text{ W} \cdot \mu\text{m}^4/\text{m}^2$$

Similarly,

$$\frac{hc}{k} = \frac{(6.626 \times 10^{-34})(3 \times 10^8)}{1.38066 \times 10^{-2}} = 1.44 \times 10^{-2}\text{m} \cdot {}^\circ\text{K} = 1.44 \times 10^4 \ \mu\text{m} \cdot {}^\circ\text{K}$$

Expressing λ in micrometers, we write Eq. (10-1) as

$$\frac{dE_b}{d\lambda} = \frac{3.746 \times 10^8 \lambda^{-5}}{e^{\frac{1.44 \times 10^4}{\lambda T}} - 1}$$

We now construct the table

λ (μm)	$dE_b/d\lambda$ (W/m$^2 \cdot$ m)	
	$T = 2000°$K	$T = 1400°$K
0	0	0
1	2.8×10^5	1.3×10^4
1.5	4.1×10^5	5.2×10^4
2	3.3×10^5	6.9×10^4
2.5	2.3×10^5	6.4×10^4
3	1.5×10^5	5.2×10^4
4	7.2×10^4	3.0×10^4
5	3.7×10^4	1.8×10^4

We observe that the peak energy density occurs at a shorter wavelength at the higher temperature. This movement of the peak toward shorter wavelengths with increasing temperature is what causes bodies to become "red hot" as the thermal radiation enters the visible light region from the red side, and ultimately to become "white hot." By spreadsheet on a personal computer, we would use the following:

	A	B
1	PI	3.14159
2	PLANCK	6.625 E − 34
3	LIGHT SPEED	3E8
4	CONSTANT 1	+B1*B2*(B3 ∧ 2)
5	BOLTZMANN	1.38066 E − 23
6	CONSTANT 2	+B2*B3 / B5
7	WAVELENGTH	ENTER VALUE
8	NUMERATOR	+B4 / (B7 ∧ 5)
9	TEMPERATURE	ENTER VALUE
10	ARGUMENT	+B6 / (B7*B9)
11	EXPONENTIAL	@ exp (B10)
12	DENOMINATOR	+B11 − 1
13	ENERGY DENSITY	+B8 / B12

After doing this for one value of wavelength and temperature, we can repeat by replacing entered values in Column B or by replicating Column B into Column C and then making changes in wavelength and/or temperature.

10-3. SHAPE FACTORS

Consider two black bodies, as illustrated in Fig. 10-3-1. Some fraction of energy emitted by body 1 strikes body 2. Call that fraction F_{12}. The energy intercepted by body 2 is absorbed by body 2, since, as a black body, it absorbs all incident energy. The heat transmitted from body 1 to body 2 is

$$q_{1 \to 2} = E_{b1} A_1 F_{12} \tag{10-7}$$

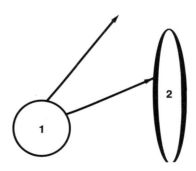

FIGURE 10-3-1. Rays from Body 1 may strike or miss Body 2.

The quantity F_{12} is called a shape factor or view factor. We can also write

$$q_{2 \to 1} = E_{b2} A_2 F_{21} \qquad (\text{10-8})$$

The net heat transfer between the two bodies is

$$q_{12} = q_{1 \to 2} - q_{2 \to 1} = E_{b1} A_1 F_{12} - E_{b2} A_2 F_{21} \qquad (\text{10-9})$$

which can also be written

$$q_{12} = \sigma A_1 F_{12} T_1^4 - \sigma A_2 F_{21} T_2^4 \qquad (\text{10-10})$$

For net heat transfer to take place only when there is a difference in temperature, it must be true that

$$A_1 F_{12} = A_2 F_{21} \qquad (\text{10-11})$$

This is called a reciprocity relationship. We shall see shortly that this reciprocity relationship can be derived on mathematical grounds.

Energy radiated diffusely moves out equally in all solid angles. To find the shape factor, we first find the emitted intensity per unit solid angle $dE_b/d\Omega$. Since any point on a surface radiates outward into a hemisphere, it should be true that if we integrate over a hemisphere, we obtain the total black body radiation; that is,

$$\int d\Omega \frac{dE_b}{d\Omega} \cos \phi = E_b \qquad (\text{10-12})$$

where $\cos \phi$ denotes the outward component of the energy emitted. Using the spherical geometry notation of Fig. 10-3-2, we see that

$$d\Omega = \frac{dA}{r^2} = \frac{(r \sin \phi \, d\theta) r \, d\phi}{r^2} \qquad (\text{10-13})$$

$$E_b = \frac{dE_b}{d\Omega} \int_0^{2\pi} d\theta \int_0^{\pi/2} d\theta \sin \phi \cos \phi \qquad (\text{10-14})$$

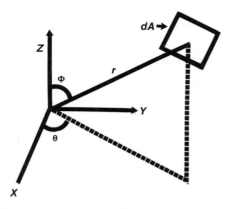

FIGURE 10-3-2. Spherical geometry for element of area on a hemisphere.

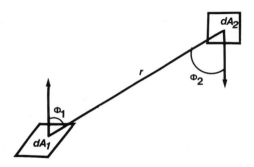

FIGURE 10-3-3. Orientation of differential surfaces on Bodies 1 and 2.

Evaluating the integrals yields

$$\frac{dE_b}{d\Omega} = \frac{1}{\pi} E_b \qquad (10\text{-}15)$$

Now let us consider two arbitrary black bodies with differential surfaces dA_1 and dA_2 as shown in Fig. 10-3-3. The orientation of surface 2 to surface 1 is such that

$$d\Omega = \frac{dA_2\cos\phi_2}{r^2} \qquad (10\text{-}16)$$

(In the hemisphere case above, $\cos\phi_2$ was 1.) The amount leaving surface 1 is obtained as before, so the total energy from dA_1 that reaches dA_2 is

$$dq_{1\rightarrow2} = \frac{E_{b1}}{\pi} dA_1\cos\phi_1\frac{dA_2\cos\phi_2}{r^2} \qquad (10\text{-}17)$$

The total energy leaving all of body 1 that reaches any part of body 2 is then

$$q_{12} = \frac{E_{b1}}{\pi}\int dA_1\int dA_2\frac{\cos\phi_1\cos\phi_2}{r^2} = E_{b1}A_1F_{12} \qquad (10\text{-}18)$$

We therefore may identify the shape factor by

$$A_1F_{12} = \frac{1}{\pi}\int dA_1\int dA_2\frac{\cos\phi_1\cos\phi_2}{r^2} \qquad (10\text{-}19)$$

We may observe that the reciprocity relationship holds because the right-hand side of this equation is unaffected by interchanging the indices 1 and 2.

Shape factor have been evaluated by carrying out the indicated integrations. Table 10-3-1 contains formulas for several commonly encountered configurations. Figure 10-3-4–10-3-7 provide graphs of these formulas for selected cases.

TABLE 10-3-1
Shape Factors for Common Configurations

1. Infinite parallel planes

$$F_{12} = \sqrt{\frac{1}{4}\left(\frac{a_2}{a_1}+1\right)^2 + \left(\frac{h}{a_1}\right)^2} - \sqrt{\frac{1}{4}\left(\frac{a_2}{a_1}-1\right)^2 + \left(\frac{h}{a_1}\right)^2}$$

2. Parallel rectangles

$$F_{12} = \frac{2}{\pi}\left[\frac{1}{a}\sqrt{a^2+h^2}\tan^{-1}\frac{b}{\sqrt{a^2+h^2}} + \frac{\sqrt{b^2+h^2}}{b}\tan^{-1}\frac{a}{\sqrt{b^2+h^2}} \right.$$
$$\left. -\frac{h}{a}\tan^{-1}\frac{b}{h} - \frac{h}{b}\tan^{-1}\frac{a}{h} + \frac{h^2}{2ab}\ln\frac{(a^2+h^2)(b^2+h^2)}{(a^2+b^2+h^2)h^2}\right]$$

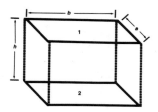

3. Touching perpendicular rectangles

$$F_{12} = \frac{1}{\pi}\left[\tan^{-1}\frac{a}{b} + \frac{c}{b}\tan^{-1}\frac{a}{c} - \sqrt{\left(\frac{c}{b}\right)^2+1}\tan^{-1}\frac{a}{\sqrt{b^2+c^2}}\right.$$
$$+\frac{c^2}{4ab}\ln\frac{(a^2+b^2+c^2)c^2}{(a^2+c^2)(b^2+c^2)} + \frac{b}{4a}\ln\frac{(a^2+b^2+c^2)b^2}{(a^2+b^2)(b^2+c^2)}$$
$$\left.-\frac{a}{4b}\ln\frac{(a^2+b^2+c^2)a^2}{(a^2+b^2)(a^2+c^2)}\right]$$

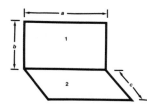

(continued)

4. Parallel disks, common axis

$$F_{12} = \left[\sqrt{\left(\frac{D_2}{2D_1} + \frac{1}{2} \right)^2 + \left(\frac{h}{D_1} \right)^2} - \sqrt{\left(\frac{D_2}{2D_1} - \frac{1}{2} \right)^2 + \left(\frac{h}{D_1} \right)^2} \right]^2$$

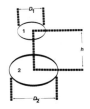

5. Row of tubes along wall

$$F_{12} = 1 - \sqrt{1 - \left(\frac{D}{S} \right)^2} + \frac{D}{S} \tan^{-1} \sqrt{\left(\frac{S}{D} \right)^2 - 1}$$

6. Parallel cylinders

$$F_{12} = \frac{1}{\pi} \left[\sqrt{\left(\frac{s+D}{D} \right)^2 - 1} + \sin^{-1} \left(\frac{D}{s+D} \right) - \frac{s+D}{D} \right]$$

$$s = h - D$$

Not all cases can be expressed directly in terms of the shape factors given. However, quite often we can express nonstandard surfaces in terms of standard surfaces. An example is given in Fig. 10-3-8. In this case, we would like to obtain the shape factor F_{12}. We may proceed by considering surface 1 to be part of a large surface 3 having an additional surface 4. It must be true that

$$A_3 F_{32} = A_1 F_{12} + A_4 F_{42}$$

The shape factors F_{32} and F_{42} are for standard configurations. We therefore may evaluate F_{12} indirectly by first evaluating F_{32} and F_{42}.

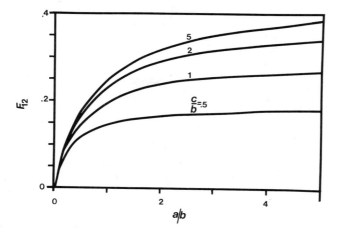

FIGURE 10-3-4. Shape factors for touching perpendicular rectangles. Notation from Table 10-3-1.

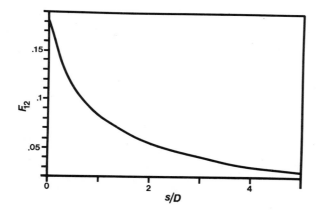

FIGURE 10-3-5. Shape factors for parallel cylinders.

The manipulation of shape factors of standard configurations to obtain a shape factor of a nonstandard configuration is called shape factor algebra. Another item of information useful in dealing with shape factors is that all shape factors for radiation leaving a surface must add to 1 (i.e., all radiation must go somewhere). Thus, if there are N surfaces,

$$\sum_{n=1}^{n} F_{in} = 1 \tag{10-20}$$

Note that for some surfaces, the shape factor to itself is nonzero. Figure 10-3-9 shows concave and convex surfaces, that is, surfaces that can and cannot see themselves.

EXAMPLE 10-2. Find the shape factors between two infinite concentric cylinders.

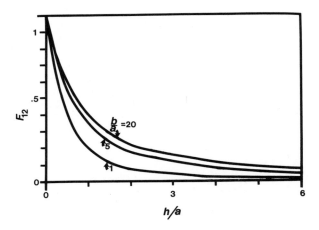

FIGURE 10-3-6. Shape factors for parallel rectangles. Notation from Table 10-3-1.

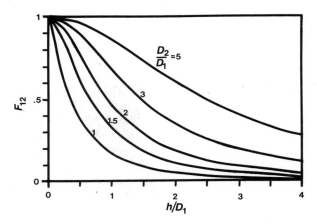

FIGURE 10-3-7. Shape factors for parallel disks on a common axis. Notation from Table 10-3-1.

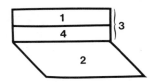

FIGURE 10-3-8. Configuration of surfaces for shape factor algebra.

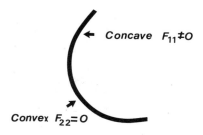

FIGURE 10-3-9. Convex and concave surfaces.

SOLUTION. Denote the inner and outer cylinder surfaces as 1 and 2. Note that the inner surface is convex so

$$F_{11} = 0$$

Since all radiation must go somewhere

$$F_{11} + F_{12} = 1$$

so the shape factor from inside to outside F_{12} is 1. By reciprocity, we find

$$A_2 F_{21} = A_1 F_{12} = A_1$$

Then we find the shape factor of surface 2 to itself by

$$F_{22} = 1 - F_{21}$$

Charts for shape factors, as in Figs. 10-3-4–10-3-7, are commonly available in texts and handbooks in the literature. They generally provide convenience and acceptable accuracy for standard configurations. However, if a significant amount of algebra is involved in obtaining values indirectly, problems of small differences of large numbers, combined with the modest accuracy to which graphs can be read, can lead to results of little value. The formulas for actual configurations, for the most part, are such that they can be evaluated readily with personal computing or, with a bit more effort, with electronic calculators.

The curves of shape factors are best used to gain insight into the nature of the variation of these factors. For example, consider Fig. 10-3-7 for a disk radiating to a parallel larger disk on a common axis. (To treat a large disk to a small disk, apply reciprocity.) Note that as the separation goes to 0, the shape factor goes to 1, since all radiations from the small disk must strike the larger disk. (For a large disk to a small disk then, reciprocity would tell us that the shape factor would go to the ratio of areas as the separation goes to 0.) As the separation increases, the shape factor declines, with the rate of decline slower for large ratios of diameters.

COMPUTER PROJECT 10-1. Write a computer program that will evaluate on request any of the standard shape factor formulas of Table 10-3-1.

The discussion thus far has been in terms of two black bodies. It is possible to have interchange among several black bodies. Let us consider the case of three black bodies as an illustration. The amount of energy received and absorbed by body 1 is

$$G_1 A_1 = E_{b2} A_2 F_{21} + E_{b3} A_3 F_{31} \tag{10-21}$$

The energy emitted by body 1 is E_{b1}. If there is an energy source Q_1 inside body 1, then

$$Q_1 + G_1 A_1 = E_{b1} A_1 \qquad (10\text{-}22)$$

The temperature of body 1 will be such that it can emit the energy generated internally and reradiate the energy absorbed radiation from other bodies. For a general number N of bodies, we may write

$$E_{bi} A_i = Q_i + \sum_{j=1}^{N} E_{bj} A_j F_{ji} \qquad (10\text{-}23)$$

It is tacitly assumed that any radiation not intercepted by one of the bodies may be considered as emitted to space. Space may be thought of as a black body (it absorbs all energy sent to it) of zero temperature (it emits no energy) and infinite area (so that absorbing a finite amount of energy does not affect its temperature).

Problems may be posed in alternate ways. One way is to stipulate the heat sources Q_i and the geometry. Then a linear set of simultaneous algebraic equations may be solved for the E_{bi} and, thereafter, for the temperature. It is also possible that the heat source in one body may be regulated so as to maintain its temperature at a given value. For that body, we would replace the corresponding equation in the above set with the equation

$$E_{bi} = \sigma T_i^4 \qquad (10\text{-}24)$$

where T_i is known. After the algebraic equations are solved, we could then evaluate the Q_i needed.

It is sometimes desirable to introduce geometric construction to place configurations in standard form. Consider the surfaces denoted by solid lines in Fig. 10-3-10. Let the line associated with surface A_2 be extended, and let a line be drawn from the left end of surface A_1 perpendicular to the extended line. We now have a right triangle. From this right triangle, we may use our standard perpendicular surface formulas and our knowledge that all radiation must go somewhere to evaluate the shape factors F_{12} and F_{21}. The task is left as an exercise. In general, when faced with a novel configuration, it is prudent to see if constructions can be made which incorporate known situations, like perpendicular surfaces or fully enclosed surfaces.

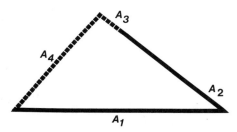

FIGURE 10-3-10. Geometric construction for shape factor calculation.

EXAMPLE 10-3.   Two square black bodies with insulated backs, 1 m sides, and a common edge are at right angles to one another. One body is maintained at a temperature of 1000°K. No heat is supplied to or removed from the second body.

a. What is the temperature of the second body?
b. How much heat must be supplied to the first body?

SOLUTION.   (a) For the two bodies, Eq. (10-23) yields two equations

$$E_{b1} A_1 = Q_1 + E_{b2} A_2 F_{21}$$

$$E_{b2} A_2 = Q_2 + E_{b1} A_1 F_{12}$$

Let us consider the second equation. We know that

$$Q_2 = 0$$

and we can evaluate

$$E_{b1} = \sigma T_1^4$$

since we know T_1. Actually, we need not multiply out all terms because some will cancel. The equation becomes

$$\sigma T_2^4 A_2 = \sigma T_1^4 A_1 F_{12}$$

The σ terms cancel, and we note that the areas are equal. It follows that

$$T_2 = T_1 F_{12}^{1/4}$$

For squares, the formula in Table 10-3-1 simplifies considerably. We have

$$\frac{a}{b} = \frac{a}{c} = 1 \qquad \tan^{-1}(1) = \frac{\pi}{4}$$

$$\frac{a}{\sqrt{b^2 + c^2}} = \frac{a^2}{b^2 + c^2} = \frac{1}{\sqrt{2}} = 0.707$$

$$\tan^{-1}(0.707) = 0.615$$

$$\frac{c^2}{ab} = \frac{b}{a} = \frac{a}{b} = 1$$

All the logarithmic terms become

$$\ln \frac{3a^2 a^2}{(2a^2)(2a^2)} = \ln \frac{3}{4} = -0.288$$

$$F_{12} = \frac{1}{\pi} \left[\frac{\pi}{4} + \frac{\pi}{4} - (1.414)(0.615) - 0.288 \left(\frac{1}{4} + \frac{1}{4} - \frac{1}{4} \right) \right]$$

$$= 0.2$$

and we find

$$T_2 = 669°K$$

(b) We now return to the first equation generated by Eq. (10-23) to find

$$Q_1 = E_{b1}A_1 - E_{b2}A_2 F_{21}$$
$$= \sigma T_1^4 A_1 - \sigma T_2^4 A_2 F_{21}$$

Noting that F_{21} is the same as F_{12} and that areas are equal,

$$Q_1 - \sigma A\left(T_1^4 - 0.2 T_2^4\right) = 5.669 \times 10^{-8}\left[1000^4 - 0.2(669)^4\right]$$
$$= 5.44 \times 10^4 \text{ W}$$

EXAMPLE 10-3 (Spreadsheet). We proceed by first preparing in Columns A–D an evaluation of the general touching perpendicular rectangles shape factor. We could choose just to evaluate the special case for the square, but this way, we have a procedure to use with more general problems. Then, in subsequent columns, we perform the analysis for this particular problem. The spreadsheet follows below for the first four columns to evaluate the shape factor, followed by additional columns to treat evaluation of temperature and heat supply. When done this way, whenever a problem requiring shape factors for teaching perpendicular rectangles is called for, this spreadsheet can be retrieved, Columns E and F may be deleted, and the details of the new problem can be treated.

	A	B	C	D
1	TOUCHING	PERPENDICULAR	RECTANGLES	
2	SHAPE	FACTOR		
3	A EDGE	1	ATAN A / B	@ATAN(B6)
4	B FACE 1	1	C / B ATAN A / C	+B8*@ATAN(B7)
5	C FACE 2	1	TERM 3	-B9*@ ATAN (B11)
6	A / B	+B3 / B4	TERM 4	+(B8 / (4*B7))*B16
7	A / C	+B3 / B5	TERM 5	+B18 / (4*B6)
8	C / B	+B5 / B4	TERM 6	-B20*B6 / 4
9	ROOT	@ SQRT(1 + (B8 ∧ (2))	SUM	@ SUM(D3...D8)
10	B² ∧ 2 + C ∧ 2	+(B4 ∧ 2) + (B5 ∧ 2)	F12	+D9 / @PI
11	RATIO 1˙	+B3 / SQRT(B10)		
12	A ∧ 2 + B ∧ 2 + C ∧ 2	+(B3 ∧ 2) + B10		
13	A ∧ 2 + C ∧ 2	+(B3 ∧ 2) + (B5 ∧ 2)		
14	RATIO 2	+B12 / B13		
15	C ∧ 2 / (B ∧ 2 + C ∧ 2)	+(B5 ∧ 2) / B10		
16	FIRST LN	@ LN (B14*B15)		
17	B ∧ 2 / (B ∧ 2 + C ∧ 2)	+(B4 ∧ 2) / B10		
18	SECOND LN	@ LN (B14*B17*B13 /(B3 ∧ 2 + B4 ∧ L 2)		
19	A ∧ 2 / (A ∧ 2 + B ∧ 2)	+(B3 ∧ 2) / ((B3 ∧ 2) +(B4 ∧ 2))		
20	THIRD LN	@ LN(B14*B19)		

	E	F
1	EVALUATE	DETAILS
2	SPECIFIC	PROBLEM
3	HOT TEMP	1000
4	TEMP 2	+F3*(D10 ∧ .25)
5	SIGMA	5.669E - 8
6	HEAT	+B5*(F3 ∧ 4 - D10 *(F4 ∧ 4))

A body that is not black will absorb a fraction α of radiation incident on it. Equilibrium considerations require that such a body emit less than the black body radiation

$$E = \varepsilon E_b \qquad (10\text{-}25)$$

and that the fractions ε and α must be the same. If not all the incident energy is absorbed, then the remainder must be either reflected or transmitted. In this section, we assume that there is no transmission.

The equilibrium argument is straightforward as to why ε and α must be the same. Suppose a small body is in thermal equilibrium (i.e, same temperature) with a large surrounding body, as shown in Fig. 10-4-1. The small body receives an incident energy density E_I. If the small body is a black body, then at equilibrium it must emit as much energy as it absorbs, so

$$E_b A = E_I A \qquad (10\text{-}26)$$

If the body is not black, and if the surrounding body is so large that it cannot be affected by the small one, then the incident energy E_I is the same. However, the absorbed energy is only $\alpha E_I A$. At equilibrium, the same amount of energy must be emitted as absorbed, so the emitted energy is

$$E = \varepsilon E_b = \alpha E_b \qquad (10\text{-}27)$$

The absorption and the emission can deviate nonuniformly from 1 as a function of wavelength. The average emissivity is given by

$$\varepsilon E_b = \int_0^\infty d\lambda \, \varepsilon(\lambda) \frac{dE_b}{d\lambda} \qquad (10\text{-}28)$$

If $\varepsilon(\lambda)$ is constant, then the body is called a gray body. In this section, we are not concerned with wave length dependence of emissivity.

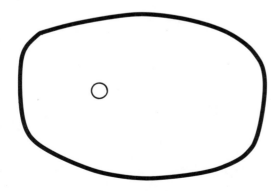

FIGURE 10-4-1. Small body contained within a large body.

If we "look" at a gray body, we "see" thermal radiation of two types—the emitted radiation and the reflected radiation (what is not absorbed is reflected). Let us define the total amount of radiation leaving the surface as the radiosity J:

$$J = \varepsilon E_b + \rho G = \varepsilon E_b + (1 - \varepsilon)G \qquad \text{(10-29)}$$

where G, as in Section 10-3, is the total incident radiation. Reflection ρ is equal to $1 - \varepsilon$ when there is no transmission. As noted in Section 10-3, if there is a source of heat Q within the body, there is a net transfer of heat from the body,

$$\frac{Q}{A} = J - G \qquad \text{(10-30)}$$

Note that it is radiosity, not emitted radiation, that describes departing energy. We may eliminate G from the last two equations to obtain

$$Q = \frac{E_b - J}{(1 - \varepsilon)/\varepsilon A} \qquad \text{(10-31)}$$

This last relationship suggests an electrical analogy as illustrated in Fig. 10-4-2. The quantity $(1 - \varepsilon)/\varepsilon A$ may be thought of as an electrical resistance. Voltages E_b and J exist at opposite ends of the resistor, so a current Q flows. This analogy may be extended to deal with interactions with other bodies.

The amount of energy incident on body j that left body i is

$$q_{i \to j} = J_i A_i F_{ij} \qquad \text{(10-32)}$$

Similarly, the energy going from body j to body i is

$$q_{j \to i} = J_j A_j F_{ji} \qquad \text{(10-33)}$$

The net energy flow between these bodies is

$$q_{i \to j} - q_{j \to i} = q_{ij}^{\text{net}} = \frac{J_i - J_j}{1/A_i F_{ij}} \qquad \text{(10-34)}$$

where we have made use of the reciprocity relationship for shape factors.

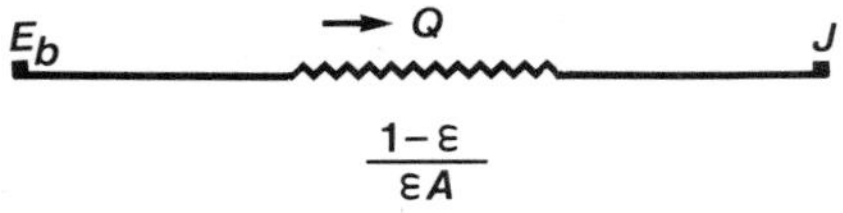

FIGURE 10-4-2. Electrical analog with current Q flowing between voltages E_b and J.

Again, heat flow looks like a current. The radiosities look like potentials. The quantity $1/A_i F_{ij}$ looks like an electrical resistance. Figure 10-4-3 shows the electrical network description of the interaction between two bodies. Similarly, Fig. 10-4-4 shows the electrical network for interactions among three bodies.

Let us note some special cases. First, let us ask what happens if we have a black body. (Since $\varepsilon = 1$ for that body, the "internal resistance" $(1 - \varepsilon)/\varepsilon A$ goes to 0. Since there is no resistance between E_b and J, these two voltages must be the same. Thus, J can be replaced by E_b for a black body. This is to be expected since for a black body emission and radiosity are the same.

Another special case is free space. Suppose that surface 3 in Fig. 10-4-4 is free space. Since free space can be considered a black body at zero temperature, its internal resistance is zero and the node J_3 can be replaced by E_{b3} which is a zero potential. This is like an electrical ground.

A special case encountered commonly is that of a reradiating surface. This is a surface that is insulated so that it neither receives nor provides heat on a net basis. Such a surface was involved in Example 10-3. Referring to Fig. 10-4-2, we observe that $Q = 0$ at such a surface. Therefore, E_b and J must be the same.

Finally, we consider a special case where one surface is very large. For example, suppose we have two small hot places facing into a very large room. Figure 10-4-4 still applies with surfaces 1 and 2 denoting the plates and surface 3 denoting the room. Since the room is postulated to be very large, the area A_3 may be presumed to be essentially infinite for the evaluation of the internal resistance $(1 - \varepsilon_3)/\varepsilon_3 A_3$. Thus, even though the room is not a black

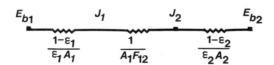

FIGURE 10-4-3. Electrical analog for interaction between two bodies.

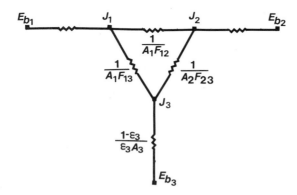

FIGURE 10-4-4. Electrical analog for interaction between three bodies.

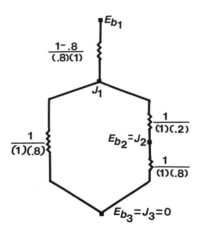

FIGURE 10-4-5. Simplified network for Example 10-4.

body, it appears to be, for practical purposes, a black body from the perspectives of the small plates.

Interactions with the atmosphere provide interaction with a large surface. During a clear day, solar radiation yields a source of about 700 W/m² at the surface of the earth. This energy density results from attenuation of the sunshine by the atmosphere (as per Section 10-8). The sunlight may be taken to have a temperature of 5800°K for the purpose of assigning wavelength distribution (of concern when we deal later with transmission), but *not* for the purpose of calculating energy density. In other words, a black body at 5800°K will not emit 700 W/m². The difference for evaluation of properties is important because absorptivities tend to be significantly different for solar radiation than for radiation at lower temperatures.

Just as the atmosphere does not permit all sunshine to reach the surface of the earth, it does not permit all radiation from the earth to leave to free space. A rough estimate of the influence of the atmosphere on a clear night, according to a suggestion by Swinbank (1963), is to treat it as a black body at a temperature of $0.0552\, T_{\text{air}}^{1.5}$. Cloud cover increases the effective black body temperature of the atmosphere.

The equations for the case of Fig. 10-4-4 may be obtained by balance considerations at each node. Just as current must be conserved in electrical circuits, heat is conserved at nodes using the electrical analogy. Thus, we obtain, at the node where "voltage" is J_1,

$$\frac{E_{b1} - J_1}{(1 - \varepsilon_1)/\varepsilon_i A_i} + \frac{J_2 - J_1}{1/A_1 F_{12}} + \frac{J_3 - J_1}{1/A_1 F_{13}} = 0 \qquad (\mathbf{10\text{-}35})$$

and similar equations at the other radiosity nodes. At the black body radiation nodes, we get

$$Q_i = \frac{E_{bi} - J_i}{(1 - \varepsilon_i)/\varepsilon_i A_i} \qquad (10\text{-}36)$$

We thus have six equations corresponding to the six nodes.

How we proceed depends on what is known and what is not known, as in Section 10-3 with black bodies. If the heat sources are given, we can solve the six equations for the three radiosities and three black body radiations, from which the temperatures may be obtained. Alternately, for a particular body, the temperature may be specified, and we may replace the last equations by

$$E_{bi} = \sigma T_i^4 \qquad (10\text{-}37)$$

as in Section 10-3 on black bodies.

While the electrical analogy is often convenient, we also may proceed with direct balance equations. At any one of an array of N bodies, the energy emitted must be equal to the source of the heat within the body plus the energy absorbed:

$$\varepsilon_i E_{bi} A_i = Q_i + \alpha_i \sum_{j=1}^{n} J_j A_j F_{ji} \qquad (10\text{-}38)$$

In addition, we may say that the total energy leaving must be balanced by the total energy arriving and the source,

$$J_i A_i = Q_i + \sum_{j=1}^{n} J_j A_j F_{ji} \qquad (10\text{-}39)$$

If the heat sources are specified, then the last equation may be solved for the radiosities, which may then be substituted into the previous equation to obtain black body radiation E_{bi} and thereby temperature. If the temperatures (and therefore E_{bi}) are specified, the two equation sets may be subtracted to yield (noting $\alpha_i = \varepsilon_i$)

$$\varepsilon_i E_{bi} A_i = J_i A_i - (1 - \varepsilon_i) \sum_{j=1}^{n} j_j A_j F_{ji} \qquad (10\text{-}40)$$

which may be solved for the radiosities, after which the Q_i may be obtained.

Reciprocity relationships may be used to simplify equations. The last equation, for example, may be written in the form

$$\varepsilon_i E_{bi} = J_i - (1 - \varepsilon_i) \sum_{j=1}^{n} J_j F_{ij} \qquad (10\text{-}41)$$

The area A_i cancels out after application of reciprocity.

Each of the two approaches (electrical network, balance equations) works, so we should consider the circumstances under which we might choose one option or the other. The electrical network approach is simple and usually aids in the visualization of the problem. (This is a subjective statement that applies to different degrees to different people.) However, the simplicity of the electrical analogy tends to be lost when the number of interacting surfaces is large and the circuit diagram contains multiple overlaps and crossings. For large numbers of surfaces, the system of balance equations can be of greater use.

The circuit diagram approach can be followed by a set of rules, as follows:

1. Pick a surface i.
2. Start with its "internal voltage" E_{bi}.
3. Connect the appropriate internal resistance $R_{int, i}$.
4. Connect $R_{int, i}$ to a radiosity quantity.
5. From the radiosity quantity, draw connecting lines to other surfaces with which it interacts.
6. Place geometric resistances along those lines.
7. Repeat for other surfaces until the network is complete.
8. Write heat current balance equations at all nodes.

It may not always be necessary to go through all the steps, since insight into the problem (e.g., recognition of one of the special cases like reradiating surfaces cited above) may make it easy to write a network directly. However, the listed procedure is a useful guide when it is not clear how to approach the problem. In addition, the procedure will generalize into other situations later in this chapter involving transmission and specular reflection.

EXAMPLE 10-4. Repeat Example 10-3 with the surfaces taken as gray bodies with emissivities of 0.8. Use both the network and the balance equation approaches.

SOLUTION. Let us first use the network approach to the problem. Following the stipulated procedure, we obtain the three-body network of Fig. 10-4-4 with surface 3 denoting free space. We note that surface 2 is reradiating, so E_{b2} and J_2 are the same. We could have obtained this information by applying a current balance (Step 8 of our procedures) at node E_{b2}, and noting that since the heat source and current there are 0, E_{b2} and J_2 are the same. We also note that free space is a black body at zero temperature. We may draw a simplified circuit diagram as shown in Fig. 10-4-5. This diagram notes that the areas A_1 and A_2 are each 1, as per Example 10-3, that $F_{12} = 0.2$ as per Example 10-3, and therefore that F_{13} (and therefore F_{23}) is $1 - 0.2 = 0.8$.

Note that this simplified diagram has parallel paths of resistance between nodes J_1 and E_{b3}. The left path resistance is

$$R_L = \frac{1}{A_1 F_{13}} = \frac{1}{(1)(0.8)} = 1.25$$

The right path resistance is

$$R_R = \frac{1}{A_1 F_{12}} + \frac{1}{A_2 F_{23}} = \frac{1}{(1)(0.2)} + \frac{1}{(1)(0.8)} = 6.25$$

The equivalent resistance R_{eq} of resistors in parallel is

$$\frac{1}{R_{eq}} = \frac{1}{R_L} + \frac{1}{R_R} = 0.8 + 0.16 = 0.96$$

$$R_{eq} = 1.042$$

The total resistance between E_{b1} and E_{b3} is

$$R_{tot} = \frac{1 - 0.8}{0.8} + 1.042 + 0.250 + 1.042 = 1.292$$

The total heat flow and therefore the heat supplied to surface 1 is

$$Q_1 = \frac{E_{b1} - E_{b3}}{R_{tot}} = \frac{\sigma T_1^4 - 0}{1.292} = \frac{5.669 \times 10^4}{1.292} = 4.39 \times 10^4 \text{ W}$$

We next find J_1 from

$$Q_1 = \frac{E_{b1} - J_1}{0.25}$$

$$J_1 = E_{b1} - 0.25 Q_1 = 4.57 \times 10^4 \text{ W/m}^2$$

The current through the right side path is

$$Q_R = \frac{J_1 - E_{b3}}{R_R} = \frac{4.57 \times 10^4 - 0}{6.25} = 7.31 \times 10^3$$

To get the voltage E_{b2}, we note that

$$Q_R = \frac{J_1 - E_{b2}}{5}$$

$$E_{b2} = 4.57 \times 10^4 - (5)(7.31 \times 10^3) = 9.14 \times 10^3 \text{ W/m}^2$$

The temperature at the reradiating surface is

$$T_2 = \left(\frac{E_{b2}}{\sigma} \right)^{1/4} = \left(\frac{9.15 \times 10^3}{5.669 \times 10^{-8}} \right)^{1/4} = 634^\circ \text{ K}$$

Let us next use the balance equation approach. At surface 1, temperature is specified, so we write Eq. (10-40).

$$\varepsilon_1 E_{b1} A_1 = J_1 A_1 - (1 - \varepsilon_1) J_2 A_2 F_{21}$$

We note that because of the reradiating surface, we can replace J_2 by E_{b2} to yield

$$G_1 E_{b1} A_1 = J_1 A_1 - (1 - \varepsilon_1) E_{b2} A_2 F_{21}$$

Had we not recognized that $J_2 = E_b$, we would obtain this by multiplying Eq. (10-39) for surface 2 by ε_2, subtracting from Eq. (10-38), and noting that $Q_2 = 0$.

We next write Eq. (10-38) for surface 2:

$$\varepsilon_2 E_{b2} A_2 = \varepsilon_2 J_1 A_1 F_{12}$$

We now have two equations in the two unknowns J_1, E_{b2}. Solving for $J_1 A_1$ we obtain

$$J_1 A_1 = \frac{E_{b2} A_2}{F_{12}}$$

and substituting yields

$$\varepsilon_1 E_{b1} A_1 = \frac{E_{b2} A_2}{F_{12}} - (1 - \varepsilon_1) E_{b2} A_2 F_{21}$$

The equal areas cancel and

$$E_{b2} = \frac{\varepsilon_1 E_{b1}}{1/F_{12} - (1 - \varepsilon_1) F_{21}} = \frac{(0.8)(5.669 \times 10^4)}{5 - (0.2)(0.2)} = 9.15 \times 10^3$$

in agreement with the other evaluation of this quantity. We now apply Eq. (10-38) for surface 1

$$\varepsilon_1 E_{b1} A_1 = Q_1 + \varepsilon_1 E_{b2} A_2 F_{21}$$

All quantities are known but Q_1, which is evaluated as

$$Q_1 = (0.8)(5.669 \times 10^4)(1) - (0.8)(9.15 \times 10^3)(1)(0.2) = 4.39 \times 10^4$$

10-5. RADIATION SHIELDS

It may be desirable to try to retard radiation heat transfer from a high-temperature surface. This may be done to reduce heat loss and/or to prevent an excessive temperature at another surface (or to reduce a cooling requirement at this other surface). One way to accomplish this reduction is to place a shield in the path of the radiation.

Consider two infinite parallel planes. One is maintained at a temperature. T_1. The second surface is maintained at temperature T_2 which is less than T_1. The electric circuit for this case is given in Fig. 10-5-1. The heat transfer

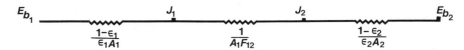

FIGURE 10-5-1. Circuit for two surfaces without shield.

"current" between voltages E_{b1} and E_{b2} is

$$q = \frac{E_{b1} - E_{b2}}{(1 - \varepsilon_1)/\varepsilon_1 A_1 + 1/A_1 F_{12} + (1 - \varepsilon_2)/\varepsilon_2 A_2} \qquad (10\text{-}42)$$

For infinite parallel planes, the areas are equal and $F_{12} = 1$. Thus, we may write

$$\frac{q}{A} = \frac{E_{b1} - E_{b2}}{1/\varepsilon_1 + 1/\varepsilon_2 - 1} \qquad (10\text{-}43)$$

Let us now interject a shield between these planes. The electric circuit is shown in Fig. 10-5-2. Again noting that shape factors are 1, we find that

$$\left(\frac{q}{A}\right)_{\text{shielded}} = \frac{E_{b1} - E_{b2}}{1/\varepsilon_1 + 2/\varepsilon_s + 1/\varepsilon_2 - 2} \qquad (10\text{-}44)$$

The ratio of heat transferred in two cases is

$$\frac{(q/A)_{\text{shielded}}}{q/A} = \frac{1/\varepsilon_1 + 1/\varepsilon_2 - 1}{1/\varepsilon_1 + 2/\varepsilon_s + 1/\varepsilon_2 - 2} \qquad (10\text{-}45)$$

Note that the percentage effect on heat transfer is independent of the temperatures involved. Suppose $\varepsilon_1 = 0.8$, $\varepsilon_2 = 0.2$, and the shield is a polished surface with $\varepsilon_s = 0.04$. Then the ratio of shielded to unshielded heat transfer is 0.097. The shield reduces heat transfer by 90%.

While a highly polished surface is particularly effective, any surface would have a significant effect. Suppose we have two surfaces of equal ε and add a shield with the same value of ε. The ratio would become

$$\frac{\left(\frac{q}{A}\right)_{\text{shielded}}}{q/A} = \frac{2/\varepsilon - 1}{4/\varepsilon - 2} = \frac{1}{2} \qquad (10\text{-}46)$$

for a 50% reduction.

In the examples above, we specified temperatures T_1 and T_2. The shield reduced heat transfer and thereby made it possible to reduce the cooling required to maintain temperature T_2. Suppose, on the other hand, that the heat source at plane 1 is fixed, so heat transfer is the same in each case. We then

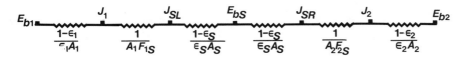

FIGURE 10-5-2. Circuit for surfaces with radiation shield.

obtain

$$\frac{(E_{b1} - E_{b2})_{\text{shielded}}}{(E_{b1} - E_{b2})_{\text{unshielded}}} = \frac{1/\varepsilon_1 + 2/\varepsilon_s + 1/\varepsilon_2 - 2}{1/\varepsilon_1 + 1/\varepsilon_2 - 1} = \frac{(T_1^4 - T_2^4)_{\text{shielded}}}{(T_1^4 - T_2^4)_{\text{unshielded}}}$$

$$(\textbf{10-47})$$

Even with the same amount of heat transferred, if T_1 is kept the same in both cases, then T_2 will be lower in the shielded case.

Adding a radiation shield to a radiation heat transfer situation can be considered an analogue of adding an insulator to a conduction heat transfer situation. Each of these actions can be looked on as adding a thermal resistance to reduce the current (heat flow). For each of these actions, a particular class of material is best suited for providing resistance—low thermal conductivity in the case of an insulator, low emissivity in the case of a radiation shield.

Shields can be considered in other geometries. Suppose we have a cylinder containing a very hot or very cold fluid within an inner surface, and we have a vacuum between that inner surface and the outer surface. In the absence of the shield, we may express the heat transfer by

$$q = \frac{E_{bi} - E_{bo}}{(1 - \varepsilon_i)/\varepsilon_i A_i + 1/A_i F_{io} + (1 - \varepsilon_o)/\varepsilon_o A_o} \qquad (\textbf{10-48})$$

Noting that the shape factor is 1 (all heat leaving the inner cylinder reaches the outer cylinder), we obtain

$$\frac{q}{A_i} = \frac{E_{bi} - E_{bo}}{1/\varepsilon_i + (A_i/A_o)(1/\varepsilon_o - 1)} \qquad (\textbf{10-49})$$

If we place a shield between these cylinders, we obtain

$$q = \frac{E_{bi} - E_{bo}}{(1 - \varepsilon_i)/\varepsilon_i A_i + 1/A_i F_{is} + 2(1 - \varepsilon_s)/\varepsilon_s A_s + 1/A_s F_{so} + (1 - \varepsilon_o)/\varepsilon_o A_o}$$

$$(\textbf{10-50})$$

Noting that the shape factors are 1, we obtain

$$\left(\frac{q}{A_i}\right)_{\text{shielded}} = \frac{E_{bi} - E_{bo}}{1/\varepsilon_i + (A_i/A_s)(2/\varepsilon_s - 1) + (A_i/A_o)(1/\varepsilon_o - 1)} \qquad (\textbf{10-51})$$

The ratio of shielded to unshielded heat transfer is

$$\frac{(q/A_i)_{\text{shielded}}}{q/A_i} = \frac{1/\varepsilon_i + (A_i/A_o)(1/\varepsilon_o - 1)}{1/\varepsilon_i + (A_i/A_s)(2/\varepsilon_s - 1) + (A_i/A_o)(1/\varepsilon_o - 1)} \qquad (\textbf{10-52})$$

Because of cylindrical geometry, results depend on area ratios. If the evacuated regions, however, are very thin in comparison to the radius of the inside cylinder, the area ratios will be about 1 and the ratios will be similar to those obtained in plane geometry.

Radiation shielding can be improved by adding a second shield. By the same procedures used above, we may show (as is called for in a problem at the end of the chapter) that

$$\frac{(q/A)_{2 \text{ shields}}}{(q/A)_{1 \text{ shield}}} = \frac{1/\varepsilon_1 + 2/\varepsilon_s + 1/\varepsilon_2 - 2}{1/\varepsilon_1 + 4/\varepsilon_s + 1/\varepsilon_2 - 3} \qquad (10\text{-}53)$$

The benefit of the second shield approaches 50% when ε_s is much smaller than both ε_1 and ε_2. With the sample values of ε_1, ε_2, and ε_s used earlier in the section, the benefit of the second shield is 47.5% of the heat loss with one shield. Recall that the first shield yielded a benefit of 90% of the original heat transfer. Thus, the second shield yields a benefit of about 5% of the original heat transfer.

10-6. INTERACTIONS INVOLVING TRANSMISSION

Radiation incident on a surface may be transmitted through the body. Since all radiation must go somewhere, absorption α, reflection ρ, and transmission τ must satisfy

$$\alpha + \rho + \tau = 1$$

Note that equilibrium still requires α to equal ε. For materials of interest, transmission may be highly nonuniform. Substantial transmission may occur over a particular range of wavelengths, while the body may appear opaque for other wavelengths.

Let us assume that the radiation incident on the plate has the wavelength distribution of black body radiation at some temperature; that is, let us assume that the radiation comes from a black or gray body. The wavelength distribution of black body radiation may be expressed in the form

$$\frac{dE_b}{d\lambda} = \frac{C_1 T^5}{(\lambda T)^5 (e^{C_2/\lambda T} - 1)} \qquad (10\text{-}55)$$

$$C_1 = 3.74 \times 10^8 \text{ W} \cdot \mu\text{m}^4/\text{m}^2 \qquad (10\text{-}56)$$

$$C_2 = 1.4387 \times 10^4 \, \mu\text{m} \cdot {}^\circ\text{K} \qquad (10\text{-}57)$$

Because of the sizes of wavelengths of thermal radiation, it is common to have them specified in micrometers ($\mu\text{m} = 10^{-6}$ m). The fraction of total black body radiation occurring in the range $\lambda_1 - \lambda_2$ is

$$f = \frac{1}{\sigma T^4} \int_{\lambda_1}^{\lambda_2} d\lambda \frac{C_1 T^5}{(\lambda T)^5 (e^{C_2/\lambda T} - 1)} = \frac{C_1}{\sigma} \int_{\lambda_1 T}^{\lambda_2 T} \frac{dx}{x^5 (e^{C_2/x} - 1)} \quad (10\text{-}58)$$

This fraction is a function of λT and may be considered to be $f(\lambda_1 T, \lambda_2 T)$. Table 10-6-1 provides a table of $f(0, \lambda T)$. From the table, we may evaluate

$$f(\lambda_1 T, \lambda_2 T) = f(0, \lambda_2 T) - f(0, \lambda_1 T) \quad (10\text{-}59)$$

To find the energy transmitted through a "window," we first find the energy incident on the window. We then find, using Table 10-6-1, the fraction of energy in a particular range of wavelengths. We then apply the transmissivity τ to find the fraction of that energy that goes through the material.

Let us now consider the influence of transmission on radiating surfaces. Consider the three finite planes of Fig. 10-6-1. Plane 2 can transmit a portion of the radiation incident on it. We consider two possibilities—transparent and diffuse transmissions.

Radiation passes through a transparent medium as if that medium were not there. When we see an accurate image through a window, we consider the window to be transparent to the transmitted light. Transparent transmission

TABLE 10-6-1
Radiation Fraction f

λT	f	λT	f
700	10^{-5}	5,400	0.680
1,000	3.21×10^{-4}	5,600	0.701
1,200	2.13×10^{-3}	5,800	0.720
1,400	7.79×10^{-3}	6,000	0.738
1,600	1.97×10^{-3}	6,400	0.769
1,800	3.93×10^{-2}	6,800	0.796
2,000	6.67×10^{-2}	7,200	0.819
2,200	0.101	7,600	0.839
2,400	0.140	8,200	0.856
2,600	0.183	8,500	0.875
2,800	0.228	9,000	0.890
3,000	0.273	9,500	0.903
3,200	0.318	10,000	0.914
3,400	0.362	11,000	0.932
3,600	0.404	12,000	0.945
3,800	0.443	13,000	0.955
4,000	0.481	14,000	0.963
4,200	0.516	15,000	0.970
4,400	0.549	16,000	0.974
4,600	0.579	18,000	0.981
4,800	0.608	20,000	0.986
5,000	0.634	25,000	0.992
5,200	0.659	30,000	0.995

FIGURE 10-6-1. Three radiating planes. Center plane transmits radiation.

does not interfere with the directions of passage of radiation. For transparent transmission τ_{2t}

$$q_{1 \rightarrow 3} = J_1 A_1 F_{13} \tau_{2t} \qquad (10\text{-}60)$$

This also may be referred to in the literature as specular transmission.

In diffuse transmission τ_{2d}, radiation interacts with the medium so that it emerges with random orientation; that is, it emerges with the same distribution as we assume for emission. This process may be considered analogous to transmission of light through certain glass plates (as in some offices, doors for shower rooms, etc.) through which you cannot see. Under circumstances, we have

$$q_{1 \rightarrow 3} = J_1 A_1 F_{12} \tau_{2d} F_{23} \qquad (10\text{-}61)$$

It is possible, of course, to have both kinds of transmission take place simultaneously.

The electric circuit for the problem of Fig. 10-6-1 is shown in Fig. 10-6-2. Note that the transmission introduces a direct path of current between radiosities 1 and 3.

Let us consider the energy balance at surface 2. Since transmitted energy has been treated separately, let us define a modified radiosity excluding transmitted energy.

$$J_2' = \varepsilon_2 E_{b2} + \rho_2 G_2 \qquad (10\text{-}62)$$

The net heat balance for surface 2 may be written

$$Q_2 = \varepsilon_2 E_{b2} A_2 - \alpha_2 G_2 A_2 = \varepsilon_2 E_{b2} A_2 - \varepsilon_2 G_2 A_2 \qquad (10\text{-}63)$$

The existence of transmission does not change the requirement that the difference between emission and absorption be the internal source. If we multiply the preceding two equations by $\varepsilon_2 A_2$ and by ρ_2, respectively, and then add, we get, upon noting that,

$$\varepsilon_2 + \rho_2 = 1 - \tau_2 = 1 - \tau_{2t} - \tau_{2d} \qquad (10\text{-}64)$$

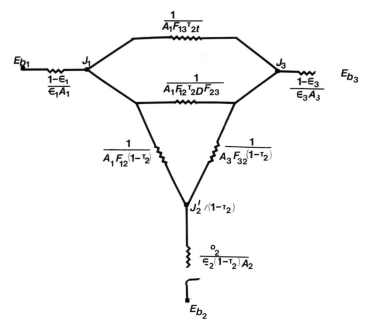

FIGURE 10-6-2. Electric circuit with transmission.

The equation

$$\rho_2 Q_2 = \varepsilon_2(1 - \tau_2) A_2 E_b - \varepsilon_2 J_2'$$ (10-65)

This may be rearranged into the electric circuit form

$$Q_2 = \frac{E_{b2} - J_2'/(1 - \tau_2)}{\rho_2/\varepsilon_2 A_2(1 - \tau_2)}$$ (10-66)

Consequently, there is a node $J_2'/(1 - \tau2)$ associated with surface 2 in Fig. 10-6-2. Note that when τ_2 goes to 0, ρ_2 goes to $1 - \varepsilon_2$, and the result reduces to the case with no transmission.

Introducing an alternate radiosity is a procedure that aids in treating certain problems. It also will prove useful in dealing with specular reflection. Note that the set of steps outlined in Section 10-4 for setting up a problem in electric circuit form still applies. The internal resistance [denominator of Eq. (10-66)] and the radiosity term $J_2'/(1 - \tau_2)$ differ from the analogous terms in Section 10-4 but still permit use of the procedures.

While body 2 transmits radiation, it also has the ability to absorb and emit radiation (assuming τ_2 is not 1). The reduced (by the amount that will be transmitted) heat flows are

$$q_{1 \rightarrow 2} = J_1 A_1 F_{12}(1 - \tau_2)$$

$$q_{2 \rightarrow 1} = J_2' A_2 F_{21} = \frac{J_2'}{1 - \tau_2} A_1 F_{12}(1 - \tau_2)$$

The net heat flow is then

$$q_{12} = \frac{J_1 - J_2'/(1 - \tau_2)}{1/A_1 F_{12}(1 - \tau_2)}$$

EXAMPLE 10-5.   A window of area 0.5 m² has a transmissivity of 0.9 for wavelengths less that 2.5 μm and zero elsewhere. The window is in a large room whose walls are at 20°C. How much heat is lost by radiation through the window? Sunshine provides a heat flux of 700 W/m² to the window. How much gets inside.

SOLUTION.   Radiations from the walls to the window is characterized by 20°C = 293°K. The energy radiated through the window to the outside will be

$$q_{io} = \left(\sigma T_{\text{wall}}^4 \right) \left(A_{\text{window}} F_{\text{window,wall}} \right) f(0, \lambda T) \tau$$

We have used reciprocity to replace $A_{\text{wall}} F_{\text{wall, window}}$ with $A_{\text{window}} F_{\text{window, wall}}$. This is because we know very little about the room–only that it is large. Since radiation from the window goes into the room, the shape factor $F_{\text{window, wall}} = 1$. We do not need to know the details of the room. Note that relative to the window, the large areas of the walls make the walls appear as a black body. For $\lambda = 2.5$ μm, Table 10-6-1 yields f on the order of 10^{-5}, so we obtain

$$q_{io} = (5.669 \times 10^{-8})(293)^4 (0.5)(1)(10^{-5})(0.9) = 1.88 \times 10^{-3} \text{ W}$$

Note that f is very small and the energy radiated through the window is very small. For sunlight, the characteristic temperature is 5800°K. At this value of τ, Table 10-6-1 yields $f = 0.966$. The heat transmitted from outside to inside is

$$q_{oi} = (700 \text{ W/m}^2)(0.5 \text{ m}^2)(0.966)(0.9) = 304 \text{ W}$$

Note that while only a minute fraction of energy from the inside was transmitted to the outside, most of the energy from the outside was transmitted to the inside. This sensitivity of transmission to wavelength is an important factor for greenhouses.

10-7. SPECULAR REFLECTION

In earlier sections, we treated reflected radiation as diffuse, that is, emerging from the surface randomly with no "recollection" of where it came from originally. With visible light, this is the type of reflection we see from most surfaces. However, for some surfaces, which we call mirrors, radiation is reflected so that angle of reflection equals angle of incidence. This type of reflection is called specular reflection. Analysis of this type of reflection can be complicated, especially when dealing with curved surfaces. This section reviews elementary principles of specular reflection.

Let us consider separately the diffuse and specular portions of radiosity. Define the diffuse radiosity as

$$J_D \varepsilon E_b + \rho_D G \tag{10-67}$$

where ρ_D is the diffuse reflection (and ρ_s will be the specular reflection). If there is a source of heat Q at the body that must be dissipated, then

$$Q = A(\varepsilon E_b - \alpha G) \qquad (10\text{-}68)$$

Eliminating G from these two equations and noting that

$$\varepsilon + \rho_D + \rho_s = 1 \qquad (10\text{-}69)$$

we find, with a little algebra, that

$$Q = \frac{E_b - J_D/(1 - \rho_s)}{\rho_D/\varepsilon A(1 - \rho_s)} \qquad (10\text{-}70)$$

This relationship is in electric circuit form, as illustrated in Fig. 10-7-1. Note that the existence of specular reflection is accounted for, even though the node is in terms of diffuse radiosity.

Consider the interaction illustrated in Fig. 10-7-2 between a specularly reflecting plane surface and another diffusely reflecting plane surface. Let us first consider diffuse interaction. The diffuse energy from surface 1 to surface 2 is

$$q^D_{1 \to 2} = J_{1D} A_1 F_{12} \qquad (10\text{-}71)$$

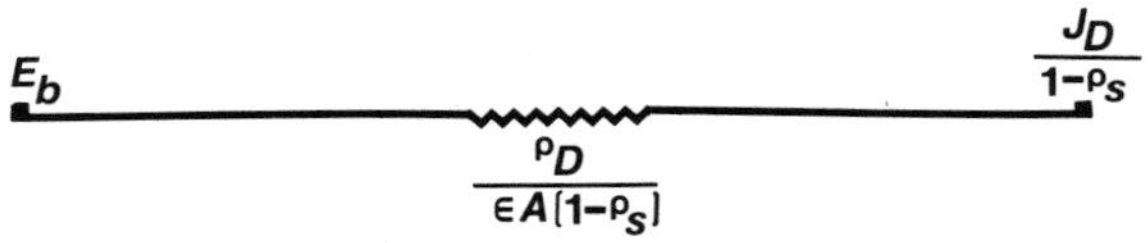

$$E_b \underbrace{\overset{\dfrac{J_D}{1-\rho_s}}{\wedge\!\wedge\!\wedge}}_{\dfrac{\rho_D}{\varepsilon A(1-\rho_s)}}$$

FIGURE 10-7-1. Electric analogy at a specular surface.

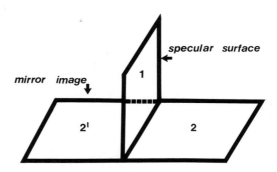

FIGURE 10-7-2. A specularly reflecting surface (1), a diffuse surface (2), and the mirror image of the diffuse surface (2′).

while the diffuse interaction from 2 to 1 is

$$q_{2\to 1}^D = J_2 A_2 F_{21}(1 - \rho_{1s})\qquad(10\text{-}72)$$

The net diffuse interaction is

$$q_{12}^{\text{net}, D} = q_{1\to 2}^D - q_{2\to 1}^D = \frac{J_{1D}/(1 - \rho_{1s}) - J_2}{1/A_1 F_{12}(1 - \rho_{1s})}\qquad(10\text{-}73)$$

where we have noted shape factor reciprocity and divided by $(1 - \rho_{1s})$ to place the equation in electric circuit form with a $J_{1D}/(1 - \rho_{1s})$ node.

Let us consider free space as a third node which is a black body of zero temperature. The energy lost to space from the surfaces may expressed as

$$q_{13} = J_{1D} A_1 F_{13} + J_{1s} A_1 F_{13}^s\qquad(10\text{-}74)$$

$$q_{23} = J_2 A_2 F_{23}\qquad(10\text{-}75)$$

We note that the specular reflection from surface 1 is equivalent to diffuse radiation from the mirror image 2' of surface 2, or

$$J_{1s} A_1 F_{13} = J_2 A_{2'} \rho_{1s} F_{2'3}\qquad(10\text{-}76)$$

If we reorganize our energy flows, we can define

$$q_{13}' = \frac{J_{1D}}{1 - \rho_{1s}} A_1 F_{13}(1 - \rho_{1s})\qquad(10\text{-}77)$$

$$q_{23}' = J_2(A_2 F_{23} + A_{2'} \rho_{1s} F_{2'3})\qquad(10\text{-}78)$$

where total heat transfer is preserved:

$$q_{13} + q_{23} = q_{13}' + q_{23}'\qquad(10\text{-}79)$$

In effect, we are considering heat that leaves surface 2 and is specularly reflected from surface 1 in the direction of surface 3 (space) to go directly from surface 2 to surface 3. The electric circuit corresponding to this situation is given Fig. 10-7-3

By treating the net diffuse exchange between surfaces 1 and 2 and the total exchange (diffuse and specular) with free space, we have treated the problem completely. The specular reflection from surface 1 to surface 2 is a circular loop from node J_2 back to itself and thus does not introduce any current to the problem.

When dealing only with diffuse reflection, it was not necessary to introduce free space in a problem involving two plane surfaces. The losses to space were implicit in the problem. With specular reflection, however, more information is needed.

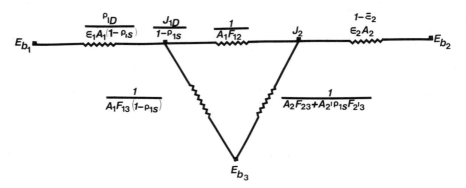

FIGURE 10-7-3. Electric circuit diagram for a plane specularly reflecting surface interacting with a plane diffuse surface and free space.

Consider two rectangles of equal dimensions at right angles to one another, as in Configuration 3 of Table 10-3-1(see p. 267). Now let one of the rectangles be moved such that it is parallel to the first, as in Configuration 2 of Table 10-3-1. In the case of Configuration 3, all specular reflection goes to free space, while in the case of Configuration 2, some specular reflection will return to the original surface. The difference between these two situations is represented in Fig. 10-7-3 by the shape factor $F_{2'3}$.

Note that the set of steps outlined in Section 10-4 for setting up a problem in electric circuit form still applies. The internal resistance [denominator of Eq. (10-70)] and the radiosity term $J_D/(1 - \rho_s)$ differ from the analogous terms in Section 10-4 but still permit use of the procedures.

EXAMPLE 10-6.   Repeat Example 10-4 for total heat loss assuming that the reflection at the reradiating surface is specular.

SOLUTION.   The circuit is analogous to that of Example 10-4. Since $\varepsilon = 0.8$, reflection is $1 - 0.8 = 0.2$. Note that the geometric resistances are modified relative to Example 10-4, so now we have

$$\frac{1}{A_1 F_{12}(1 - \rho_{2s})} = \frac{1}{(1)(0.2)(1 - 0.2)} = 6.26$$

$$\frac{1}{A_2 F_{23}(1 - \rho_{2s})} = \frac{1}{(1)(0.8)(1 - 0.2)} = 1.5625$$

In addition, the mirror image of surface 1 appears to radiate to space so in the left path we have

$$\frac{1}{A_1 F_{13} + A_1' \rho_{2s} F_{13}'} = \frac{1}{(1)(0.8) + (1)(0.2)(0.2)} = \frac{1}{0.84} = 1.19$$

The left path resistance is thus

$$R_L = 1.19$$

The right path resistance is

$$R_R = 6.25 + 1.56 = 7.81$$

The equivalent resistance is given by

$$\frac{1}{R_{eq}} = \frac{1}{R_L} + \frac{1}{R_R} = 0.84 + 0.13 = 0.97$$

$$R_{eq} = 1.03$$

The total resistance between E_{b1} and E_{b3} is then

$$R_{tot} = \frac{1 - 0.8}{0.8} + 1.03 = 1.28$$

Note that the total resistance has declined relative Example 10-4. Because of the specular reflection from surface 2, radiation that returned to surface 1 in Example 10-4 now goes directly to space. The total heat flow is then

$$Q_1 = \frac{E_{b1} - E_{b3}}{R_{tot}} = 4.43 \times 10^4 \text{ W}$$

10-8.   GASES

In previous sections, we assumed that the space between surfaces did not influence radiation heat transfer. Thus, any gas, such as air, implicitly was assumed to be "nonparticipating." For many commonly occurring gases such as oxygen and nitrogen (the principal constituents of air), this assumption is reasonable. In general, symmetric nonpolar diatomic molecules (like O_2) are expected to be nonparticipating, as are monatomic gases (like argon). On the other hand, more complicated molecules of a nonsymmetric polar type are expected to be participating. Included among these molecules are products of combustion—carbon dioxide, carbon monoxide, water vapor, and sulfer dioxide. Also included are hydrocarbon gases and ammonia.

The products of combustion, being hot, can emit radiation which will come into contact with nearby surfaces. Reflections and emissions from these surfaces will, in turn, cause radiation to be absorbed by the gas. Analysis of such situations, which are important, involves recognition of volumetric absorption and of wavelength dependence.

Thus far, we have considered radiation heat transfer in terms of surfaces. Gas molecules provide a very large number of very small surfaces, so treatment as a surface phenomenon is not usually practical. Accordingly, we define a quantity k as the absorption per unit length within the gas. For radiation moving in the x direction with intensity I, the absorption in interval dx is (Fig. 10-8-1)

$$dA = -dI = kI\,dx \tag{10-80}$$

If a beam of radiation is incident normal to a slab (plane) region containing a gas, then integration yields

$$I = I_0 e^{-kL} \tag{10-81}$$

The gas appears to have an absorptivity

$$\alpha = 1 - e^{-kL} \tag{10-82}$$

Actually, radiation can enter a region at any angle, and gas-containing volumes may have a variety of shapes. In principle, we may follow the procedure used in Section 10-3 for evaluating shape factors and introduce a gas attenuation factor f_g to modify Eq. (10-19) to

$$A_1 F_{12} f_g = \frac{1}{\pi} \int dA_1 \int dA_2 \frac{\cos\phi_1 \cos\phi_2}{r^2} e^{-kr} \tag{10-83}$$

An alternate approach has been to define a mean beam length

$$L_e = \frac{1}{A_1 F_{12}} \int dA_1 \int dA_2 r \frac{\cos\phi_1 \cos\phi_2}{\pi r^2} \tag{10-84}$$

and use this to evaluate absorptivity and therefore also emissivity according to

$$\alpha = \varepsilon = 1 - e^{-kL_e} \tag{10-85}$$

Mean beam lengths for selected configurations are given in Table 10-8-1.

In dealing with volumetric absorption, we have used thus far a single coefficient k to characterize absorption per unit length. We have not yet

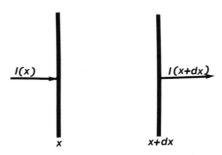

FIGURE 10-8-1. Entering and exiting radiation from interval dx where $\kappa I\,dx$ is absorbed.

TABLE 10-8-1
Mean Beam Lengths

Geometry	Mean beam length
Cylinders, diameter D, length L	
$\quad$ L semi-infinite, radiate to base	$0.65D$
$\quad$ L infinite, radiate to surface	$0.95D$
$\quad$ $L/D = 1$, radiate to surface	$0.60D$
$\quad$ $L/D = 1$, radiate to center of base	$0.71D$
Sphere, diameter D	$0.65D$
Hemisphere, diameter D,	
Radiate to center of base	$0.50D$
Infinite plane geometry, thickness L	$1.8L$
Cube, edge L, radiate to face	$0.60L$
Exterior of array of tubes, diameter D	
$\quad$ Square array, pitch P	$3.5(P - D)$
$\quad$ Triangular array, pitch $P = 2D$	$3.0(P - D)$
General body, volume V, area A,	
radiate to surface	$3.6V/A$

considered the second feature noted earlier, namely, wavelength dependence. Gases emit and absorb radiation in discrete bands of wavelengths. For engineering purposes, one frequently (as shall be done here) seeks effective overall average properties by which to characterize the gas. It should be noted, however, that there can be situations where explicit wavelength dependence is important, and more elaborate references (such as those cited at the end of the chapter) should then be consulted.

The average absorption or emission in a particular band i may be expressed as

$$A_i = \frac{1}{\lambda_{iU} - \lambda_{iL}} \int_{\lambda_{iL}}^{\lambda_{iU}} \alpha(\lambda) d\lambda \qquad (10\text{-}86)$$

This average quantity per band tends to be measured experimentally, but the detailed wavelength dependence within the band may not be well known. The total emission from this band is then, using Eq. (10-59),

$$E_i = A_i f(\lambda_{iL}T, \lambda_{iU}T)\sigma T^4 \qquad (10\text{-}87)$$

The total gas emission E_g is the sum of the emissions in all bands

$$E_g = \sum_i E_i \qquad (10\text{-}88)$$

and the effective emissivity of the gas E_g is

$$\varepsilon_g = \frac{E_g}{\sigma T^4} \qquad (10\text{-}89)$$

Effective emissivities as functions of temperature for water vapor and carbon dioxide are given in Fig. 10-8-2 and 10-8-3. These figures apply at 1 atm, with parametric curves given in terms of the partial pressure of the participating component (i.e., H_2O or CO_2) and the mean beam length. Corrections for other pressures are given Figs. 10-8-4 and 10-8-5.

When more than one participating gas is present (as in the case for combustion products, when H_2O and CO_2 are both expected), there is interaction among these gases. With two participating gases

$$\varepsilon_g = \varepsilon_{g1} + \varepsilon_{g2} - \Delta\varepsilon_{12} \qquad (10\text{-}90)$$

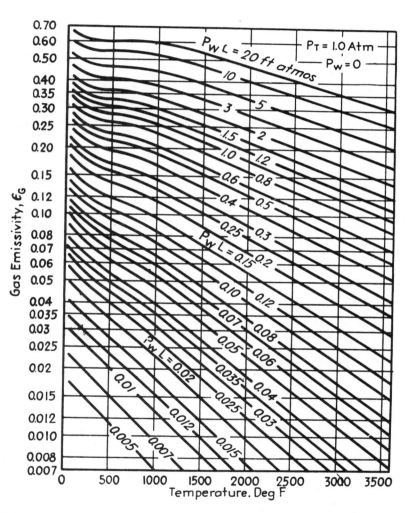

FIGURE 10-8-2. Effective emissivity of water vapor. From H. C. Hottel and R. B. Egbert, *Trans. AIChE* **38**, 531 (1942). Reproduced with permission.

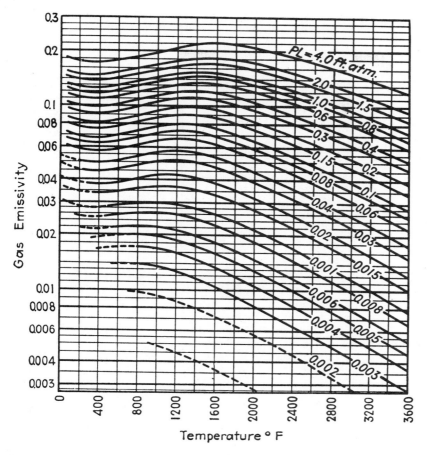

FIGURE 10-8-3. Effective emissivity of carbon dioxide. From H. C. Hottel and R. B. Egbert, *Trans. AIChE* **38**, 531 (1942). Reproduced with permission.

The correction factor for water vapor and carbon dioxide is given in Fig. 10-8-6.

For an interchange of radiation between a surface and surrounding black walls, the net energy flux is

$$q'' \varepsilon_g(T_g) \sigma T_g^4 - \alpha_g(T_w) \sigma T_w^4 \qquad (10\text{-}91)$$

where g and w subscripts denote gas and wall. The absorption by the gas is evaluated at the temperature of the wall. The absorption for a mixture is evaluated in the following empirical manner. The absorption is again

$$\alpha_g = \alpha_{CO_2} + \alpha_{H_2O} - \Delta\alpha \qquad (10\text{-}92)$$

EFFECT OF TOTAL PRESSURE AND PARTIAL PRES-
SURE ON WATER VAPOR RADIATION

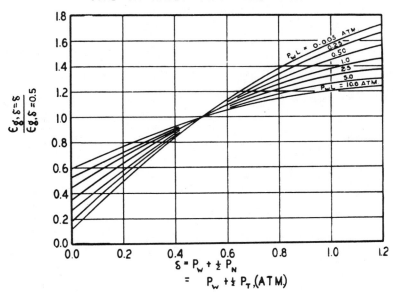

FIGURE 10-8-4. Pressure correction for water vapor. From H. C. Hottel and R. B. Egbert, *Trans. AIChE* **38**, 531 (1942). Reproduced with permission.

The individual absorptions are evaluated by

$$\alpha_{CO_2} = \varepsilon_{CO_2}\left(T_w P_{CO_2} L_e \frac{T_w}{T_g}\right)\left(\frac{T_g}{T_w}\right)^{0.65} \tag{10-93}$$

$$\alpha_{H_2O} = \varepsilon_{H_2O}\left(T_w P_{H_2O} L_e \frac{T_w}{T_g}\right)\left(\frac{T_g}{T_w}\right)^{0.45} \tag{10-94}$$

$$\Delta\alpha = \Delta\varepsilon(T_w) \tag{10-95}$$

If the surface wall is not black but has a high emissivity (above 0.8), then the net heat transfer may be evaluated by

$$\frac{q_{net}}{q_{net,\text{black wall}}} = \frac{1 + \varepsilon_w}{2} \tag{10-96}$$

The above empirically based approach may be used for estimating effects involving hot gases. For more elaborate treatments and for cases that do not fit the conditions of the prescriptions used, the reader should consult the

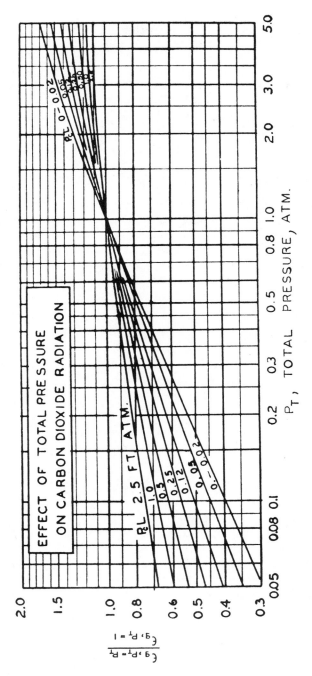

FIGURE 10-8-5. Pressure correction for carbon dioxide. From H. C. Hottel, R. B. Egbert, *Trans. AIChE* **38**, 531 (1942). Reproduced with permission.

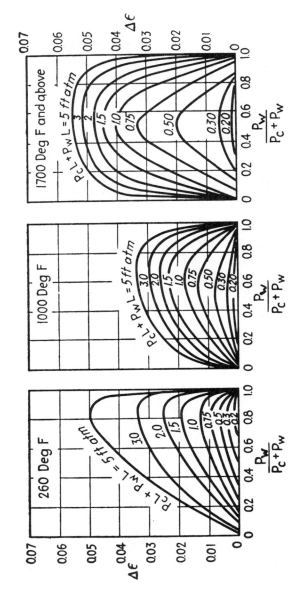

FIGURE 10-8-6. Correction factors for mixtures of water vapor and carbon dioxide. From H. C. Hottel, R. B. Egbert, *Trans. AIChE* **38**, 531 (1942). Reproduced with permission.

specialized references on radiation heat transfer provided at the end of the chapter.

301

Radiation

PROBLEMS

10-1. A cylinder (body 1) of height H_1 and diameter D_1 is contained within a cylinder (body 2) of height H_2 and diameter D_2. Both cylinders have a common axis and center. What are F_{11}, F_{12}, F_{21}, and F_{22}?

10-2. A cavity is in the shape of an equilateral triangle. What are the shape factors between surfaces in the cavity?

10-3. A structure is made of a level floor over which a hemisphere-shaped dome is placed. What is the shape factor from the interior of the hemisphere to the floor?

10-4. Configuration 2 in Table 10-3-1 is for parallel rectangles of equal dimension. Find the shape factor if one of the dimensions of rectangle 2 differs from the corresponding dimension of rectangle 1.

10-5. Find a simplified expression for shape factor for touching perpendicular rectangles when the common edge (a in Table 10-3-1) is infinite.

10-6. Show that the formula for parallel rectangles goes to the formula for infinite parallel planes of equal dimension when one of the dimensions of the rectangle goes to infinity.

10-7. Find the shape factors associated with Fig. 10-3-10 in terms of standard shape factors.

10-8. Show how the spreadsheet of Example 10-3 may be adapted to evaluate F_{12} of Fig. 10-3-8.

10-9. Suppose the two square black bodies of Example 10-3 were opposite one another with a separation of 1 m. Find the amount of heat that has to be supplied to the $1000°$K body and the temperature of the other body.

10-10. Suppose the two square bodies of Problem 10-9 were replaced by circular bodies of the same area. Find the amount of heat that has to be supplied to the $1000°$K body and the temperature of the other body.

10-11. Suppose you have to store a cryogenic liquid in a cylindrical container and you wish to have a small heat transfer rate into the liquid. The diameter of the container is 20 cm and the temperature is $20°$K. The container wall is 1 cm thick with a high thermal conductivity. Outside the wall is a 1 cm vacuum gap to a second 1 cm wall that is at $20°$C.

    a. If the walls have $\varepsilon = 0.8$, what is the rate per unit length of cylinder at which heat is transferred to the liquid?

    b. If the walls are covered with a thin foil with a polished surface where $\varepsilon = 0.04$, what is the rate at which heat is transferred to the liquid?

10-12. You wish to improve on the situation of Problem 10-11 by adding a radiation shield. Outside the second 1 cm wall you add another 1 cm vacuum gap followed by a third 1 cm wall. This third wall is at $20°$C.

    a. What is the heat transfer if all three walls have $\varepsilon = 0.8$?

    b. What is the heat transfer if the middle wall, that is, the shield, has $\varepsilon = 0.04$ on each face while the other walls have $\varepsilon = 0.8$?

    c. What is the heat transfer rate if all walls have polished surfaces with $\varepsilon = 0.04$?

10-13. You wish to improve further on the situation of Problems 10-11 and 10-12 by adding a layer of insulating material ($k = 0.04$ W/m $\cdot °$ K) to the outside wall. The outside of the insulating material will be at $20°$C. For each of the cases considered in Problems 10-11 and 10-12, make tables or plots as functions of insulation thickness of the heat transfer and the temperature at the inside of the insulation.

10-14. Consider again the situations in Problems 10-11–10-13. You would like to keep the lowest temperature of the insulator above 200° K and you would like to keep the overall diameter of the container below 40 cm. Select from the results of Problem 10-13 designs that are acceptable.

10-15. Given the constraints of Problem 10-14, would you be better off adding more radiation shields (assuming each gap to be 1 cm and each wall to be 1 cm thick) instead of insulation while staying within the 40 cm diameter limit?

10-16. A long duct that has a cross-section of an equilateral triangle has one wall maintained at 1000° K, one wall at 500° K, and one wall insulated. All walls have $\varepsilon = 0.8$. How much heat must be supplied per unit area of the 1000° K wall?

10-17. A window 10 cm square is used to view a large furnace area whose walls are maintained at 1000° K. The temperature of the walls in the outside room are at 293° K. The window has a transparent transmissivity of 0.9 for wavelengths less than 2.5 μm and zero elsewhere. How much net energy is transmitted through the window from the furnace to the outside room?

10-18. The possibility is raised of using different window materials for the configuration of Problem 10-17. Tabulate or plot the net energy transmission as a function of the upper limit of the wavelength interval in which transmission takes place?

10-19. Consider the configuration of Problem 10-9. Suppose that each of the bodies is a gray body with $\varepsilon = 0.8$. Find the amount of heat that has to be supplied to the 1000° K body and the temperature of the other body.

10-20. Find the heat that has to be supplied to the 1000° K body of Problem 10-19 if the other body's reflection is specular.

10-21. Repeat Problem 10-10 for the case where each body is a gray body with $\varepsilon = 0.8$.

10-22. Repeat Problem 10-21 if the reflection of the cooler body is specular.

10-23. Find the heat that has to be supplied to the 1000° K body of Problem 10-19 if the reflection of that 1000° K body is specular.

10-24. Show that when transmission occurs only in an interval $\lambda_A < \lambda < \lambda_B$, in which the transmission factor is τ_0, the effective average transmission for radiation of all wavelengths originating from a source at temperature T is $\tau = \tau_0[f(0, \lambda_B T) - f(0, \lambda_A T)]$.

10-25. Show that when transmission occurs only in an interval $\lambda_A < \lambda < \lambda_B$, in which the transmission factor is τ_0, the interchange of radiation for the configuration of Fig. 10-6-1 is, for transparent transmission,

$$q_{13} = A_1 F_{13} \tau_0 \{ J_1[f(0_1\lambda_B T_1) - f(0_1\lambda_A T_1)]$$
$$- J_3[f(0_1\lambda_B T_3) - f(0_1\lambda_A T_3)] \}$$

10-26. Generalizing from Problem 10-25, show that energy interchanges in the interval $\lambda_A < \lambda < \lambda_B$ can be expressed as in Section 10-6 with radiosities and black body radiations modified with wavelength interval fractions.

10-27. For the configuration of Fig. 10-6-1, if transparent transmission is τ_0 in the interval $\lambda_A < \lambda < \lambda_B$, prepare sets of equations for radiation heat interchanges between surfaces in the intervals
a. $\lambda A < \lambda < \lambda_B$; and
b. $\lambda < \lambda_A$, $\lambda < \lambda_B$.

10-28. Air is blown over the window of Problem 10-17 at a rate such the window temperature is maintained at 300° K. What is the total radiation heat transfer between the furnace and the outside room? The ε for the window is 0.8 outside the wavelength interval $\lambda_A < \lambda < \lambda_B$ and is 0.05 in that interval.

10-29. With the energy interchanges of Problem 10-27 and the total energy sources at each surface, prepare a complete set of equations to characterize the configuration of Fig. 10-6-1.

10-30. The blower of Problem 10-28 fails so that there is no significant heat removal from the window other than by radiation.

 a. What is the total radiation heat transfer between the furnace and the outside room?

 b. What is the temperature of the window?

REFERENCES

1. J. A. Duffie and W. A. Beckman, *Solar Energy Thermal Processes*, Wiley, New York, 1974.
2. H. C. Hottel and R. B. Egbert, "The radiation of Furnace Gases," *Trans. ASME*, **63**, 297 (1941).
3. H. C. Hottel and R. B. Egbert, "Radiant Heat Transmission from Water Vapor," *Trans. AIChE*, **38**, 531 (1942).
4. H. C. Hottel and A. F. Sarofim, *Radiative Transfer*, McGraw-Hill, New York, 1967.
5. J. R. Howell, A Catalog of Radiation Configuration Factors, McGraw-Hill, New York, 1982.
6. S. S. Kutateladze and V. M. Borishanskii, *A Concise Encyclopedia of Heat Transfer*, Pergamon, New York, 1966.
7. T. J. Love, *Radiative Heat Transfer*, Merrill, Columbus, OH, 1968.
8. A. K. Oppenheim, "Radiation Analysis by the Network Method," *Trans. ASME*, **78**, 725 (1956).
9. W. M. Rohsenow and J. P. Hartnett, *Handbook of Heat Transfer*, McGraw-Hill, New York, 1973.
10. R. Siegel and J. R. Howell, *Thermal Radiation Heat Transfer*, 2nd ed., McGraw-Hill, New York, 1980.
11. E. M. Sparrow and R. D. Cess, *Radiation Heat Transfer*, Hemisphere, Washington D.C., 1978.
12. W. C. Swinbank, "Long-Wave Radiation from Clear Skies," *Q. J. R. Meteorol. Soc.*, **89** (1963).
13. J. A. Wiebelt, *Engineering Radiation Heat Transfer*, Holt, Rinehart and Winston, New York, 1966.

Chapter Eleven

HEAT EXCHANGERS

11-1. INTRODUCTION

Thus far, we have been concerned with heat transfer phenomena. In this chapter, we are concerned with types of equipment that are intended to accomplish the transfer of heat. General classes of heat exchangers are described, as are general considerations of heat exchanger design. Then general procedures for analyzing heat exchangers are explained.

11-2. TYPES OF HEAT EXCHANGERS

Heat exchangers have the function of transferring heat from one fluid to another. Many types of heat exchanger exist. This is because there are many design considerations involved. Some design considerations may even be in basic conflict. Several design considerations follow.

1. Resistance to heat transfer should be minimized.
2. Contingencies should be anticipated via safety margins; for, example, allowance for fouling during operation.
3. The equipment should be sturdy.
4. Costs and material requirements should be kept low.
5. Corrosion should be avoided.

305

6. Pumping costs should be kept low.
7. Space required should be kept low where applicable.
8. Required weight should be kept low.

Design involves trade-offs among factors not related directly to heat transfer. Meeting the objective of minimized thermal resistance implies thin walls separating fluids. Thin walls may not be compatible with sturdiness. Auxiliary steps may have to be taken, for example, the use of support plates for tubing, to realize sturdiness.

Use of thin walls and the presence of support plates may facilitate corrosion or other types of chemical action, ultimately leading to equipment damage. The formation and growth of a product at the location where tubes pass through a support plate have caused significant difficulty in large steam generators, for example. Potential for corrosion within the heat exchanger may lead to requirements on treating fluids prior to entry into the heat exchanger. Filters and demineralizers might be used for such purposes.

Thus, design or selection of heat exchangers would require the engineer to anticipate possible source of corrosion. These sources of corrosion may originate outside the process of interest. If a heat exchanger uses air, the engineer would consider what may be present in the air where the equipment is to be used. If a heat exchanger uses water (e.g., a power plant condenser may use river or ocean water), the engineer should consider what is present in the water. A power plant condenser may contain different materials, depending on whether it is to be used with river or ocean water.

Impurities entering the heat exchanger can also cause deterioration of heat transfer performance. Material may settle so as to produce coatings that "foul" surfaces, leading to reduced heat transfer coefficients and possibly other problems too.

Impurities can also originate within the equipment. The heat exchanger may be part of a complex machine. A fluid may go through several other pieces of equipment in addition to the heat exchanger itself. Contaminants may enter the fluid in one of the other pieces of equipment and cause difficulty in the heat exchanger.

The prospect for deterioration of performance due to fouling implies a desirability or a safety margin in heat exchanger design so that the overall industrial process can function after some time. However, the friction–heat transfer analogy implies that the heat transfer margin may require increased pumping costs, which leads to a conflict with another objective.

Particularly in mobile equipment, space and weight are likely to be important factors, possibly even overriding other objectives. Automobile radiators, for example, are quite thin.

Heat exchangers differ considerably depending on which objectives are most important. Let us examine some considerations in heat exchanger design, and see how they relate to classes of heat exchangers.

A simple type of heat exchanger, which will be analyzed in Section 11-3, is the double pipe. The two fluids that are to exchange heat flow in the concentric pipes either in the same or in opposite directions, as illustrated in Fig. 11-2-1.

The heat exchanger of Fig. 11-2-1 may require a very long pipe. Instead of one long pipe, we may use a number of shorter pipes in parallel, as shown in Fig. 11-2-2. With all these individual pipes passing through a "box," we can modify the heat exchanger so as to have the outside fluid occupy the full volume between pipes, and thus not be associated with particular pipes. This modification, shown in Fig. 11-2-3, is a form of shell-and-tube heat exchanger.

A further modification to the shell-and-tube exchanger is shown in Fig. 11-2-4. Here the tube undergoes a U-bend so that fluid enters and leaves the

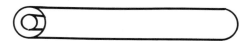

FIGURE 11-2-1. Double pipe heat exchanger. One fluid flows through central cylinder. Second fluid flows through outer annulus.

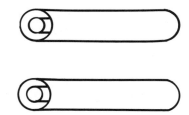

FIGURE 11-2-2. Short double pipes in parallel.

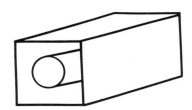

FIGURE 11-2-3. Tube within a shell.

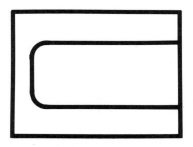

FIGURE 11-2-4. Schematic of a U-tube within a shell.

heat exchanger at the same side. This may compromise heat exchanger effectiveness which, as will be observed in Section 11-3, tends to be greatest under counterflow conditions (this statement does not apply if one fluid is changing phase). The two-pass arrangement of Fig. 11-2-4 may be preferred, however. For example, it may be convenient because of access considerations to have fluid enter and leave at the same side. Another factor that may be pertinent is that the U-tube accommodates thermal expansion (or avoids thermal stresses from differences in coefficients of expansion of shell and tube materials).

To enhance heat transfer, baffles tend to be introduced to promote circulation of the shell-side fluid, as illustrated in Fig. 11-2-5. Thus, flow tends to be both across and along the tubes. As noted in the examples of Chapter 7, heat transfer tends to be higher in cross-flow that in axial flow. Baffles that tend to convert an axial flow situation to a cross-flow situation thereby enhance heat transfer. The mixing also promotes uniformity of heat transfer. In the two-pass arrangement shown, temperatures along the first tube pass differ from temperatures along the second tube pass. If the shell-side fluid did not mix, then fluid flowing along the first pass tubing would have a larger temperature change (because of a larger temperature differential to tube-side fluid) than would fluid flowing along the second pass tubing.

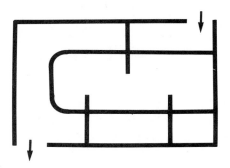

FIGURE 11-2-5. Baffles direct flow of shell-side fluid in a shell- and tube heat exchanger.

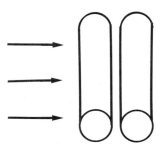

FIGURE 11-2-6. Cross flow. Flow within tubes unmixed. Flow outside tubes mixed.

In Fig. 11-2-5, fluid flows across tubes in addition to flowing along tubes. Another heat exchanger, the cross-flow heat exchanger, concentrates on the cross-flows as in Figs. 11-2-6 and 11-2-7. The two basic variants on this type of heat exchanger, as shown, are associated with mixing of fluids. Cross-flow frequently is used with open heat exchangers. Air may be drawn in by a fan and blown over a set of tubes. There may be no fixed outer dimension to the heat exchanger (as the shell of a shell-and-tube heat exchanger defines the dimensions of the device).

Unmixed cross-flow can lead to highly nonuniform heat transfer. If the tubes are arranged in a two-pass configuration, however, as illustrated in Fig. 11-2-8, then uniformity is promoted. The interference with mixing may be desired explicitly or may be due to fins for enhancing heat transfer.

The above are the heat exchanger types with which we are concerned primarily in this chapter. Other types of heat exchanger also exist. Compact heat exchangers employ various approaches to maximize surface area within small volumes. Regenerators are heat exchangers that provide for intermediate

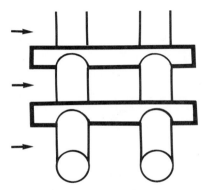

FIGURE 11-2-7. Cross flow. Both fluids unmixed.

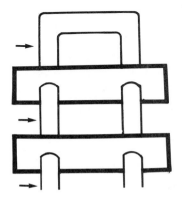

FIGURE 11-2-8. Cross flow of unmixed fluids with two-pass tubing.

storage of heat before transfer. The reader is directed to the references at the end of this chapter for further information on other types of heat exchanger.

11-3. CONCENTRIC PIPE HEAT EXCHANGER

Let us consider the simple concentric pipe heat exchanger shown in Fig 11-2-1. We assume that both fluids have properties that are constant in the heat exchanger. The fluids may be flowing in the same or in opposite directions. We assume that the hot fluid is moving in the positive direction.

Consider a differential section of length dx. The heat transferred from the hot to the cold fluid is (with P being perimeter, i.e., $P\,dx = dA$)

$$dq = UP(T_h - T_c)\,dx \qquad (11\text{-}1)$$

The quantity UP may be evaluated at either the inside or outside radius of the pipe. If the inside is used

$$UP = \frac{2\pi r_i}{1/h_i + r_i\ln(r_o/r_i)/k + (r_i/r_o)(1/h_o)} \qquad (11\text{-}2)$$

The heat transferred per unit length can be translated into temperature loss per unit length of hot fluid

$$\dot{m}_h c_h \frac{dT}{dx} = -\frac{dq}{dx} \qquad (11\text{-}3)$$

This also can be used to find the temperature gain of the cold fluid

$$\dot{m}_c c_c \frac{dT_c}{dx} = \pm\frac{dq}{dx} \qquad (11\text{-}4)$$

The plus sign applies if the cold fluid is flowing in the same direction as the hot fluid (parallel flow), and the minus sign applies if it is flowing in the opposite direction (counterflow). In either case, the cold fluid gains energy as it flows. In the counterflow case, it flows in the negative x direction.

For both hot and cold fluids, let

$$C_i = \dot{m}_i c_i \qquad (11\text{-}5)$$

for simplicity. For parallel flow, we combine equations to get

$$\frac{d}{dx}(T_h - T_c) = -\frac{dq}{dx}\left(\frac{1}{C_h} + \frac{1}{C_c}\right) = -\left(\frac{1}{C_h} + \frac{1}{C_c}\right)UP(T_h - T_c) \quad (11\text{-}6)$$

This can be integrated over the heat exchanger to yield (with A total area)

$$\ln \frac{T_{h2} - T_{c2}}{T_{hi} - T_{ci}} = -\left(\frac{1}{C_h} + \frac{1}{C_c}\right) UA \qquad (11\text{-}7)$$

The total heat transfer is

$$q = C_h(T_{h1} - T_{h2}) - C_c(T_{c2} - T_{c1}) \qquad (11\text{-}8)$$

We thus may write, by solving for $1/C_h$ and $1/C_c$ and then adding,

$$-\left(\frac{1}{C_h} + \frac{1}{C_c}\right) = \frac{1}{q}(T_{h2} - T_{h1} + T_{c1} - T_{c2}) = \frac{1}{q}[(T_{h2} - T_{c2}) - (T_{h1} - T_{c1})]$$

$$(11\text{-}9)$$

Accordingly, we find

$$q = UA\,\Delta T_m \qquad (11\text{-}10)$$

$$\Delta T_m = \frac{(T_{h2} - T_{c2}) - (T_{h1} - T_{c1})}{\ln[(T_{h2} - T_{c2})/(T_{h1} - T_{c1})]} \qquad (11\text{-}11)$$

This quantity is called the log-mean temperature difference. It is an effective average temperature difference that characterizes heat transfer in a heat exchanger, accounting for the fact that both hot and cold fluids vary in temperature as they pass through the heat exchanger.

For a counterflow heat exchanger, the same result is obtained. We note that

$$\frac{d}{dx}(T_h - T_c) = -\frac{dq}{dx}\left(\frac{1}{C_h} - \frac{1}{C_c}\right) \qquad (11\text{-}12)$$

There is a difference on the right-hand side rather than a sum as was the case for parallel flow. However, the total heat transfer is

$$q = C_h(T_{h1} - T_{h2}) = -C_c(T_{c2} - T_{c1}) \qquad (11\text{-}13)$$

which introduces a compensating minus sign for the cold fluid. Thus, the same overall result is obtained.

Let us consider further how temperatures vary in the heat exhanger. In parallel flow, we obtained

$$\frac{d}{dx}(T_h - T_c) = -\left(\frac{1}{C_c} + \frac{1}{C_h}\right) UP(T_h - T_c) \qquad (11\text{-}14)$$

If we integrate to some intermediate distance along the heat exchanger, we obtain

$$T_h - T_c = (T_{h1} - T_{c1})e^{-(1/C_c + 1/C_h)UPx} \tag{11-15}$$

The temperature difference between the hot and cold fluids is a maximum at the inlet and declines exponentially with passage through the heat exchanger, as illustrated in Fig. 11-3-1. The maximum temperature that can be attained by the cold fluid is limited by the minimum temperature that can be attained by the hot fluid.

For counterflow, we have

$$\frac{d}{dx}(T_h - T_c) = -\left(\frac{1}{C_h} - \frac{1}{C_c}\right)UP(T_h - T_c) \tag{11-16}$$

which may be integrated to yield

$$T_h - T_c = (T_h - T_c)_1 e^{-(1/C_h - 1/C_c)UPx} \tag{11-17}$$

Whether the temperature difference between the hot and cold fluids increases or decreases across the heat exchanger depends on the relative values of the heat capacities. If the heat capacities are balanced, then temperature difference will be constant along the heat exchanger. This situation is illustrated in Fig. 11-3-2. Note that the highest temperature of the cold fluid is not limited by the lowest temperature of the hot fluid.

In a counterflow heat exchanger, it is possible that the difference between hot and cold fluid temperatures may not change substantially, since heat capacities may be adjusted (via relative flows) to make heat transfer along the heat exchange nearly uniform. Under such conditions, the log-mean temperature formula can be simplified. Let us rewrite

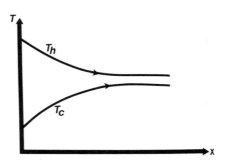

FIGURE 11-3-1. Temperature variations in a parallel flow heat exchanger. Arrows show flow direction.

$$\Delta T_m = \frac{\Delta T_2 - \Delta T_1}{\ln(\Delta T_2/\Delta T_1)} = \frac{1}{2}(\Delta T_2 + \Delta T_1)\frac{\Delta T_2 - \Delta T_1}{\frac{1}{2}(\Delta T_2 + \Delta T_1)}\frac{1}{\ln(\Delta T_2/\Delta T_1)} \quad \text{(11-18)}$$

Let us express

$$\frac{\Delta T_2}{\Delta T_1} = 1 + \varepsilon \quad \text{(11-19)}$$

so that the log-mean temperature difference may be expressed as

$$\Delta T_m = \frac{1}{2}(\Delta T_2 + \Delta T_1)\frac{\varepsilon}{(1 + \frac{1}{2}\varepsilon)\ln(1 + \varepsilon)} \quad \text{(11-20)}$$

Let us expand the logarithm up to a cubic term; that is,

$$\ln(1 + \varepsilon) = \varepsilon - \tfrac{1}{2}\varepsilon^2 + \tfrac{1}{3}\varepsilon^3 \quad \text{(11-21)}$$

We then find that keeping all terms to ε^2 results in

$$\Delta T_m = \frac{1}{2}(\Delta T_2 + \Delta T_1)\frac{1}{1 + \frac{1}{12}\varepsilon^2} \quad \text{(11-22)}$$

We thus may observe that the arithmetic average temperature difference is a very good approximation to the logarithmic mean temperature difference even when there is a substantial percentage variation of temperature difference. For example, if the difference is 50% ($\varepsilon = 0.5$), then the error is about 2%. This error is likely to be acceptable when compared with the accuracy of the overall heat transfer coefficient.

Using the linear average can be a useful convenience. When one of the temperatures is unknown, the log-mean average requires solution of a nonlinear equation by iterative means. Use of the linear average permits noniterative solution of a linear equation. When the variation in temperature difference is

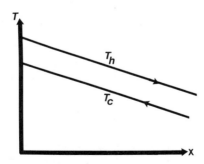

FIGURE 11-3-2. Temperature variation in a counterflow heat exchanger. Arrows show flow direction.

large enough for error to be of concern, it still may be possible to use the linear average to obtain a first guess for the iteration process.

In dealing with heat exchanger problems, it is useful to examine the information provided to see if additional information can be added easily to that which has been specified. For example, if both flow rates have been specified, and if three temperatures have been specified, then a simple heat balance yields the fourth temperature. The scope of information readily available can affect the choice of method used to analyze the heat transfer problem, that is, whether to use a method based on this section or a method based on the effectiveness concept to be introduced in Section 11-5.

EXAMPLE 11-1. It is desired to heat water from 40 to 90°C in a counterflow double pipe heat exchanger. The hot fluid has a specific heat of 2000 J/kg · °C and enters the heat exchanger at 120°C. The flow rate of the hot fluid is twice that of the water. What is the log-mean temperature difference?

SOLUTION. We obtain additional temperature information by equating heat balances

$$q = \dot{m}_w C_w \Delta T_w = \dot{m}_h C_h \Delta T_h$$

where h denotes the hot fluid and w denotes water.

$$\Delta T_h = \frac{\dot{m}_w}{\dot{m}_h} \frac{C_w}{C_h} \Delta T_w = \frac{1}{2} \frac{4183}{2000} (90 - 40) = 52.23°C$$

$$T_{h,\text{out}} = 120 - 52.3 = 67.7°C$$

The log-mean temperature difference is

$$\Delta T_m = \frac{\Delta T_2 - \Delta T_1}{\ln(\Delta T_2/\Delta T_1)}$$

It is not material which side is chosen as side 1. Let us take the side where the hot fluid enters as side 1

$$\Delta T_1 = T_{h,\text{in}} - T_{w,\text{out}} = 120 - 90 = 30$$

$$\Delta T_2 = T_{h,\text{out}} - T_{w,\text{in}} = 67.7 - 40 = 27.7$$

$$\Delta T_m = \frac{27.7 - 30}{\ln(27.7/30)} = 28.8°C$$

Note that the negative sign in the numerator is balanced by the negative sign of the logarithm. Note also that for this case, the linear average would be close to the log mean.

11-4. LOG-MEAN TEMPERATURE DIFFERENCE AND OTHER TYPES OF HEAT EXCHANGERS

Types of heat exchanger other than the concentric pipe (or a single-pass shell and tube exchanger which is equivalent to the concentric pipe) are more difficult to analyze) but still can be characterized by

$$q = UA\,\Delta T \tag{11-23}$$

where ΔT is a suitably derived average temperature difference. Common practice has been to relate this "true" temperature difference to the counter-flow log mean temperature by a correction factor F so that

$$q = UAF\Delta T_m \tag{11-24}$$

Since the counterflow configuration is the most effective of heat exchange arrangements, the correction factor is expected to be less than 1.

Consider the shell and tube configuration of Fig. 11-2-4. Over the first part of the tube flow, tube flow and shell flow are parallel. Over the second part of the tube flow, a counterflow situation exists. Thus, it is theoretically possible for tube exit temperature to exceed shell exit temperature. Temperature variation is shown in Fig. 11-4-1.

Note that, in this heat exchanger, the shell-side fluid is mixed and cools the two tube passes simultaneously. Thus, at some point x,

$$\frac{dq}{dx} = \frac{d}{dx}(UA)[2T_s - T_t(x) - T_t(L - x)] \tag{11-25}$$

where L is the length of the tubing and where s and t subscripts denote shell

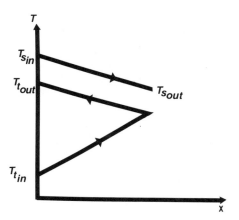

FIGURE 11-4-1. Temperature variation in a two-tube-pass shell and tube heat exchanger. Arrows show flow direction.

and tube. The analysis therefore is complicated and the result is

$$F = \frac{\sqrt{1 + R^2}\,\ln[(1 - P)/(1 - RP)]}{(R - 1)\ln\left\{\left[2 - P(R + 1 - \sqrt{1 + R^2})\right]/\left[2 - P(R + 1 + \sqrt{1 + R^2})\right]\right\}}$$

(11-26)

$$R = \frac{C_t}{C_s} = \frac{T_{\text{out},\,s} - T_{\text{in},\,s}}{T_{\text{in},\,t} - T_{\text{out},\,t}}$$

(11-27)

$$P = \frac{T_{\text{in},\,t} - T_{\text{out},\,t}}{T_{\text{in},\,t} - T_{\text{in},\,s}}$$

(11-28)

The result is plotted in Fig. 11-4-2. The quantity C is R when R is less than 1 and $1/R$ when R is greater than 1. The quantity ε, known as the heat exchanger effectiveness, is P when R is greater than 1 and RP when R is less than 1. The heat exchanger effectiveness is the ratio of the maximum fluid temperature change and the difference in inlet temperatures. The effectiveness concept will be discussed in some depth in Section 11-5.

The F factor and the ε factor can each be viewed as a figure of merit of the heat exchanger. It is desirable to have F high, since when F is high there will be minimal departure from what can be achieved with the best theoretical heat exchanger, the counterflow exchanger. It is desirable to have ε high, since when ε is high the heat exchange is close to the maximum that is possible thermodynamically. Clearly, it is possible to have a high F factor with a low ε and to have a high ε with a low F factor.

It may be observed that the F factor approaches 1 for all values of ε when C approaches 0. From Eqs. (11-27) and (11-28), it may be seen that C is the

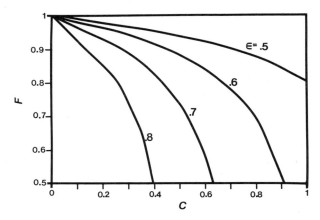

FIGURE 11-4-2. F-factor of a shell-and-tube heat exchanger.

ratio of the minimum heat capacity to the maximum heat capacity. A $C = 0$ implies that one fluid has an infinite heat capacity, that is, undergoes a change in phase. In such a case, there is no temperature change for that fluid, and the direction of flow is not important. All heat exchangers are then equivalent to counterflow, so $F = 1$.

It also may be observed that the F factor declines gradually from 1 initially as C increases from zero and then curves over to drop steeply. It is desirable to be at the upper, gradual portion of an F-factor curve not only because the F-factor figure of merit is high there, but also because it is good practice to avoid a region of steep variation. Steep variation implies that potentially large uncertainty in performance can be associated with small uncertainties in design parameters. This question of sensitivity when steepness or saturation effects are observed will arise again when effectiveness is discussed in Section 11-5.

For cross-flow, solutions are more complicated but still readily evaluated even on a small computer. Define temperature ratios relative to the difference of inlet temperatures

$$K = \frac{T_1 - T_2}{T_1 - t_1} \tag{11-29}$$

$$S = \frac{t_2 - t_1}{T_1 - t_1} \tag{11-30}$$

$$V = \frac{\Delta T}{T_1 - t_1} \tag{11-31}$$

where T and t refer to the two fluids. The F factor can be expressed as

$$F = \frac{V}{V_{\text{counterflow}}} \tag{11-32}$$

$$V_{\text{counterflow}} = \frac{K - S}{\ln(1 - S)/(1 - K)} \tag{11-33}$$

Solving the differential equation for heat transfer when both fluids are unmixed leads to the implicit equation

$$V = \sum_{n=0}^{\infty} \sum_{m=0}^{\infty} A_{nm} \left(\frac{K}{V}\right)^n \left(\frac{S}{V}\right)^m \tag{11-34}$$

$$A_{nm} = (-1)^{n+1} \frac{(n + m)!}{n!(n + 1)!m!(m + 1)!} \tag{11-35}$$

The equation for V may be solved by trial and error or by iteration. A guess is made for V on the right-hand side and used to evaluate V on the left-hand side. For a reasonably designed heat exchanger, the F factor should be high (e.g., 0.8) so that a first guess for V that is at or slightly below the counterflow value is usually prudent.

For those who would program this iteration on a personal computer, it is suggested that the coefficients A_{nm} be evaluated once and stored (rather than evaluated again at each iteration). Evaluation of a large number of factorials many times can be time consuming. A sample subroutine is provided in Appendix C. The resulting F factor is plotted in Fig. 11-4-3. Again the F vs. C presentation is used. Recall that ε is that ratio of maximum temperature change to difference in inlet temperatures and that C is the ratio of minimum to maximum heat capacity (or ratio of minimum to maximum temperature change).

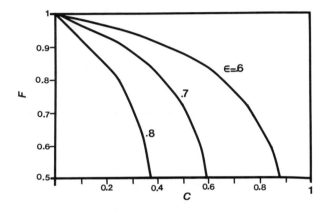

FIGURE 11-4-3. F-factor for a crossflow heat exchanger with both fluids unmixed.

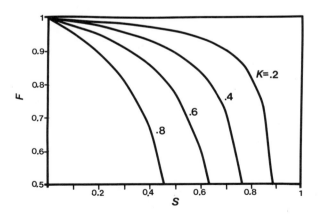

FIGURE 11-4-4. F-factor for a crossflow heat exchanger with one fluid mixed.

When one fluid is mixed (denoted by T), we obtain

$$V = \frac{S}{\ln \dfrac{1}{1 - (S/K)\ln[1/(1 - K)]}} \qquad (11\text{-}36)$$

This expression is easy to evaluate, and the F factor is plotted in Fig. 11-4-4. The presentation of Fig. 11-4-4 is somewhat different from the other F-factor curves, since the display in ε–C form would be somewhat more complicated than the form given. Note that both S and K in this plot are effectiveness-like quantities (fluid temperature change divided by inlet temperature difference). Note that effectiveness is the larger of S and K and that C is the ratio of the smaller of S and K to the larger. When both fluids are mixed, we again have an iterative situation:

$$V\left[\frac{K/V}{1 - \exp(-K/V)} + \frac{S/V}{1 - \exp(-S/V)}\right] = 1 + V \qquad (11\text{-}37)$$

A sample subroutine is provided in Appendix C.

In the various equations for F factors and V, it might appear that for certain values the expression becomes infinite or undefined. This actually is not the case because finite limits exist. However, care must be taken in computer evaluation. The fact that a ratio has a finite limit will not prevent error if numerator and denominator are evaluated separately and a ratio then taken.

On the other hand, it is possible for formulas to yield complex numbers, as when a logarithm of a negative number is required. This type of result may be interpreted as implying that the specifications cannot be realized physically; that is, the combination of inlet and outlet temperatures given cannot be achieved even with an infinite heat transfer area.

It was noted in Section 11-3 that using the arithmetic average temperature difference instead of the log-mean difference can simplify certain calculations. The situation is more complicated with the more elaborate heat exchangers because the correction factor F also depends on temperature. Thus, the approximation may not be sufficient for avoiding iteration. Some problems that require iteration may be treated more efficiently by working with the heat exchanger effectiveness approach discussed in Section 11-5.

EXAMPLE 11-2. It is desired to use a cross-flow heat exchanger to accomplish the heating of Example 11-1. The water is to flow inside the tubes and the hot fluid is to flow across the tubes. The hot fluid mixes. By what factor would the heat transfer area have to be increased?

SOLUTION. Since the heat exchange is the same, we must have

$$q = UA_0\Delta T_m = UAF\Delta T_m$$

where A_0 is the area for the counterflow exchanger. Thus,

$$\frac{A}{A_0} = \frac{1}{F}$$

Let us evaluate F by Eq. (11-36) for one fluid mixed and one fluid unmixed (the water in the tubes is unmixed since the tubes are separated).

$$F = \frac{V}{V_{cF}}$$

$$V = \frac{S}{\ln \dfrac{1}{1 - (S/K)\ln[1/(1 - K)]}}$$

$$K = \frac{T_1 - T_2}{T_1 - t_1} = \frac{120 - 67.7}{120 - 40} = \frac{52.3}{80} = 0.654$$

$$S = \frac{t_2 - t_1}{T_1 - t_1} = \frac{90 - 40}{120 - 40} = \frac{50}{80} = 0.625$$

$$V = \frac{0.625}{\ln \dfrac{1}{1 - (0.625/0.654)\ln[1/(1 - 0.654)]}} = \frac{0.625}{\ln(-70.2)}$$

The negative argument of the logarithm indicates that the problem posed is not feasible. A cross-flow heat exchanger even with infinite area cannot produce the combination of temperatures specified.

To obtain a cross-flow heat exchanger, it is necessary to modify some aspect of the problem. Suppose we modify the flow rate of the water so that the water is heated to 80°C instead of 90°C. Then

$$S = \frac{80 - 40}{120 - 40} = \frac{40}{80} = 0.5$$

$$V = \frac{0.5}{\ln \dfrac{1}{1 - (0.5/0.654)\ln[1/(1 - 0.654)]}} = 0.30$$

so cross-flow heat exchange is possible. The counterflow case for the new temperature is

$$V_{cF} = \frac{K - S}{\ln[(1 - S)/(1 - K)]} = \frac{0.654 - 0.5}{\ln[(1 - 0.5)/(1 - 0.654)]} = 0.418$$

The F factor is

$$F = \frac{V}{V_{cF}} = \frac{0.30}{0.418} = 0.72$$

The log-mean temperature difference for the new case is

$$\Delta T_m = \Delta T_{in} V = (0.418)(80) = 33.4$$

If the heat transfer coefficient is unchanged (e.g., if we modified flow rate by changing the number of tubes keeping flow per tube constant), then a counterflow heat exchanger with the new temperatures would require 28.8/33.4 or 86.2% of the reference heat transfer area. The cross-flow heat exchanger would require 1/0.72 or 39% more area than its corresponding counterflow exchanger, or 20% more area than the original counterflow exchanger.

Whether our resolution of difficulty encountered is appropriate depends on the application. The application would determine whether it is acceptable to have the heated water delivered at 80°C instead of 90°C.

11-5. HEAT EXCHANGER EFFECTIVENESS

In this section, we deal with a measure of effectiveness of heat exchangers, which also provides a basis for dealing with problems that are difficult to address with the methods of Sections 11-3 and 11-4. In particular, the log-mean temperature approach is easiest to apply when all the temperatures are specified, and we wish to find the area required to yield a specified heat exchange. We observed, however, than when some temperatures are not known (e.g., if only inlet temperatures are specified) iterative solution of nonlinear equation may be required. The approach of this section deals more readily with this case.

A parameter effectiveness is defined as

$$\varepsilon = \frac{\text{Actual heat transfer}}{\text{Theoretical maximum heat transfer}} \tag{11-38}$$

The theoretical limit to heat transfer is determined by the maximum and minimum temperatures encountered. The maximum possible heat loss by hot fluid would occur if the hot fluid were cooled to the inlet temperature of the cold fluid,

$$q_{h,\max} = C_h(T_{h,in} - T_{c,in}) \tag{11-39}$$

Similarly, the maximum possible heat gain by cold fluid would occur if it were heated to the inlet temperature of the hot fluid

$$q_{c,\max} = C_c(T_{h\,in} - T_{c,in}) \tag{11-40}$$

Since the heat transferred from the hot fluid is the same as the heat transferred to the cold fluid, the maximum possible heat transfer is the smaller of the two theoretical maxima, or

$$q_{\max} = \min(q_{h,\max} \, q_{c,\max}) \tag{11-41}$$

This implies that

$$q_{max} = C_{min}(T_{h,in} - T_{c,in}) \qquad (11\text{-}42)$$

We refer to the fluid with the lower heat capacity as the minimum fluid. It then follows that

$$\varepsilon = \frac{\Delta T_{min.fl.}}{\Delta T_{in}} \qquad (11\text{-}43)$$

where $\Delta T_{min.fl.}$ is the temperature change of the minimum fluid (which yields the maximum temperature change) in the heat exchanger and ΔT_{in} is the difference between inlet temperatures. This ratio is the form in which effectiveness was encountered in Section 11-4.

We may obtain expression for effectiveness by rearranging the equations of Sections 11-3 and 11-4. For parallel flow, we previously obtained

$$\ln \frac{T_{h2} - T_{c2}}{T_{h1} - T_{c1}} = -\left(\frac{1}{C_h} + \frac{1}{C_c}\right)UA \qquad (11\text{-}44)$$

Assume that the cold fluid is the minimum fluid (the same final expression will result if the hot fluid is the minimum fluid). Note that

$$T_{h2} - T_{c2} = T_{h2} - T_{h1} + T_{h1} - T_{c1} + T_{c1} - T_{c2} \qquad (11\text{-}45)$$

$$T_{h1} - T_{h2} = C(T_{c2} - T_{c1}) \qquad (11\text{-}46)$$

$$C = \frac{C_{min}}{C_{max}} = \frac{C_c}{C_h} \qquad (11\text{-}47)$$

We thus obtain

$$\ln[1 - (1 + C)\varepsilon] = -\frac{UA}{C_{min}}(1 + C) \qquad (11\text{-}48)$$

We can then rearrange the equation to

$$\varepsilon = \frac{1 - e^{-N(1+C)}}{1 + C} \qquad (11\text{-}49)$$

$$N = \frac{UA}{C_{min}} \qquad (11\text{-}50)$$

The quantity N (sometimes called NTU or number of transfer units) is a measure of the heat transfer area. Table 11-5-1 lists expressions for ε for common heat exchange types. Table 11-5-2 provides alternate formula for N in terms of ε. The formulas of Table 11-5-1 are plotted in Figs. 11-5-1–11-5-5.

TABLE 11-5-1
Heat Exchanger Effectiveness Formulas

Exchanger type	Formula
Double pipe	
Parallel flow	$\varepsilon = \dfrac{1 - \exp[-N(1 + C)]}{1 + C}$
Counterflow	$\varepsilon = \dfrac{1 - \exp[-N(1 - C)]}{1 - C\exp[-N(1 - C)]}$
Shell and tube One shell pass Even number of tube passes	$\varepsilon = 2\left[1 + C + \sqrt{1 + c^2}\,\dfrac{1 + \exp\left(-N\sqrt{1 + c^2}\right)}{1 - \exp\left(-N\sqrt{1 + c^2}\right)}\right]^{-1}$
Cross-flow C_{max} mixed, C_{min} unmixed	$\varepsilon = \dfrac{1 - \exp\{-C[1 - \exp(-N)]\}}{C}$
C_{max} unmixed, C_{min} mixed	$\varepsilon = 1 - \exp\left[\dfrac{1 - \exp(-NC)}{-C}\right]$
Both fluids mixed	$\dfrac{1}{\varepsilon} = \dfrac{1}{1 - \exp(-N)} + \dfrac{C}{1 - \exp(-NC)} - \dfrac{1}{N}$
Both fluids unmixed	$\varepsilon = 1 - \exp\left\{\dfrac{N^{0.22}}{C}\left[\exp(-CN^{0.78}) - 1\right]\right\}$

TABLE 11-5-2
Heat Exchanger NTU Formulas

Exchanger type	Formula
Double pipe	
Parallel flow	$N = \dfrac{-\ln[1 - \varepsilon(1 + C)]}{1 + C}$
Counterflow	$N = \dfrac{-\ln[(\varepsilon - 1)/(\varepsilon C - 1)]}{1 - C}$
Shell and tube One shell pass Even number of tube passes	$N = \dfrac{1}{\sqrt{1 + c^2}}\ln\left[\dfrac{2 - \varepsilon(1 + C + \sqrt{1 + c^2})}{2 - \varepsilon(1 + C - \sqrt{1 + c^2})}\right]$
Cross-flow C_{max} mixed, C_{min} unmixed	$N = -\ln\left[1 + \dfrac{\ln(1 - \varepsilon C)}{C}\right]$
C_{max} unmixed, C_{min} mixed	$N = \dfrac{1 + C\ln(1 - \varepsilon)}{C}$

The formulas for heat exchanger effectiveness are directly applicable to situations where flow rates and heat transfer area are specified, but limited specifications have been made regarding temperatures. This is opposed to the log-mean formula which is directly applicable to a case where all the temperatures are specified and the heat transfer area remains to be determined.

It may be observed that effectiveness for all heat exchangers becomes

$$\varepsilon = 1 - e^{-N} \tag{11-51}$$

when C goes to zero. While there is no real material that has zero or infinite heat capacity, the infinite capacity situation can be used to describe the case where one fluid is changing phase, that is, either boiling or condensing. The heat transferred is associated with latent heat of vaporization and is not

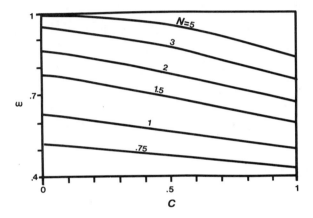

FIGURE 11-5-1. Effectiveness of a counterflow heat exchanger.

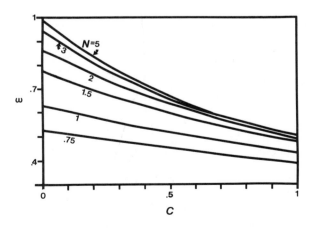

FIGURE 11-5-2. Effectiveness of a parallel flow heat exchanger.

associated with a temperature change. If temperature does not change, then direction of flow does not influence effectiveness.

In keeping with Eq. (11-51), the corresponding (i.e., for the same N) curves in Figs. 11-5-1–11-5-5 start at the same points for $C = 0$. Effectiveness declines with C since, for a given C_{min}, lowering C_{max} (and thereby increasing $C = C_{min}/C_{max}$) lowers the average temperature difference across the heat exchanger. The average temperature difference is lowered the least for the counterflow heat exchanger, the most effective type, so the decline is slowest with this heat exchanger (Fig. 11-5-1). By contrast, the average temperature difference is lowered most for the parallel flow heat exchanger, for which the effectiveness saturates rapidly with NTU. Note from the formulas of Table 11-5-1 that when NTU (or heat transfer area) goes to infinity, the effectiveness

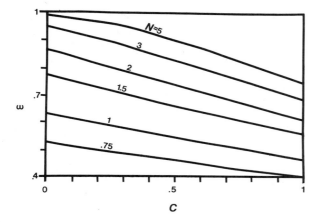

FIGURE 11-5-3. Effectiveness of a crossflow heat exchanger with both fluids unmixed.

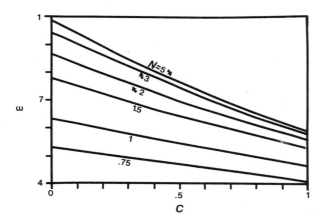

FIGURE 11-5-4. Effectiveness of a shell-and-tube heat exchanger.

of a counterflow heat exchanger goes to 1 for any value of C, but goes to $1/(1 + C)$ for a parallel flow exchanger. This explains the saturation effect observed in Fig. 11-5-2.

It is interesting to observe that the cross-flow heat exchanger with both fluids unmixed provides the next highest effectiveness of the exchangers shown after the counterflow exchanger. Again, when NTU goes to infinity, ε goes to 1 for all C. On the other hand, cross-flow with both fluids mixed provides a saturation effect, so one should be cautious about generalizing with regard to cross-flow. It is interesting to note that in the case of both fluids mixed, there is a departure from monotonic variation with NTU (i.e., curves with different values of NTU cross).

The shell-and-tube heat exchanger with two tube passes has aspects of both parallel flow and counterflow. With this exchanger, effectiveness saturates with increasing NTU as may be verified from the formulas of Table 11-5-1.

Saturation effects are important to recognize in the design process. For example, in evaluating heat transfer area to obtain a specified effectiveness goal, there may be great sensitivity to small variations in parameters. Thus, a large saving in cost of material could be obtained for a minor reduction in effectiveness.

While the effectiveness approach tends to be easier to apply than the log-mean approach in situations where limited specifications are made for temperatures, it may not avoid the need for iteration altogether. To evaluate NTU, we first must evaluate the overall heat transfer coefficient. To evaluate the convection coefficients for the two fluids, we must, for each fluid, assume a temperature at which to evaluate properties. Since we do not know all the temperatures (e.g., we may know just the two inlet temperatures), we have to assume temperatures for property evaluation and then check, at the end of the problem, to see if the assumed temperatures were reasonable.

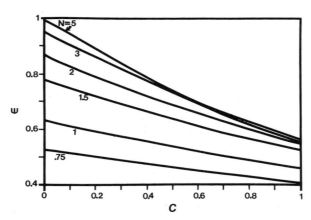

FIGURE 11-5-5. Effectiveness of a crossflow heat exchanger with both fluids mixed.

One way of making an initial estimate is to average over temperatures that bound the true average. For example, if we know the inlet temperatures for both fluids, we know that the average temperature must be between these limiting values. Thus, the average of the inlet temperatures might be a basis for an initial guess for the purposes of property evaluation.

Another potential requirement for assumption requiring checking is that the specifications of the problem may not make it obvious which fluid will be the minimum fluid. Therefore, it may be necessary to assume which will be the minimum fluid and to check upon carrying out the solution whether the assumption was correct. This will be illustrated in Example 11-4.

A word of caution is appropriate in regard to how to interpret the concept of effectiveness. We may interpret the observation that one heat exchanger is more effective than another one having the same inlet temperatures, heat capacities, and heat transfer area to mean that more heat is exchanged in the more effective heat exchanger. Suppose we reduce the flow rate of the minimum fluid, reducing C_{min}. The value of $N = UA/C_{min}$ will increase. (The value of U will decrease, since heat transfer depends on Reynolds number, but to a power less than 1. The heat transfer coefficient for the second fluid does not change except through temperature effects on properties.) For all heat exchanger types, Figs. 11-5-1–11-5-5 show that effectiveness generally increases with N. Thus, although the amount of heat transferred is reduced when the flow rate is lowered, the effectiveness increases. By changing flow rates, we change the basis for comparison. A lower flow rate of fluid can result in a larger temperature change (thereby yielding a higher effectiveness) while giving less heat transfer (which is based on the product of flow rate and temperature change).

It is possible to relate the concepts of effectiveness and F factor. It may be shown (the proof is left as a problem at the end of the chapter) that if a counterflow heat exchanger and another type of heat exchanger operate with the same inlet and outlet temperatures, the same fluids, and the same flow rates, then each of these two heat exchangers will have the same effectiveness, the same value of C, and the F factor will be the ratio of the NTU of the counterflow exchanger to the NTU of the other exchanger. Thus, it is possible to evaluate F factors by using the explicit formulas for NTU given in Table 11-5-2.

EXAMPLE 11-3. Let us modify the statement of Example 11-1. We now specify that we know the inlet temperatures of the two fluids to be 40 and 120°C. We specify the water to flow at 0.4 kg/sec, the heat transfer area to be 15 m², and the heat transfer coefficient to be 193.66 W/m² · °C. The hot fluid flow rate is 0.8 kg/sec. We do not know the exit temperature of either fluid. Find the heat transfer and the exit temperatures.

SOLUTION. This is a problem that is suited to application of the effectiveness approach, since we do not have an easy way at the outset to find the log-mean temperature difference. We note that for the hot and cold fluids

$$C_c = \dot{m}_c c_{pc} = (0.4)(4183) = 1673.2$$

$$C_h = \dot{m}_h c_{ph} = (0.8)(2000) = 1600$$

$$C = \frac{1600}{1673} = 0.9564$$

so the hot fluid is the minimum fluid. We evaluate

$$N = \frac{UA}{C_{min}} = \frac{(15)(193.7)}{1600} = 1.816$$

For a counterflow double pipe exchanger

$$\varepsilon = \frac{1 - \exp[-N(1 - C)]}{1 - C\exp[-N(1 - C)]} = \frac{1 - \exp[-(1.816)(0.0437)]}{1 - 0.956\exp[-(1.816)(0.0437)]}$$

$$\varepsilon = 0.654$$

$$q = \varepsilon C_{min}\,\Delta T_{in} = (0.654)(1600)(80) = 83{,}660 \text{ W}$$

We can now get the outlet temperatures:

$$q = C_c \Delta T_c = 1673.2(T_{co} - 40) = 83{,}660$$

$$T_{co} = 40 + \frac{83{,}660}{1673.2} = 40 + 50 = 90°C$$

$$q = C_h \Delta T_n = 1600(120 - T_{ho}) = 83{,}660$$

$$T_{ho} = 120 - \frac{83{,}660}{1600} = 67.7°C$$

EXAMPLE 11-4.   Suppose, in Example 11-3, we were to be given that the exit water temperature had to be 90°C, but that the flow rate of the hot fluid is not known. Find the heat transfer and the flow rate and exit temperature of the hot fluid.

SOLUTION.   This problem may be simple or complicated depending on which fluid is the minimum fluid. Since we do not know in advance the flow rate of the hot fluid, we do not know in advance which fluid is the minimum fluid.

The total heat transfer may be obtained first without knowing which is the minimum fluid from the heat balance

$$q = C_c \Delta T_c = (1673)(50) = 83{,}660$$

If the cold fluid were the minimum fluid, we could find ε by

$$\varepsilon = \frac{C_c \Delta T_c}{C_c \Delta T_{in}} = \frac{50}{80} = 0.625$$

We could evaluate N by

$$N = \frac{UA}{C_c} = \frac{(193.7)(15)}{1673} = 1.737$$

The minimum value of ε occurs when $C = 1$. The counterflow formula goes, upon expanding the formula, near $C = 1$ to

$$\varepsilon = \frac{N}{N+1} = \frac{1.737}{2.737} = 0.634$$

Thus, there is no value of C for which, at $N = 1.737$, ε could be as low as 0.625 and we have a contradiction. Therefore, the hot fluid must be the minimum fluid.

We could proceed iteratively with the ε–NTU method. We know that

$$q = \varepsilon C_{min} \Delta T_{in}$$

so we know that

$$\varepsilon C_{min} = \frac{q}{\Delta T_{in}} = \frac{83,660}{80} = 1046$$

We could, for several values of C_{min} evaluate ε from $\varepsilon = 1046/C_{min}$ and by evaluating $N = UA/C_{min}$ and evaluating ε from Table 11-5-1. At the C_{min} for which the two values of ε match, the value of C_{min} is appropriate. Analyzing the problem this way is well suited to the what-if usage of repetitive spreadsheet calculations. A spreadsheet that may be used is given below. One proceeds by changing C_{min} in B7 until B8 and B14 agree.

An alternative that may be simpler, or that may give a good first guess, would be to apply the log-mean temperature formula with the linear average approximation:

$$\Delta T_{lm} = \frac{q}{UA} \frac{83,660}{(193.66)(15)} = 28.8$$

	A	B
1	EFFECTIVENESS	ITERATION
2	DELTA T	80
3	Q	83660
4	U	193.7
5	A	15
6	CMAX	1673.2
7	CMIN GUESS	ENTER VALUE
8	FIRST EPS	+ B3/(B2 * B7)
9	N	+ B4 * B5/B7
10	N(1 - C)	+ B9 * [1 - (B7/B6)]
11	EXP	@ EXP (-B10)
12	NUMERATOR	1 - B11
13	DENOM	1 - [(B7/B6) * B11]
14	SECOND EPS	+ B12/B13

Next let

$$\Delta T_{lm} \sim \tfrac{1}{2}(\Delta T_1 + \Delta T_2) \approx \tfrac{1}{2}(T_{h,\text{in}} - T_{c,\text{out}} + T_{h,\text{out}} - T_{c,\text{in}})$$

$$28.8 = \tfrac{1}{2}(120 - 90 + T_{h,\text{out}} - 40)$$

$$T_{h,\text{out}} = 67.7$$

This turns out to be very close to the true answer. We may then estimate C_{min} by

$$C_{\text{min}}(120 - 67.7) = 83{,}660$$

or $C_{\text{min}} = 1599.6$. We could accept this answer or iterate further in our spreadsheet.

This example serves to illustrate that not all problems can be solved easily. It demonstrates that the most effective way to approach a heat exchanger problem may not be evident at the outset and that the most effective way can involve a combination of the log-mean temperature and effectiveness approaches.

11-6. PRACTICAL OPERATING CONSIDERATIONS— FOULING FACTORS

Equipment operating in realistic situations is subject to potential deterioration of performance. In a heat exchanger, for example, surfaces may corrode or deposits may form on surfaces. This deterioration may be progressive or it may reach a limiting value.

To deal with the prospect of deterioration, we define a "fouling factor"

$$R_f = \frac{1}{U} - \frac{1}{U_N} \tag{11-52}$$

where U_N is the overall heat transfer coefficient when the equipment was new. Factors suggested for use with particular fluids are listed in Table 11-6-1.

Fouling causes overdesign of equipment so that acceptable performance can be obtained even in a degraded condition. Incentive therefore exists to try to prevent and/or correct fouling. In addition, the fouling may foretell problems to be expected from corrosion.

TABLE 11-6-1

Fluid	Approximate fouling factor ($\times 10^4$)
Seawater, boiler feedwater	
$\quad T < 50°C$	1
$\quad T > 50°C$	2
River water	1.5
Refrigerant liquids	2
Steam (free of oil)	9
Fuel oil	9

Another practical operating consideration is variability in operating conditions. A given heat exchanger may be required to provide satisfactory performance in various circumstances. For example, the heat exchanger may use ambient air or water from a river as the cold fluid. The temperature of the cold fluid thus may vary considerably over a year and even from day to day. The engineer must anticipate the most limiting conditions that the equipment will encounter and provide for meeting heat removal requirements under those conditions.

Still another practical operating consideration is the likelihood of wear. For example, leaks may develop in heat exchanger tubes. In a heat exchanger with a large number of tubes, this wear may be dealt with by plugging tubes with leaks. In effect, the number of tubes in the heat exchanger is reduced. If it is anticipated that wear will take place, then the heat exchanger should be designed to remove the required heat even if the number of tubes in the heat exchanger were reduced by some specified value.

Thus, the engineer must deal with "design basis" conditions and nominal conditions. The design basis conditions include provision for fouling, variation in temperature of available cooling fluid, plugging of tubes, and so on. Given a heat exchanger suitable for meeting design basis conditions, the engineer must determine how the equipment will behave under nominal or typical conditions.

11-7. VARIABLE PROPERTIES

If the temperature change of one of the fluids in a heat exchanger is large, and if the properties of the fluid vary significantly with temperature, it may not be advisable to use an overall heat exchanger formula based on average temperature. Under such circumstances, one may divide the heat exchanger into sections and evaluate the properties separately in each section. Some ambiguities can exist in evaluating properties as we shall discuss later.

Consider a parallel flow concentric pipe heat exchanger. Consider the first section near the inlet. Treating this section as an entire heat exchanger, we may write

$$\ln\left(\frac{T_{h2} - T_{c2}}{T_{h1} - T_{c1}}\right) = -(UA)_1\left(\frac{1}{\dot{m}_h C_{h1}} + \frac{1}{\dot{m}_c C_{c1}}\right) \qquad (11\text{-}53)$$

The properties for the section 1 area are used for evaluation of U and of specific heats. If the inlet temperatures are known, then the temperature difference $T_{h2} - T_{c2}$ may be evaluated. We then proceed to the second section and evaluate

$$\ln\left(\frac{T_{h3} - T_{c3}}{T_{h2} - T_{c2}}\right) = -(UA)_2\left(\frac{1}{\dot{m}_h C_{h2}} + \frac{1}{\dot{m}_c C_{c2}}\right) \qquad (11\text{-}54)$$

and continue from section to section.

The heat transferred in the first section is

$$q_1 = (UA)_1 \Delta T_{ml} = (UA)_1 \frac{\Delta T_2 - \Delta T_1}{\ln(\Delta T_2/\Delta T_1)} \qquad (11\text{-}55)$$

Since this heat is removed from the hot fluid and added to the cold,

$$q_1 = \dot{m}(T_{h2} - T_{h1}) = \dot{m}_c(T_{c1} - T_{c2}) \qquad (11\text{-}56)$$

Thus, as we march out through the heat exchanger, we obtain the increments of heat transferred, and the increments in temperature in addition to the change in temperature difference between the hot and cold fluids.

The above process yields the total heat transferred in a heat exchanger of specified area with specified inlet temperatures and flow rates. The process gets more complicated when other, more elaborate heat exchangers are considered. For a counterflow heat exchanger, for example, the inlet temperatures are on opposite sides of the exchanger. Thus, if inlet temperatures are specified, the process becomes iterative. Complications increase in dealing with shell-and-tube and cross-flow heat exchangers.

One question of interest is whether the variable property analysis is justified by the accuracy with which the heat transfer coefficient is known. Heat transfer correlations in relatively well-defined geometries can have considerable uncertainty, and particular heat exchanger conditions, including baffles for mixing of shell-side fluids, can introduce additional uncertainty.

11-8. AVERAGING AND EVALUATING PROPERTIES

The analysis performed thus far for parallel flow and counterflow heat exchangers assumed uniform parameters. There are two sources of nonuniformity which are considered here—nonuniformity of the heat transfer coefficient and nonuniformity of the temperature (and therefore of physical properties).

As found above, for a parallel flow heat exchanger, the temperature difference between hot and cold fluids is greatest at the inlet and tends to decline exponentially with distance along the heat exchanger. Therefore, heat transfer tends to be higher near the inlet than near the outlet. In addition, because of entrance effects, we expect the heat transfer coefficients on both hot and cold sides to be higher near the inlet than near the outlet. Thus, the nonuniformity of the heat transfer coefficient is significant.

Let us recall the equation for temperature difference

$$\frac{d}{dx}(T_h - T_c) = -\left(\frac{1}{C_c} + \frac{1}{C_h}\right) PU(x)(T_h - T_c) \qquad (11\text{-}57)$$

We now note that the overall heat transfer coefficient can be a function of position. We may integrate this equation to obtain

$$\ln\left(\frac{T_{h2} - T_{c2}}{T_{h1} - T_{c1}}\right) = -\left(\frac{1}{C_c} + \frac{1}{C_h}\right)PL\left(\frac{1}{L}\right)\int_0^L U(x)\,dx \qquad \text{(11-58)}$$

Note that we have multiplied and divided by L on the right-hand side. If we define the average overall heat transfer coefficient by

$$\bar{U} = \frac{1}{L}\int_0^L U(x)\,dx \qquad \text{(11-59)}$$

then the results are obtained for the parallel flow heat exchanger (and, by a similar development, for the counterflow case) if U is replaced by $\bar{U}$.

Note, however, that heat transfer correlations tend to provide us with average coefficients on each side separately. Use of these coefficients does not yield the average $\bar{U}$. For the concentric pipe,

$$U(x)P = \frac{2\pi r_i}{1/h_i(x) + r_i\ln(r_o/r_i)/k + (r_i/r_o)[1/h_o(x)]} \qquad \text{(11-60)}$$

In principle, evaluation of $\bar{U}$ should involve integration of the right-hand side with explicit local dependence of $h_i(x)$ and $h_o(x)$. In practice, however $\bar{U}$ frequently is evaluated by using $\bar{h}_i$ and $\bar{h}_o$.

Whether this assumption is reasonable depends on the circumstances. In counterflow heat exchangers, for example, flow rates may be selected so that temperature difference and therefore heat transfer are nearly uniform, thereby reducing the relative influence of the zone near the entrance compared with parallel flow. In addition, the tube may have a high ratio of length-to-diameter, again reducing the influence of the zone near entry.

Evaluation of the heat transfer coefficients h_i and h_o requires knowledge of the average bulk temperatures of the hot and cold fluids. In earlier chapters, we assumed for simplicity that the average bulk temperature is simply the average of inlet and outlet temperatures. However, because temperature variation may be highly nonuniform, particularly in parallel flow, some consideration of this question may be in order.

The hot fluid will drop in temperature (assuming no change in phase) according to

$$C_h\frac{dT_h}{dx} = -UP(T_h - T_c) \qquad \text{(11-61)}$$

A similar expression can be written for the cold fluid. We have obtained the position-dependent solution for $T_h - T_c$ for the case of U being constant, which yields

$$\frac{dT_h}{dx} = -\frac{U}{C_h}\frac{dA}{dx}(T_h - T_c)_1 e^{-(1/C_h \pm 1/C_c)UPx} \qquad \text{(11-62)}$$

where the plus and minus signs in the exponential correspond to parallel flow and counterflow. The case of nonuniform $U(x)$ is left as a problem at the end of the chapter. This equation is of the form

$$\frac{dT_h}{dx} = -K_1 e^{-K_2 x} \tag{11-63}$$

which may be integrated to yield

$$T_h = T_{h1} - \frac{K_1}{K_2}(1 - e^{-K_2 x}) \tag{11-64}$$

The average temperature may be obtained by

$$\overline{T}_h = \frac{1}{L}\int_0^L dx\, T_h(x) \tag{11-65}$$

$$\overline{T}_h = T_{h1} - \frac{K_1}{K_2}\left[1 - \frac{1}{K_2 L}(1 - e^{-K_2 L})\right] \tag{11-66}$$

If we were to evaluate average temperature by a linear average of inlet and outlet, we would get

$$\overline{T}_h' = \tfrac{1}{2}(T_{h1} + T_{h2}) = T_{h1} - \frac{1}{2}\frac{K_1}{K_2}(1 - e^{-K_2 L}) \tag{11-67}$$

These two averages are different. For counterflow, where K_2 may be small, the linear average may be a good approximation. By expending exponentials, we may obtain

$$1 + \frac{1}{K_2 L}(1 - e^{-K_2 L}) \approx 1 \frac{1 - [1 - K_2 L + \tfrac{1}{2}(K_2 L)^2]}{K_2 L} = \frac{1}{2}K_2 L \tag{11-68}$$

$$\overline{T}_h \approx T_{h1} - \frac{K_1}{2K_2}(K_2 L) \approx T_{h1} - \frac{1}{2}\frac{K_1}{K_2}(1 - e^{-K_2 L}) = \overline{T}_h' \tag{11-69}$$

On the other hand, there are circumstances where K_2 would not be expected to be small. For example, if one fluid were undergoing a phase change, that temperature would stay constant while the temperature of the other fluid changed exponentially. Then the linear average might not be appropriate.

Another consideration is that the conditions of flow may not correspond precisely to the conditions under which the standard correlations were obtained. As noted earlier in this chapter, baffles may be introduced to enhance shell-side heat transfer in a shell-and-tube heat exchanger. Heat transfer will then be influenced by baffle configuration and spacing.

Shell-side heat transfer calculations are complicated by the fact that ideal conditions may not prevail. While the baffles may be designed to promote cross-flow, pure cross-flow is not achieved. One way of approaching heat exchanger analysis is to evaluate the heat transfer coefficient according to

$$h_{\text{shell-side}} = h_{\text{ideal}} f_c f_L f_B f_s f_E \qquad (11\text{-}70)$$

The correction f_c relates to baffle cut and spacing. The baffles will occupy a certain fraction of the cross-section of the shell and will be a certain distance apart. A well-designed heat exchanger may achieve a value of f_c near 1.

The correction f_L relates to leakage. There are holes in the baffle plates through which the tubes pass, for example. Some of the shell-side flow will pass through the clearances of these holes. The value of f_L is typically in the range 0.7–0.8.

The correction f_B relates to bypass flow. Some shell-side fluid may flow around the outside of the tube bundle while contributing effectively to heat transfer from the tubes. This correction is typically in the range of 0.7–0.9.

The correction f_s applies if baffle spacings near the inlet and outlet differ from spacings in the remainder of the heat exchanger. This can come about from practical design considerations. Thus, the effective Reynolds number for cross-flow in the end sections may be lower than in the remainder of the exchanger, with an adverse effect on average heat transfer coefficient. This factor usually is in the range 0.85–1.

The correction f_E relates to entrance effects when low Reynolds number laminar flow exists in cross-flow over a bank of tubes; f_E will be near 1 for most applications.

Procedures for estimating these correction factors have been developed in the heat exchanger literature. The overall product of these correction factors is typically about 0.6. The reader is referred to the literature on heat exchanger design for more details on this subject. For the purpose of this text, we assume, for the shell-side of a baffled shell-and-tube heat exchanger,

$$h = 0.6 h_{\text{ideal}} \qquad (11\text{-}71)$$

The ideal heat transfer coefficient is based on the flow being cross-flow.

In calculating the ideal cross-flow heat transfer, the flow area should be based on the baffle spacing when calculating the Reynolds number to use for a bank of tubes. This is because each cross-flow pass between baffles may be looked on as a separate heat transfer problem for flow over a bank of tubes.

PROBLEMS

11-1. Consider Example 11-1. Suppose that the outlet temperature of the hot fluid were specified to be 70°C. What would the flow rate be relative to that of the cold fluid?

11-2. Suppose that you wish to study your design options relative to Example 11-1 given a fixed objective (flow rate, temperature rise) for the water. Tabulate or plot the log-mean temperature difference and the hot fluid outlet temperature versus flow rate (relative to the water flow rate).

11-3. Relative to the pipe length of the heat exchanger in Example 11-1, what would the pipe lengths be for the cases considered in the table or plot of Problem 11-2, assuming that the heat transfer coefficient U does not change (a significant assumption).

11-4. Suppose that you wish to use parallel flow in Example 11-1. The hot fluid inlet temperature is still 120°C. Tabulate or plot, as a function of flow rate of hot fluid, the outlet temperature of the hot fluid, the log-mean temperature difference, and the pipe length (relative to that of Example 11-1). Note any restriction on the allowed range of flow rates.

11-5. In Example 11-2, it was found necessary to modify specifications in order to use a cross-flow heat exchanger. Suppose that instead of changing the outlet temperature of the water to obtain a feasible heat exchanger, you were to change the outlet temperature of the hot fluid. Tabulate or plot the F factor as a function of hot fluid outlet temperature. For each outlet temperature, determine the log-mean temperature and the fluid flow rate relative to that of the cold fluid. Examine your results and suggest a range of values of hot fluid outlet temperature within which you would be likely to select a design point.

11-6. Suppose that you wish to use a shell-and-tube heat exchanger to heat the water of Example 11-1. The hot fluid temperature is fixed at 120°C, but you may vary flow rate and therefore outlet temperature of the hot fluid. Tabulate or plot the F factor and the hot fluid flow rate as a function of the hot fluid outlet temperature. Also tabulate or plot the log-mean temperature and the hot fluid flow rate. Examine your results and suggest a range of values within which you would be likely to select a design point.

11-7. Consider Example 11-1.
a. Which fluid is the minimum fluid?
b. What is the effectiveness of the heat exchanger?
c. What is the NTU value of the heat exchanger?

11-8. Consider the heat exchanger obtained in Example 11-2.
a. Which fluid is the minimum fluid?
b. What is the effectiveness of the heat exchanger?
c. What is the NTU value of the heat exchanger?

11-9. Show that each of the effectiveness formulas in Table 11-5-1 reduces to Eq. (11-51) when C goes to 0.

11-10. Show that for a counterflow heat exchanger

$$\lim_{C \to 1} \varepsilon = \frac{N}{N+1}$$

11-11. For a heat exchanger of interest and a counterflow heat exchanger operating with the same fluids, flow rates, and inlet and outlet temperatures, show the following:
a. The effectiveness is the same for each heat exchanger.
b. C is the same for each heat exchanger.
c. F is given by the ratio of the NTU of the heat exchanger of interest.

11-12. Use the formulas of Table 11-5-2 to evaluate the F factor for several points in the plot of Fig. 11-4-4. Verify that the results are the same.

11-13. Suppose, in examining the results in Example 11-3, that you wish to improve on the effectiveness of the heat exchanger by increasing heat transfer area. Tabulate or plot the effectiveness with increasing area.

11-14. Suppose in Problem 11-13 that you select an exchanger with a heat transfer area double that of Example 11-3.
   a. What is the effectiveness?
   b. How much heat will be transferred?
   c. What will be the outlet temperatures of the hot and cold fluids?

337

Heat Exchangers

11-15. Suppose, as in Problem 11-14, that you double the area to increase effectiveness, but you wish to keep the heat transferred to the water fixed. By how much could you change the inlet temperature of the hot fluid?

11-16. Repeat Example 11-4 with the heat transfer area doubled. Assume that the heat transfer coefficient is the same.

11-17. For specified flow rates and specific heats, find the maximum possible effectiveness for the heat exchangers of Table 11-5-1.

11-18. Assume for Example 11-3 that the tube is very thin and that the heat transfer coefficients for the inner and outer pipes are equal, that is Eq. (11-2) can be written

$$UP = \frac{2\pi r}{1/h_i + 1/h_o} \qquad h_i = h_o$$

Suppose that the flow rate of the cold fluid were doubled. Repeat Example 11-3 for this case. Assume that the reference flow is turbulent.

11-19. Starting from the same assumption as in Problem 11-18, suppose that the flow rate of the hot fluid were doubled. Repeat Example 11-3 for this case. Assume that the reference flow is turbulent.

11-20. Assuming properties and the heat transfer coefficient to be uniform derive an expression for the cumulative heat transferred as a function of distance from the inlet of a parallel flow heat exchanger. Also obtain an expression for the fraction of the total heat transfer as a function of position.

11-21. Repeat Problem 11-20 for a counterflow heat exchanger.

11-22. Using the results of Problem 11-20, find the fraction of the total energy transfer that takes place in the first 25%, 50%, and 75% of the parallel flow heat exchangers considered in Problem 11-4.

11-23. Using the results of Problem 11-21, find the fraction of the total energy transfer that takes place in the first 25%, 50%, and 75% of the counterflow heat exchanger of Examples 11-1 and 11-3.

11-24. The general form of the solution for the hot fluid temperature in a double pipe heat exchanger is given in Eq. (11-64). Obtain explicit expressions for hot fluid temperature for
   a. a counterflow heat exchanger; and
   b. a parallel flow heat exchanger.

11-25. Extend the result of Problem 11-24 to obtain explicit expressions for cold fluid temperature in counterflow and parallel flow heat exchangers.

11-26. Apply the expressions obtained in Problems 11-24 and 11-25 to find the hot and cold fluid temperatures as functions of the fraction of total length for the counterflow heat exchanger of Examples 11-1 and 11-3.

11-27. Apply the expressions obtained in Problems 11-24 and 11-25 to find the hot and cold fluid temperatures as functions of the fraction of total length for one of the parallel flow heat exchangers considered in Problem 11-4.

11-28. Using the same type of heat transfer coefficient assumption as in Problem 11-18, repeat Problem 11-4.

11-29. Using the same type of heat transfer coefficient assumption as in Problem 11-18, repeat Problem 11-5.

11-30. Using the same type of heat transfer coefficient assumption as in Problem 11-18, repeat Problem 11-6.

REFERENCES

1. Anonymous, *Standards of the Tubular Exchanger Manufacturers Association*, 6th ed., TEMA, New York, 1978.

2. J. M. Chenworth and M. Impagliazzo (eds), *Fouling in Heat Exchange Equipment*, ASME Symposium Vol. HTD-17, ASME, New York, 1981.

3. J. E. Coppage and A. L. London, "The Periodic Flow Regenerator: A Summary of Design Theory," *Trans. ASME*, **75**, 779 (1953).

4. A. P. Fraas and M. N. Ozisik, *Heat Exchanger Design*, Wiley, New York, 1964.

5. H. Hansen, *Heat Transfer in Counterflow, Parallel Flow and Crossflow*, McGraw-Hill, New York, 1983.

6. J. P. Holman, *Heat Transfer*, 5th ed., McGraw-Hill, New York, 1981.

7. S. Kakac, A. E. Bergles, and F. Mayinger (eds.), *Heat Exchangers*, Hemisphere, Washington, DC, 1981.

8. W. M. Kays and A. L. London, *Compact Heat Exchangers*, 3rd ed., McGraw-Hill, New York, 1980.

9. D. Q. Kern, *Process Heat Transfer*, McGraw-Hill, New York, 1950.

10. S. S. Kutateladze and V. M. Borishanskii, *A Concise Encyclopedia of Heat Transfer*, Pergamon, New York, 1966.

11. E. U. Schlunder (ed.), *Heat Exchanger Design Handbook*, 5 volumes, Hemisphere, Washington, DC, 1982.

12. F. W. Schmidt and A. J. Willmott, *Thermal Energy Storage and Regeneration*, Hemisphere, Washington, DC, 1981.

13. E. F. C. Somerscales and J. G. Knudsen, *Fouling of Heat Transfer Equipment*, Hemisphere, Washington, DC, 1980.

14. G. Walker, *Industrial Heat Exchangers: A Basic Guide*, Hemisphere, Washington, DC, 1982.

Chapter Twelve

HEAT TRANSFER ANALYSIS AND DESIGN PROBLEMS

12-1. INTRODUCTION

In the bulk of this text, emphasis has been on individual mechanisms of heat transfer. Where multiple mechanisms were encountered, they were typically in sequence, that is, a conduction problem with a convection boundary condition. It was always clear what mechanism of heat transfer applied. In real situations, it is not always obvious which mechanism will be predominant, and, indeed, more than one mechanism may be pertinent.

12-2. HEAT LOSS FROM BUILDINGS

Heat transfer problems with conduction were considered in which layered walls characterized the exterior of a building. In particular, an example was considered in which the wall was represented as an inner layer (e.g., paneling), a middle layer of insulation, and an outer layer (e.g., brick). This is a simplification in several respects. One simplification is that we neglected the fact that air enters and leaves the building because doors open and close (see Fig. 12-2-1).

Suppose that there is an average flow rate W kg/sec entering the house bringing air at a temperature of T_∞, and an equal average flow W leaving the house at temperature T_i (i denoting inside). If the average specific heat is C, then the average loss by convection is $WC(T_i - T_\infty)$. The total heat loss is the sum of this and the conduction heat loss; that is,

$$q = UA(T_i - T_\infty) + WC(T_i - T_\infty) \qquad (12\text{-}1)$$

The average flow rate is sometimes characterized in terms of air change (AC) per unit time (e.g., 1 air change per hour). To relate air change per unit time (AC) to mass flow per unit time, we convert via volume and density. Air change per hour would convert to flow rate by

$$W = \text{AC } V(\text{m}^3) \times \frac{\text{hr}}{3600 \text{ sec}} \times \rho(\text{kg/m}^3) \qquad (12\text{-}2)$$

Thus, the total heat loss would be

$$q = UA(T_i - T_\infty) + \frac{\text{AC}}{3600}\rho V(T_i - T_\infty) \qquad (12\text{-}3)$$

We may define an effective overall heat loss coefficient by

$$U' = U + \frac{\text{AC}}{3600}\rho C\frac{V}{A} \qquad (12\text{-}4)$$

In Chapter 2, Example 2-2, we obtained a $U = (1/2.44) = .41$. Consider a structure with a flat roof and no attic that is 10 m wide, 20 m long, and 8 m high. Assume that the floor is insulated. The surface area is

$$A = (10)(20) + 2(8)(10) + 2(8)(20) = 680 \text{ m}^2 \qquad (12\text{-}5)$$

The volume is

$$V = (10)(20)(8) = 1600 \text{ m}^3 \qquad (12\text{-}6)$$

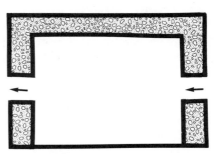

FIGURE 12-2-1.

Take the density of air as about 1.2 kg/m³ and the specific heat as 1000. We then get, for 1 air change per hour,

$$U' = .41 + \left(\frac{1}{3600}\right)(1.2)(1000)\left(\frac{1600}{680}\right) = 1.20 \qquad (12\text{-}7)$$

Thus, there is a significant component of heat loss associated with infiltration of air.

Attempts to reduce heat loss from buildings usually involve both insulation and weatherization. Insulation attempts to reduce U by increasing thermal resistance. Weatherization attempts to reduce AC by eliminating drafts by weatherstripping.

A second simplification in earlier analysis is that we have not considered radiation heat transfer between the building and its surroundings. As noted in Chapter 10, sunshine may bring an average of about 700 W/m² on a clear day (the amount would vary over the year), while on a very clear night, the building may see the atmosphere as a black body at about $0.0552\, T_{air}^{1.5}$. In determining the required capacities of heating and cooling systems, we usually are concerned with incident heat load during the day for cooling purposes and the heat loss at night for heating purposes. Let us consider the night heat loss rate.

Relative to our discussion in Chapter 2, we should change our boundary condition at the wall to

$$q = -kA\frac{dT}{dx}\bigg|_w = hA(T_w - T_\infty) + \frac{E_{bw} - E_{ba}}{R} \qquad (12\text{-}8)$$

where E_{bw} and E_{ba} are the black body radiations associated with outside wall and with atmosphere and R is the resistance for radiation heat transfer,

$$R = \frac{1 - \varepsilon_w}{\varepsilon_w A_w} + \frac{1}{A_w F_{wa}} + \frac{1 - \varepsilon_a}{\varepsilon_a A_a} \qquad (12\text{-}9)$$

We assume the building to be convex and isolated so that the shape factor to the atmosphere is 1. The atmosphere is large so that A_a is essentially infinite. If the outside material is brick, then ε_w is about 0.95 and

$$q = hA_w(T_w - T_\infty) + \frac{(E_{bw} - E_{ba})A_w}{1 + 0.05/0.95} \qquad (12\text{-}10)$$

In the conditions of Example 2-2, where $h = 10$, $T_\infty = -15°C$, and q/A was found to be 14.3 W/m², T_w was (although not evaluated in that example)

$$T_w = T_\infty + \left(\frac{q}{A}\right)\left(\frac{1}{h}\right) = -13.6°C \qquad (12\text{-}11)$$

With radiation, the estimate of the wall temperature will change, but let us use

−13.6°C as a first guess to evaluate E_{bw}

$$E_{bw} = \sigma(273 - 13.6)^4 = 257 \qquad (12\text{-}12)$$

We also find, by estimating the atmosphere temperature to be −45°C,

$$E_{ba} = \sigma(273 - 45)^4 = 153 \qquad (12\text{-}13)$$

This leads to a radiation loss of 98.6 W/m² and implies that on a clear night radiation cooling can be a prominent mode for losing heat from the surface of the building.

With a larger heat loss, it would not be possible for T_w to be −13.6°C. Indeed, the heat loss estimated would imply an unrealistically high temperature drop across the wall. Iteration is needed to arrive at consistency. We will have a situation where a greater heat loss implies a lower wall temperature. The wall temperature actually can get lower than the temperature of the outside air, which, therefore, would provide heat to the wall by convection while the wall is radiating to the atmosphere. (This is why a car parked outdoors at night feels very cold when you get into it.)

We proceed by iteration with the following steps:

1. Guess T_w.
2. Calculate E_{bw} and the radiant heat loss.
3. Calculate the heat loss from the building.
4. Calculate the heat provided to the building by convection.
5. See if Steps 2, 3, and 4 balance.
6. If not, adjust T_w and repeat.

This may be performed with the spreadsheet below.

	A	B
1	BUILDING	HEATLOSS
2	ORIGINAL Q/A	14.3
3	ORIGINAL TW	259.4
4	ORIGINAL TI	293
5	ORIGINAL DELT	+B4 − B3
6	T INFINITY	258
7	H	10
8	SIGMA	5.669X10^{-8}
9	TA	203
10	EBA	(B9 ∧ 4)*B8
11	RESISTANCE	1.053
12	TW	ENTER VALUE
13	EBW	(B12 ∧ 4)*B8
14	Q RAD	(B13 − B10) / B11
15	BUILDING LOSS	+B2*(B4 − B12) / B5
16	CONVECTION IN	(B6 − B12)*B7
17	CHECK	B15 − B16 − B14

If, on a particular guess of T_w, entry B17 is negative, then the wall temperature is not low enough and should be reduced. Similarly, if B17 is positive, the wall temperature is too low. For our problem, iteration on the spreadsheet leads quickly to $T_w = 252.4°K$ and a building heat loss of 17.3 W/m². The heat loss is 20% greater than would have been predicted without considering radiation.

It is interesting to note that while the changes in wall temperature and building heat loss are moderately significant, the actual absolute transfers by radiation and convection can be fairly large. With the assumed parameters for this case, the radiation is about 93 W/m² and the convection gain is about 76 W/m², both larger than the net building loss of about 17 W/m².

The existence of surface temperatures below ambient air temperatures because of radiation cooling is a phenomenon of practical importance in other connections. In agriculture, for example, this phenomenon can lead to freezing of crops in the fall on nights when the air temperature does not drop to the freezing temperature of water.

There are other simplifications that we have made. The walls of the building are not likely to be uniform. In particular, there are likely to be windows or glass doors covering some fraction of the surface. These windows will provide a parallel heat transfer path between the inside and the outside. In cases where the window is a single pane of glass, the thermal resistance provided can be small and a moderate fraction of the surface area of the building could provide for a substantial fraction of total heat loss. On the other hand, the window can be more complex involving two or more layers of glass (either as a thermopane window or by use of a storm window). Again, there will be the possibility of multiple modes of heat transfer. There can be conduction across the air gap between the panes and radiation across the gap.

This section is meant to illustrate that in heat transfer in general, as in convection problems in particular, it is important to ask what kind of problem we have. It should not necessarily be taken for granted that a particular mode of heat transfer will predominate. In the experience of this author, the relative roles of convection (via air changes) and conduction and the relative roles of convection and radiation on the outside are not found to be intuitively obvious. The reader would be well advised, when in actual engineering situations, to test for the relative influence of the possible routes of heat transfer.

12-3. HEAT LOSS FROM PIPING

In an industrial plant, hot fluid may be transported via piping from one area to another. For example, steam may be taken from a boiler to a turbine. We pose the following problem. We have in an area with ambient temperature of 300°K a long 15 cm inner diameter pipe made of 5 cm thick steel ($k = 30$) containing steam at 800°K. We have available pipe wrapping insulation

($k = 0.07$) in 5 cm thick layers. We wish to evaluate the maximum heat loss per unit length of pipe for various thicknesses of insulation. We also wish to know the maximum surface temperature at the outside of the insulation since we wish to avoid hazards to personnel in the plant. Because personnel protection is involved, we would like to have some conservative assumptions made in the analysis. We have the option of placing thin aluminum foil ($\varepsilon = 0.04$) around the outside of the insulation, so we wish to investigate the merits of this possibility. The ε of the insulating material itself is 0.9. (See Fig. 12-3-1.)

To allow for uncertainty, especially in the estimation of convection heat transfer coefficients, we assume that the inside convection coefficient is infinite; that is, the inside pipe wall temperature is 800°K, the maximum steam temperature. This assumption is conservative, since the actual inside temperature will be lower. As a problem at the end of the chapter, you will be asked to inquire into this assumption. In addition, we neglect the temperature drop in the steel pipe and assume that the 800°K temperature prevails at the inside of the insulating layer. This assumption is also conservative, although the temperature drop in the steel is expected to be small ($\Delta r/k$ for the steel layer will be much smaller than $\Delta r/k$ for the insulation).

At the surface of the insulated pipe, heat is lost according to

$$q = \pi DLh(T_w - T_A) + \frac{E_{bw} - E_{bA}}{R} \tag{12-14}$$

where w and A denote wall and ambient conditions. The radiation resistance is

$$R = \frac{1 - \varepsilon_w}{\varepsilon_w A_w} + \frac{1}{A_w F_{wA}} + \frac{1 - \varepsilon_A}{\varepsilon_A A_A} \tag{12-15}$$

The room is assumed to be large (A_A essentially infinite) and the shape factor F_{wA} is taken as 1, so

$$q = \pi DLh(T_w - T_s) + \pi DL \frac{\varepsilon_w \sigma (T_w^4 - T_A^4)}{\varepsilon_w} \tag{12-16}$$

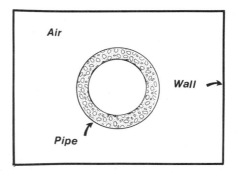

FIGURE 12-3-1.

and the heat loss per unit length is

$$\frac{q}{L} = \pi D \left[h(T_w - T_A) + \varepsilon\sigma(T_w^4 - T_A^4) \right] \qquad \text{(12-17)}$$

In addition, from Chapter 2, we note that for the conduction across the insulator

$$\frac{q}{L} = \frac{2\pi k}{\ln(D/D_i)}(T_i - T_w) \qquad \text{(12-18)}$$

The two expressions for q/L yield a single nonlinear equation for T_w:

$$D\left[h(T_w - T_A) + \varepsilon_w\sigma(T_w^4 - T_A^4) \right] = \frac{2\pi k}{\ln(D/D_i)}(T_i - T_w) \qquad \text{(12-19)}$$

In this equation the diameter D is given by

$$D = D_i + 2n\Delta \qquad \text{(12-20)}$$

where n is the number of insulating strips of thickness Δ that are used.

We have a parametric design survey to conduct. We wish to select several values of n. For each value of n, we consider two values of ε_w (depending on whether aluminum foil is used). For each case, we have to solve a nonlinear equation for T_w.

Note that in this problem the ambient air temperature and the ambient temperature for the room surfaces are the same. Thus, this problem differs from the one in Section 12-2. We know that T_w must be between T_i and T_A. We can set the problem up on the spreadsheet below and adjust T_w until the left-hand side and right-hand side balance. The spreadsheet below is for one layer without foil, but modification for the other options is straightforward.

The heat transfer coefficient depends on wall temperature. We use the general correlation

$$\text{Nu} = \left\{ 0.60 + \frac{0.387\, \text{Ra}^{1/6}}{\left[1 + (0.559/\text{Pr})^{9/16} \right]^{8/27}} \right\}^2 \qquad \text{(12-21)}$$

We previously (Chapter 8) prepared a spreadsheet that treats this correlation, so we incorporate it directly into the spreadsheet we construct. We may rewrite our heat flow equation in the form

$$\text{Nu}(T_w - T_A) + \frac{\varepsilon_w\sigma D}{k_A}(T_w^4 - T_A^4) = \frac{2k}{\ln(D/D_i)k_A}(T_i - T_w) \qquad \text{(12-22)}$$

We begin with the spreadsheet Columns A and B containing the problem data. We then use Columns C, D, E, and F for the Nusselt number spreadsheet of Example 8-4. We then follow with Columns G and H.

TABLE 12-3-1
Pipe Design Results

Number of layers	1	1	2	2
Foil	No	Yes	No	Yes
Radiation loss (W/m)	210	17	140	12
Convection loss (W/m)	186	357	109	228
Total loss (W/m)	396	374	249	240
Surface temperature (°K)	340	366	321	337

In performing the iteration with this spreadsheet, time may be saved by evaluating properties at the film temperature associated with the initial guess, and keeping those properties until a close to satisfactory answer has been obtained and then reevaluating the properties. This was found to be satisfactory in this problem. A more general approach would be to enter a table of key properties, and use the @ LOOKUP function and interpolation in the spreadsheet itself to upgrade the properties. This is left as an exercise.

We find from our results that radiation from the surface in the absence of foil for one strip of insulation yields 53% (i.e., slightly over half) of the heat loss. Thus, when dealing with natural convection problems, we should, as a matter of course, test for the influence of radiation. It is interesting to note that when we place foil over the insulation, the radiation loss drops by about 90%, but the convection loss almost doubles. The total reduction in heat loss from 396 to 374 W/m is a moderate percentage of the heat loss without the foil. In addition, the surface temperature increases from 340 to 366°K. Thus, the measure taken to reduce heat loss conflicts with the objective for worker safety of lowering surface temperature.

The design problem called for considering more than one strip of insulation. Repeating with a second strip yields a reduction of surface temperature to 321°K without foil and to 337°K with foil. Heat loss is reduced to 249 W/m without foil and to 240 W/m with foil. (See Table 12-3-1.)

	A	B
1	PIPING	PROBLEM
2	HEAT	LOSS
3	DI	.15
4	N	1
5	DELTA	.05
6	D	(2*B5*B4) + B3
7	EPSILON	.9
8	SIGMA	5.669 E - 8
9	K	.029
10	D / DI	+B6 / B3
11	LN (D/DI)	@ LN (B10)
12	K INSULATE	.07

```
13      AMBIENT T    300
14      INSIDE T     800
15      RAD COEFF    +B7*B8*B6 / B9
16      TA ∧ 4       +B13 ∧ 4


        G                H
1       HEAT FLOW     EQUATION
2       NUSSELT       +F11
3       TW
4       CONV          (H3 − B13)*H2
5       TW ∧ 4        +H3 ∧ 4
6       RAD TERM      (H5 − B16)*B15
7       COND TERM     (B14 − H3)*2 / B11
8       NET           +H4 + H6 − H7
9       HEAT LOSS     +H7*B9* @ PI
10      CONV. LOSS    +H4*B9* @ PI
11      RAD. LOSS     +H6*B9* @ PI
```

12-4. FLAT PLATE SOLAR COLLECTOR

In Chapters 8 and 10, examples were considered that could be related to flat plate solar energy collection. In Chapter 8, Example 8-5, heat transfer by natural convection was considered for the enclosure between the hot black surface and the glass. In Chapter 10, Example 10-5, net heat transmission by radiation through a window was considered.

In an actual situation, even more combinations of phenomena are involved. While the glass plate (as per Example 10-5) transmits very little radiation originating at the hot black surface, it will absorb and re-emit radiation. There will be a radiation heat loss from the glass to the atmosphere. The magnitude of this loss will depend on atmospheric conditions, that is, on what the effective atmospheric temperature appears to be. That effective temperature may range from ambient air temperature to about −45°C which is characteristic of clear conditions. (The incident solar energy will also be affected by atmospheric conditions.)

Heat loss from the glass to the atmosphere will take place by convection. This heat loss will depend on the atmospheric conditions in regard to air temperature and wind speed. The outside glass surface temperature will be affected by conduction thermal resistance of the glass.

The temperature of the hot black surface may be regulated to stay at a certain value, for example, 340°K. The temperature must be high enough to supply heat at a desired temperature to a circulating fluid (e.g., ethylene glycol). Depending on the incident solar radiation and the heat losses, the average flow rate may be varied. If the solar collection does not provide sufficient heat for internal purposes, supplementary heating (gas, electric, etc.) may be provided.

While most of the incident solar energy is transmitted through the glass, a significant amount will be absorbed by the glass. This energy absorption will influence the glass temperature and radiative and convective interactions.

Let us summarize the interactions that take place.

1. There is an incident energy q_s''.
2. By the procedures of Example 10-5, a fraction $f\tau$ of this energy is transmitted through the glass and absorbed on the hot black collector surface.
3. The remaining incident energy $(1 - f\tau)q_s''$ is absorbed on the outside (atmosphere side) of the glass (assuming none is reflected—another possible complication).
4. At the outside surface of the glass, energy is radiated to the atmosphere according to $\varepsilon_g(E_{bgo} - E_{ba})$. Based on Example 10-5, we may assume that the transmitting wavelength interval has only a minor effect on effective ε for the glass temperature.
5. At the outside surface of the glass there is convection to the air according to $h(T_{go} - T_\infty)$.
6. The net heat loss between the outside glass surface and the outside world is thus

$$\varepsilon_g(E_{bgo} - E_{ba}) + h_o(T_{go} - T_\infty) - (1 - f)q_s''$$

7. Heat transfer by conduction through the glass is given by $k_g(T_{gi} - T_{go})/t_g$, where t_g is the thickness of the glass.
8. Convection between the hot black surface and the inside glass surface at temperature T_{gi} is given by the procedures of Example 8-5 in the form $h_i(T_{bs} - T_{gi})$.
9. Radiation heat transfer between the hot black surface and the inside surface of the glass is given by $\varepsilon_g(E_{bs} - E_{bgi})$. (Note that the shape factor between these two surfaces is essentially 1.)
10. The energy interchange between the black surface and the inside glass surface is thus

$$h_i(T_{bs} - T_{gi}) + \varepsilon_g(E_{bc} - E_{bgi})$$

A useful design exercise is to investigate the heat exchange under alternate postulated conditions, for example, different outside temperatures, different wind conditions, different assumed atmospheric effective radiation temperatures, different specified black surface temperatures. It is also appropriate to consider variations that may affect performance, for example, the thickness of air space and glass.

Engineering design also involves economic considerations. Suppose, for example, that the solar collection system entails a certain cost per square meter of surface. The amount of energy collected per square meter will increase as the temperature of the collector surface is reduced (i.e., since losses by convection and radiation are reduced). However, the requirements of the heating system (e.g., home hot water heating, industrial drying, etc.) may entail minimum temperature requirements. Depending on the cost of supplemental energy, there may be an optimum balance.

The 10 steps discussed above provide sufficient information to set up the calculations required. The equations will be nonlinear and will require iteration to solve as in the other problems of this chapter involving both radiation and convection.

12-5. SUMMARY

This chapter illustrates the importance of understanding the nature of the problem being analyzed. It is not always obvious which heat transfer modes apply to a particular situation, and it is useful to test explicitly for applicability of specific modes. In particular, it should be clear that while radiation heat transfer is usually thought of as a high-temperature effect, it can be important at moderate and low temperatures also. It is useful to observe that measures taken to influence one mode of heat transfer may not have a major overall effect. It was found in the piping problem that putting a highly polished surface coating reduced the radiation heat loss dramatically, but because of a resulting increase in surface temperature, it actually increased the convection heat loss so that only a modest net improvement was made.

Sometimes there is a tendency to assume that a particular mode of heat transfer will dominate. This tendency is natural because mixed modes (e.g., radiation and convection) can involve nonlinear equations and iteration. With access to personal computing, however, the more elaborate equations do not carry as great a penalty in effort, as the spreadsheets of this chapter demonstrate.

PROBLEMS

12-1. Consider the building discussed in Section 12-2. Determine the heat loss per unit area accounting for the effects of both air changes and radiation.

12-2. Consider the radiation effect analysis in Section 12-2. Suppose that the parameters given prevailed during the day when, in addition, sunlight provides an effective incident heat flux of 120 W/m^2. What is the building heat loss per unit area?

12-3. Consider the summer cooling requirement of the building in Section 12-2. Find the building heat gain per unit area if the outside temperature is 32°C and the inside temperature is 20°C:
a. By the procedures of Example 2-2, Chapter 2.
b. Accounting for air changes per hour.

12-4. As in Problem 12-3, consider summer cooling, but as affected by incident radiation. Take the incident heat flux as 700 W/m^2.

12-5. Consider the combined effects of air changes per hour and radiation on building heat gain for the conditions of Problems 12-3 and 12-4.

12-6. A road surface has an emissivity of 0.9. The road is long and 5 m wide. The ambient air is at $-15°C$ on a clear winter night. Estimate the surface temperature of the road.

12-7. For the problem discussed in Section 12-3, how small would the heat transfer coefficient at the inside of the pipe have to be to have a 5% effect on the calculated heat loss?

12-8. Consider the road of Problem 12-6 on a hot summer day when the air temperature is $32°C$ and sunlight provides a heat flux of 700 W/m². Estimate the temperature of the road surface.

12-9. Relative to the situation of Problem 12-8, tabulate or plot the estimated road surface temperature with variation in ambient air temperature from 20 to 32°C, assuming a solar heat flux of 700 W/m².

12-10. Relative to the situation of Problem 12-8, tabulate or plot the estimated road surface temperature with variation in solar heat flux from 100 to 700 W/m².

12-11. Relative to the situation of Problem 12-8, tabulate or plot the estimated road surface temperature with variation in road surface emissivity between 0.5 and 0.9.

12-12. For the pipe of Section 12-3, determine the effect on the results of accounting for the conduction in the steel pipe itself.

12-13. Continue the design survey of Section 12-3 and tabulate or plot heat loss and surface temperature versus number of layers of insulation of 0–5 with and without foil. Include a case of foil around the pipe when there is no insulation.

12-14. In Table 12-3-1, it may be observed that with one layer of insulation, the convection loss increases substantially when the foil is added. Determine the extent to which this increase is due to changes in surface temperature and surface area.

12-15. Relative to the conditions of Section 12-3, consideration is given to sending through the pipe steam at various temperatures from 400 to 1000°K. Tabulate the heat loss and surface temperatures for 0–3 layers of insulation as a function of steam temperature.

12-16. Relative to the conditions of Section 12-3, consideration is given to using pipes of different diameters. Determine the variation of the results of Section 12-3 with inner diameter from 7.5 to 20 cm.

12-17. Set up a spreadsheet or other interactive computer procedure for the problem discussed in Section 12-4. Take glass properties and black surface temperature as per Example 10-5. Assume there is no wind outside and that glass thickness is 5 mm. Take the outside air temperature to be $-15°C$.

12-18. Suppose that the purpose of the solar collector of Problem 12-17 is to heat water from 10 to 60°C. Assuming no heat losses once energy is absorbed at the black surface (i.e., no losses from piping, etc.), how much water can be heated per unit time per unit area of collector?

12-19. To reduce heat losses, the solar collector of Section 12-4 is provided with a second glass cover and air space equal in dimensions to the first cover and space. Modify the analysis procedure of Section 12-4 to account for the second cover.

12-20. Extend the spreadsheet of Problem 12-17 to incorporate the model developed in Problem 12-19.

12-21. An empirical expression suggested for use for convective heat loss from a flat plate collector as a function of wind speed u is

$$h = 5.7 + 3.8u$$

Use this to extend Problem 12-17 to account for wind speed. Consider wind speeds in the range of 0–30 m/sec.

12-22. How does the heat transfer of Problem 12-17 change as the outside air temperature varies from $-15°C$ in the winter to $32°C$ in the summer. Consider a constant heat flux of 700 W/m². (Actually, this too would vary.)

12-23. Relative to Problems 12-18 and 12-22, how much water can be heated as a function of outside air temperature?

Appendix A

HEAT TRANSFER DATA

This appendix contains data for use with problems in the text. Data have been gathered from various primary sources and text compilations as listed in the references. Emphasis is on presentation of the data in a manner suitable for computerized database manipulation.

Properties of solids at room temperature are provided in a common framework. Parameters can be compared directly. Upon entrance into a database program, data can be sorted, for example, by rank order of thermal conductivity.

Gases, liquids, and liquid metals are treated in a common way. Attention is given to providing properties at common temperatures (although some materials are provided with more detail than others). In addition, where numbers are multiplied by a factor of a power of 10 for display (as with viscosity) that same power is used for all materials for ease of comparison. For gases, coefficients of expansion are taken as the reciprocal of absolute temperature in degrees kelvin. For liquids, actual values are used. For liquid metals, the first temperature entry corresponds to the melting point.

The reader should note that there can be considerable variation in properties for classes of materials, especially for commercial products that may vary in composition from vendor to vendor, and natural materials (e.g., soil) for which variation in composition is expected. In addition, the reader may note some variations in quoted properties of common materials in different compilations. Thus, at the time the reader enters into serious professional work, he or she may find it advantageous to verify that data used correspond to the specific materials being used and are up to date.

TABLE A-1
Properties of Solids

Material	ρ (kg/m^3)	C_p (J/kg · °C)	k (W/m · °C)	α (m^2/sec) $e^{+0.05}$
Metals and Alloys				
Aluminum	2,707	896	204	8.418
Beryllium	1,850	825	200	5.92
Bismuth	9,780	122	7.86	0.66
Boron	2,500	1,090	30	1.1
Brass	8,522	385	111	3.41
Bronze	8,666	343	26	0.859
Cadmium	8,650	231	96.8	4.84
Carbon steel (1%C)	7,801	473	43	1.17
Cast iron	7,272	420	51.9	1.7
Chrome steel (1%C)	7,913	448	59	1.67
Chromium	7,160	451	93.6	2.9
Cobalt	8,862	419	100	2.7
Cooper	8,954	3,831	386	11.23
Duralumin	2,787	883	164	6.676
Germanium	5,360	318	61.4	3.6
Gold	19,300	129	316	12.7
Inconel-X	8,510	439	11.6	3.1
Iron	7,870	452	81.1	2.28
Lead	11,340	129	129	2.41
Lithium	534	3,560	76	4
Magnesium	1,740	1,017	155	8.76
Manganese	7,430	477	78	2.2
Molybdenum	10,240	255	138	5.3
Nichrome	8,360	430	12.6	0.35
Nickel	8,900	442	90.5	2.3
Platinum	21,450	130	72	2.6
Silicon	2,330	691	153	9.5
Silver	10,500	235	427	17.3
Stainless steel (304)	7,900	477	14.9	3.95
Tantalum	16,600	138	57.5	2.51
Tin	5,750	227	67	5.13
Titanium	4,500	510	22	9.6
Tungsten	19,300	133	178	6.92
Uranium	19,070	116	27	1.2
Vanadium	6,100	486	33	1.1
Wrought iron	7,850	460	57.8	1.6
Zinc	7,140	388	122	4.4
Zirconium	6,570	280	22	1.2
Building Materials				
Bricks				
Chrome	3,000	840	2.2	0.087
Common	1,600	840	0.69	0.052
Fireclay	2,000	960	1	0.052

TABLE A-1 (*Continued*)

Material	ρ (kg/m^3)	C_p (J/kg · °C)	k (W/m · °C)	α (m^2/sec) $e^{+0.05}$
Masonry	1,700	837	0.65	0.046
Concrete	2,300	880	1	0.049
Cement mortar	1,860	780	0.9	0.062
Glasses				
Window	2,700	800	0.84	0.039
Pyrex	2,225	835	1.4	0.075
Plate	2,500	750	1.4	0.075
Plaster gypsum	1,440	840	0.5	0.041
Stones				
Granite	2,640	800	3	0.14
Marble	2,650	1,000	2.7	0.1
Sandstone	2,200	740	2.8	0.17
Woods				
Fir	420	2,720	0.11	0.0096
Maple	540	2,400	0.166	0.0128
Oak	540	2,400	0.166	0.0128
Yellow pine	540	2,400	0.166	0.0128
Plywood	550	1,200	0.12	0.018
Insulating Materials				
Asbestos	383	816	0.113	0.036
Corkboard	160	1,900	0.043	0.014
Glass wool	40	700	0.038	0.14
Glass fiber duct liner	32	835	0.038	0.014
Glass fiber brown	16	835	0.043	0.032
Urethane	70	1,045	0.026	0.036
Vermiculite	80	835	0.068	0.01
Polystyrene (R-12)	55	1,210	0.027	0.04
Miscellaneous Materials				
Ice	920	2,000	2.2	0.12
Soil	2,050	1,840	0.52	0.014
Rubber, hard	1,150	2,009	0.163	0.0062
Paper	930	1,340	0.18	0.014
Diamond, Type IIa	3,500	510	2500	140
Diamond, Type IIb	3,250	510	1350	81
Silicon carbide	3,180	675	490	23
Aluminum oxide	3,970	765	39.5	1.3
Clay	1,460	880	1.3	0.1
Cotton	80	1,300	0.06	0.058

TABLE A-2
Properties of Gases

Name	T (°K)	ρ (kg/m³)	C_p (J/kg · °K)	k (W/m · °K)	α (m²/sec) (×10⁵)	μ (kg/m · sec) (×10⁵)	ν (m²/sec) (×10⁵)	Pr	β
CO_2	200	2.68	759	0.0095	0.467	1.02	0.381	0.814	0.005
	250	2.15	806	0.0129	0.744	1.26	0.586	0.79	0.004
	300	1.79	852	0.0166	1.09	1.5	0.838	0.763	0.0033
	350	1.53	897	0.0205	1.49	1.73	1.13	0.755	0.0028
	400	1.34	939	0.0244	1.94	1.94	1.45	0.747	0.0025
	450	1.19	979	0.0283	2.43	2.15	1.8	0.743	0.0022
	500	1.07	1,017	0.0323	2.96	2.35	2.19	0.74	0.00202
	600	0.894	1,077	0.0403	4.18	2.72	3.04	0.727	0.0016
	700	0.767	1,126	0.0487	5.64	3.06	3.99	0.708	0.00143
	800	0.671	1,169	0.056	7.14	3.39	5.05	0.708	0.00125
	900	0.596	1,205	0.0621	8.65	3.89	6.19	0.716	0.00111
	1,000	0.537	1,235	0.068	10.3	3.97	7.4	0.721	0.001
Argon	200	2.44	524	0.0124	0.973	1.6	0.657	0.674	0.005
	250	1.95	522	0.0152	1.49	1.95	1	0.672	0.004
	300	1.62	522	0.0177	2.07	2.27	1.4	0.669	0.003333
	350	1.39	521	0.0201	2.78	2.57	1.85	0.666	0.002857
	400	1.22	521	0.0223	3.52	2.85	2.34	0.665	0.0025
	450	1.08	521	0.0244	4.33	3.12	2.88	0.665	0.002222
	500	0.974	521	0.0264	5.2	3.37	3.45	0.664	0.002
	600	0.812	521	0.0301	7.12	3.82	4.72	0.662	0.001666
	700	0.696	521	0.0336	9.68	4.25	6.11	0.658	0.001428
	800	0.609	521	0.0369	11.6	4.64	7.62	0.655	0.00125
	900	0.541	521	0.0398	14.1	5.01	9.26	0.654	0.001111
	1,000	0.487	521	0.0427	16.8	5.35	11	0.652	0.001
Ammonia	200	1.04	2,199	0.0153	0.67	0.689	0.663	0.99	0.005
	250	0.831	2,248	0.0197	1.05	0.853	1.03	0.973	0.004
	300	0.692	2,298	0.0246	1.55	1.03	1.48	0.959	0.0033
	350	0.593	2,349	0.0302	2.17	1.21	2.03	0.938	0.0028
	400	0.519	2,402	0.0364	2.92	1.39	2.68	0.917	0.0025
	450	0.461	2,455	0.0433	3.82	1.58	3.42	0.894	0.0022
	500	0.415	2,507	0.0506	4.86	1.76	4.25	0.873	0.002
	550	0.378	2,559	0.058	6	1.95	5.16	0.86	0.0018
	600	0.346	2,611	0.0656	7.26	2.14	6.18	0.852	0.0016
	700	0.297	2,710	0.0811	10.1	2.51	8.45	0.839	0.0014
	800	0.26	2,810	0.0977	13.4	2.88	11.1	0.828	0.0012
	900	0.231	2,907	0.1146	17.1	3.24	14	0.822	0.0011
	1,000	0.208	3,001	0.1317	21.1	3.59	17.3	0.818	0.001
Air	100	3.6	1,027	0.00925	0.25	0.692	0.192	0.77	0.0108
	150	2.37	1,010	0.0137	0.575	1.03	0.434	0.753	0.0066
	200	1.77	1,003	0.0181	1.02	1.34	0.757	0.74	0.005
	250	1.41	1,003	0.0223	1.57	1.61	1.14	0.724	0.004
	300	1.18	1,005	0.0261	2.21	1.85	1.57	0.712	0.00333
	350	1.01	1,008	0.0297	2.92	2.08	2.06	0.706	0.00286
	400	0.883	1,013	0.0331	3.7	2.29	2.6	0.703	0.00251
	450	0.785	1,020	0.0363	4.54	2.49	3.18	0.7	0.00225

TABLE A-2 (*Continued*)

Name	T (°K)	ρ (kg/m³)	C_p (J/kg · °K)	k (W/m · °K)	α (m²/sec) ($\times 10^5$)	μ (kg/m · sec) ($\times 10^5$)	ν (m²/sec) ($\times 10^5$)	Pr	β
Air	500	0.706	1,029	0.0395	5.44	2.68	3.8	0.699	0.002
	550	0.642	1,039	0.0426	6.39	2.86	4.45	0.698	0.00187
	600	0.589	1,051	0.0456	7.37	3.03	5.15	0.698	0.00167
	650	0.543	1,063	0.0484	8.38	3.19	5.88	0.701	0.00154
	700	0.504	1,075	0.0513	9.46	3.35	6.64	0.702	0.0014
	750	0.471	1,087	0.0541	10.6	3.5	7.43	0.703	0.0013
	800	0.441	1,099	0.0569	11.7	3.64	8.25	0.704	0.00123
	850	0.415	1,110	0.0597	13	3.78	9.11	0.704	0.00118
	900	0.392	1,120	0.0625	14.2	3.92	9.99	0.705	0.00111
	950	0.372	1,131	0.0649	15.4	4.05	10.9	0.706	0.00105
	1,000	0.353	1,141	0.0672	16.7	4.18	11.8	0.709	0.001
	1,100	0.321	1,159	0.0717	19.3	4.42	13.8	0.716	0.0009
	12,00	0.294	1,175	0.0759	22.2	4.65	15.8	0.72	0.00083
	1,300	0.272	1,189	0.0797	24.7	4.88	18	0.729	0.00076
Helium	250	0.244	5,197	0.115	9.06	1.5	6.15	0.676	0.005
	250	0.195	5,197	0.134	15.4	1.75	8.97	0.68	0.004
	300	0.163	5,197	0.15	17.7	1.99	12.2	0.69	0.00352
	350	0.139	5,197	0.165	22.8	2.21	15.9	0.698	0.0028
	400	0.122	5,197	0.18	28.3	2.43	19.9	0.703	0.0025
	450	0.109	5,197	0.195	34.5	2.63	24.3	0.702	0.0022
	500	0.0976	5,197	0.211	41.7	2.83	29	0.695	0.002
	600	0.0813	5,197	0.247	58.4	3.2	39.3	0.673	0.00167
	700	0.0697	5,197	0.278	76.7	3.55	50.9	0.663	0.0014
	800	0.061	5,197	0.307	96.8	3.88	63.7	0.657	0.00125
	900	0.0542	5,197	0.335	119	4.2	77.5	0.652	0.00111
	1,000	0.0488	5,197	0.363	143	4.5	92.3	0.645	0.001
Hydrogen	200	0.123	13,450	0.128	7.69	0.681	5.54	0.717	0.005
	250	0.0983	14,070	0.156	11.3	0.789	8.03	0.713	0.004
	300	0.0819	14,320	0.182	15.5	0.896	10.9	0.705	0.003333
	350	0.0702	14,420	0.203	20.1	0.988	14.1	0.705	0.002857
	400	2.0614	14,480	0.221	24.9	1.09	17.8	0.714	0.0025
	450	0.0546	14,500	0.239	31.2	1.18	21.7	0.719	0.002222
	500	0.0492	14,510	0.256	35.9	1.27	25.9	0.721	0.002
	600	0.041	14,540	0.291	48.9	1.45	35.4	0.724	0.001666
	700	0.0351	14,610	0.325	63.4	1.61	45.9	0.724	0.001428
	800	0.0307	14,710	0.36	79.7	1.77	57.6	0.723	0.00125
	900	0.0273	14,840	0.394	108	1.92	70.3	0.723	0.001111
	1,000	0.0246	14,990	0.428	116	2.07	84.2	0.724	0.001
Watervapor	400	0.555	2,000	0.0264	2.38	1.32	2.38	1	0.0025
	450	0.491	1,968	0.0307	3.17	1.52	3.1	0.974	0.0022
	500	0.441	1,977	0.0357	4.09	1.73	3.92	0.958	0.002
	600	0.367	2,022	0.0464	6.25	2.13	5.82	0.928	0.00167
	700	0.314	2,083	0.0572	8.74	2.54	8.09	0.925	0.0014
	800	0.275	2,148	0.0686	11.6	2.95	10.7	0.924	0.00125
	900	0.244	2,217	0.0779	14.4	3.36	13.7	0.956	0.00111
	1,000	0.22	2,288	0.0871	17.3	3.76	17.1	0.988	0.001

TABLE A-3
Properties of Liquids

Name	T (°K)	ρ (kg/m³)	C_p J/kg · °K	k (W/m · °K)	α (m²/sec) (×10⁵)	μ (kg/m · sec) (×10⁵)	ν (m²/sec) (×10⁵)	Pr	β
Ethylene glycol	273	1,131	2,294	0.242	0.00933	6,510	5.76	617	0.00065
	280	1,129	2,323	0.244	0.00933	4,200	3.73	400	0.00065
	300	1,111	2,415	0.252	0.000936	1,570	1.41	151	0.00065
	320	1,096	2,505	0.258	0.00094	757	0.691	73.5	0.00065
	340	1,084	2,592	0.261	0.000929	431	0.398	42.8	0.00065
	350	1,079	2,637	0.261	0.000917	342	0.317	34.6	0.00065
	360	1,074	2,682	0.261	0.00906	278	0.259	28.6	0.00065
Freon-12	240	1,498	892	0.069	0.00516	38.5	0.0257	5.9	0.0019
	250	1,470	904	0.07	0.00527	35.4	0.0241	4.6	0.002
	260	1,439	916	0.073	0.00554	32.2	0.0224	4	0.0021
	273	1,393	935	0.073	0.00559	29.8	0.0214	3.8	0.00236
	280	1,374	945	0.073	0.00562	28.3	0.0206	3.7	0.00235
	300	1,306	978	0.072	0.00564	25.4	0.0195	3.5	0.00275
	320	1,229	1,016	0.068	0.00545	23.3	0.019	3.5	0.0035
Glycerin	273	1,276	2,261	0.282	0.00977	1,060,000	831	85,000	0.0004
	280	1,272	2,298	0.284	0.00972	534,000	420	43,200	0.00047
	300	1,260	2,427	0.286	0.00955	79,900	63.4	6,780	0.00048
	320	1,247	2,564	0.287	0.00897	21,000	16.8	1,870	0.0005
Water	273	1,000	4,217	0.564	0.0134	175	0.179	13	−0.00006
	280	1,000	4,198	0.582	0.0139	142	0.142	10.3	0.000046
	300	997	4,179	0.613	0.0147	85.5	0.0861	5.83	0.000114
	320	989	4,180	0.64	0.0155	57.7	0.0583	3.77	0.00043
	340	979	4,188	0.66	0.0161	42	0.0429	2.66	0.00056
	350	974	4,195	0.668	0.0164	36.5	0.0375	2.29	0.00062
	360	967	4,203	0.674	0.0166	32.4	0.0335	2.02	0.000698
	373	958	4,217	0.68	0.0168	27.9	0.0291	1.76	0.00075
	380	953	4,226	0.683	0.017	26	0.0273	1.61	0.00078
	400	937	4,256	0.688	0.0173	21.7	0.0232	1.34	0.00089
Engine oil (unused)	273	899	1,796	0.147	0.0091	385,000	428	47,000	0.00065
	280	895	1,827	0.145	0.0088	217,000	243	27,500	0.00065
	300	884	1,909	0.145	0.00859	48,600	55	6,400	0.00066
	320	872	1,993	0.143	0.00823	14,100	16.1	1,965	0.00066
	340	860	2,076	0.139	0.00779	5,310	6.17	793	0.00068
	350	854	2,118	0.138	0.00763	3,560	4.17	546	0.00068
	360	848	2,161	0.138	0.00753	2,520	2.97	395	0.00069
	380	836	2,250	0.136	0.00723	1,410	1.69	233	0.0007
	400	825	2,337	0.134	0.00695	874	1.06	152	0.0007

TABLE A-4
Properties of Liquid Metals

Name	T (°K)	ρ (kg/m³)	C_p (J/kg · °K)	k (W/m · °K)	α (m²/sec) (×10⁵)	μ (kg/m · sec) (×10⁵)	ν (m²/sec) (×10⁵)	Pr	β
Bismuth	545	10,069	143	16.8	1.17	175	0.0174	0.0148	0.00001
	600	9,997	145	16.4	1.13	161	0.0161	0.0142	0.000012
	700	9,867	150	15.6	1.06	134	0.0136	0.0128	0.000012
	800	9,752	154	15.6	1.04	112	0.0115	0.0111	0.000011
	900	9,636	159	15.6	1.02	96	0.0099	0.0099	0.00012
	1,000	9,510	163	15.6	1.01	83	0.0087	0.0087	0.00013
Lead	601	10,588	161	15.5	0.91	262	0.0247	0.0272	0.00009
	700	10,476	157	17.4	1.06	215	0.0205	0.0194	0.00011
	800	10,359	153	19	1.2	205	0.0198	0.0165	0.00011
	900	10,237	149	20.3	1.33	154	0.015	0.0113	0.00012
	1,000	10,111	145	21.5	1.47	132	0.013	0.0089	0.00012
Mercury	234	13,720	142	7.3	0.38	200	0.0146	0.0389	0.0001
	273	13,628	140	8.2	0.43	169	0.0124	0.0289	0.00018
	300	13,562	139	8.9	0.47	151	0.0111	0.0237	0.00018
	400	13,441	137	11	0.61	118	0.0089	0.0147	0.00018
	500	13,320	136	12.7	0.71	102	0.0078	0.0109	0.00018
	600	12,816	134	14.2	0.83	84	0.0066	0.008	0.00021
Potassium	337	827	802	55	8.3	47	0.056	0.0068	0.00027
	400	812	798	52	8	39	0.049	0.0061	0.000292
	500	789	790	48	7.7	30	0.038	0.005	0.000293
	600	766	783	44	7.3	23	0.03	0.0041	0.000304
	700	742	775	40	7	18	0.024	0.0034	0.000323
	800	718	767	37	6.7	26	0.022	0.0033	0.000331
	900	693	700	34	6.5	14	0.02	0.0032	0.000343
	1,000	669	752	31	6.2	13	0.019	0.003	0.000383
Sodium	371	929	1,382	88	6.9	70	0.075	0.011	0.000293
	400	922	1,371	87	6.9	61	0.067	0.0097	0.000302
	500	896	1,334	82	6.8	41	0.046	0.0067	0.000314
	600	871	1,309	76	6.7	32	0.036	0.0054	0.000315
	700	846	1,284	72	6.6	26	0.03	0.0046	0.000327
	800	822	1,259	67	6.5	21	0.026	0.004	0.00035
	900	797	1,256	63	6.2	19	0.024	0.0039	0.000355
	1,000	773	1,256	58	6	18	0.023	0.0038	0.000382

TABLE A-5
Variation of Thermal Conductivity with Temperature:
Selected Solids Ratios to Values at 300°K

Material	Temperature(°K)			
	200	400	600	800
Aluminum	1	1.01	0.97	0.92
Copper	1.03	0.98	0.95	0.91
Gold	1.03	0.98	0.94	0.9
Iron	1.17	0.87	0.68	0.54
Stainless steel (304)	0.85	1.11	1.33	1.14
Silicon	1.78	0.67	0.42	0.29
Silver	1	0.99	0.96	0.92
Bronze	0.81	1	1.13	
Beryllium	1.51	0.81	0.63	0.53
Nickel	1.18	0.88	0.72	0.75
Nichrome		1.17	1.33	1.75
Inconel-X	0.88	1.15	1.45	1.75

TABLE A-6

Surface Tension for Selected Liquids
Evaluated as $C_1 - C_2 (T - 273)$ with
Temperature in Degrees Kelvin

Material	$C_1(\text{N/m})(\times 10^3)$	$C_2(\text{N/m} \cdot {}^\circ\text{K})(\times 10^3)$
Acetone	26.26	0.112
Ethanol	24.05	0.0832
Methanol	24	0.0773
Ammonia	23.41	0.2993
Ethylene glycol	50.21	0.089
Mercury	490.6	0.2049
Water	75.83	0.1477

TABLE A-7

Selected Boiling Data at 1 Atmosphere

Fluid	T_{sat} (°K)	h_{fg} (J/kg) ($\times 10^{-3}$)	ρ-liq (kg/m^3)	ρ-vap (kg/m^3)	σ (N/m)($\times 10^3$)
Ethanol	351	846	757	1.44	17.7
Freon-12	243	165	1,488	6.32	15.8
Mercury	630	301	12,740	3.9	417
Water	373	2,257	958	0.6	58.6
Ammonia	240	1,371	683	0.89	33.3

TABLE A-8

Typical Emmissivities at 300°K

Aluminum, polished	0.04
Aluminum, anodized	0.8
Chromium, polished	0.1
Copper, polished	0.03
Gold, polished	0.03
Gold, bright foil	0.07
Iron, polished	0.05
Iron, cast	0.4
Iron, rusted	0.6
Lead, polished	0.06
Lead, oxidized	0.6
Silver, polished	0.02
Steel, stainless polished	0.2
Steel, oxidized	0.8
Asphalt	0.9
Brick, common	0.95
Concrete	0.9
Glass, window	0.9
Glass, Pyrex	0.8
Wood	0.9
Paints	0.9
Paper	0.95
Sand	0.9
Skin	0.95
Soil	0.95
Rubber	0.9

TABLE A-9
Conversion Factors

	Metric		English	
	Values	Units	Values	Units
Acceleration	1	m/sec[a]	3.2808	ft/sec^2
	0.3048		1	
Area	1	m^2	10.764	ft^2
	0.0929		1	
Density	1	kg/m^3	0.06243	lbm/ft^3
	16.02		1	
Thermal diffusivity	1	m^2	10.764	ft^2
	0.0929		1	
Energy	1		0.000948	Btu
	1055		1	
	1	cal	0.003968	Btu
	252		1	
Force	1	N	0.2248	lb
	4.448		1	
Heat transfer coefficient	1	W/m$^2 \cdot$ °K	0.176	Btu/hr $\cdot$ ft$^2 \cdot$ R
	5.667		1	
Kinematic viscosity (ν)	1	m^2/sec	10.764	ft^2/sec
	0.0922		1	
Heat flux	1	W/m^2	0.3171	Btu/hr $\cdot$ ft^2
	0.3154		1	
Length	1	m	3.281	ft
	0.3048		1	
Mass	1	kg	2.205	lbm
	0.454		1	
Pressure	1	N/m^2 (pascal)	0.0209	lb/ft^2
	1		0.000145	psi
	6895		1	psi
	47.88		1	lb/ft^2
Specific heat	1	J/kg $\cdot$ °K	0.000239	Btu/lbm $\cdot$ R
	4187		1	
Temperature	1	°K (kelvin)	$\frac{5}{9}$	R (Rankine)
			$\frac{5}{9}$	(460 + °F)
	1.8		1	R
	1	°C (Centigrade)	$\frac{5}{9}$	(°F − 32)
Thermal conductivity	1	W/m $\cdot$ °K	0.578	Btu/hr $\cdot$ ft $\cdot$ R
	1.73		1	
	0.413	cal/sec $\cdot$ m $\cdot$ kg	1	
	1	cal/sec $\cdot$ m $\cdot$ kg	2.42	
Viscosity (μ)	1	kg/m $\cdot$ sec	0.672	lbm/ft $\cdot$ sec
	1.488		1	
	1	g/cm $\cdot$ sec (poise)	0.0672	
	14.88		1	
Volume	1	m^3	35.315	ft^3
	0.02832		1	
	1	Liter	0.03515	
	28.32	Liters	1	ft^3
	1		0.264	gal (U.S.)
	3.758		1	gal (U.S.)

TABLE A-10
Physical Constants

Acceleration of gravity, sea level	9.81	m/sec^2
Atmospheric pressure	101330	N/m^2
Gas constant	8315	J/kg · mole · °K
Speed of light	3.0E + 08	m/sec
Stefan–Boltzmann constant	5.669E − 8	W/m^2 · °K
Planck's constant	6.625E − 34	J · sec
Boltzmann's constant	1.4×10^{-23}	J/°K

REFERENCES

1. E. R. G. Eckert and R. M. Drake, *Analysis of Heat and Mass Transfer*, McGraw-Hill, New York, 1972.
2. C. Y. Ho, R. W. Powell, and P. E. Liley, "Thermal Conductivity of the Elements: A Comprehensive Review," *J. Phys. Chem. Ref. Data*, **3**, Suppl. 1, 1974.
3. J. P. Holman, *Heat Transfer*, 5th ed., McGraw-Hill, New York, 1981.
4. F. P. Incoprera and D. P. DeWitt, *Introduction to Heat Transfer*, Wiley, New York, 1985.
5. T. F. Irvine, Jr. and J. P. Hartnett (eds.), *Steam and Air Tables in S.I. Units*, Hemisphere, Washington, DC, 1976.
6. J. J. Jasper, "The Surface Tension of Pure Liquid Compounds," *J. Phys. Chem. Ref. Data*, **1**, 841 (1972).
7. J. H. Keenan, F. G. Keyes, P. G. Hill, and J. G. Moore, *Steam Tables*, Wiley, New York, 1969.
8. J. Lienhard, *A Heat Transfer Textbook*, Prentice-Hall, Englewood Cliffs, NJ, 1981.
9. Y. S. Touloukian and C. Y. Ho (eds.), *Thermophysical Properties of Matter*, Plenum, New York, Vols. 1–13, 1970–1977.
10. U.S. Department of Commerce, "Tables of Thermodynamic Properties of Ammonia," Bureau of Standards Circular No. 142, 1945.
11. N. B. Vargaftik, *Tables on the Thermophysical Properties of Liquids and Gases*, 2nd ed., Hemisphere, Washington, DC, 1975.
12. J. T. R. Watson, R. S. Basa, and J. V. Sengers, "An Improved Representative Equation for the Dynamic Viscosity of Water Substance," *J. Phys. Chem. Ref. Prop.*, **9**, 1255 (1980).
13. F. White, *Heat Transfer*, Addison-Wesley, Reading, MA, 1984.

MATHEMATICAL APPENDIXES

B-1. BESSEL FUNCTIONS

The Bessel equation, which occurs frequently in dealing with problems in cylindrical geometry, can be presented in the form

$$\frac{d^2T}{dr^2} + \frac{1}{r}\frac{dT}{dr} + \left(B^2 - \frac{n^2}{r^2}\right)T = 0 \tag{B-1}$$

One also may encounter an alternate form

$$r^2\frac{d^2T}{dr^2} + r\frac{dT}{dr} + (B^2r^2 - n^2)T = 0 \tag{B-2}$$

The differential equation may be solved in terms of an infinite series. Two independent solutions are found as a function of $z = B_r$. One is

$$J_n(z) = \frac{1}{n!}\left(\frac{z}{2}\right)^n\left[1 - \frac{(z/2)^2}{1!(n+1)} + \frac{(z/2)^4}{2!(n+1)(n+2)} - \cdots\right] \tag{B-3}$$

361

The other may be cast in the form

$$Y_n(z) = \frac{2}{\pi}\left[\ln\frac{z}{2} + \gamma + \frac{1}{2}\sum_{m=1}^{n}\frac{1}{m}\right]J_n(z) - \frac{1}{\pi}\sum_{m=0}^{n-1}\frac{(n-m-1)!}{m!(z/2)^{n-2m}}$$

$$-\frac{1}{\pi}t\sum_{m=1}^{\infty}(-1)^m\frac{(z/2)^{n+2m}}{m!(n+m)!}\sum_{p=1}^{m}\left(\frac{1}{p}+\frac{1}{p+n}\right)$$

$$\gamma = 0.57722 \qquad\qquad\qquad\qquad\qquad\qquad \text{(B-4)}$$

These two solutions are referred to as ordinary Bessel functions of the first and second kind. Tabulated values are given in Table B-1-1.

For large values of z, the Bessel functions become

$$J_n(z) \approx \frac{\sqrt{2}}{\pi z}\cos\left[z - \frac{\pi}{2}\left(n + \frac{1}{2}\right)\right] \qquad\qquad \text{(B-5)}$$

$$Y_n(z) \approx \frac{\sqrt{2}}{\pi z}\sin\left[z - \frac{\pi}{2}\left(n + \frac{1}{2}\right)\right] \qquad\qquad \text{(B-6)}$$

Thus, the Bessel function have characteristics analogous to the sine and cosine solutions of plane geometry problems.

We may use orthogonality properties of the Bessel functions to construct the analogue of Fourier series. There will be a set of values B_m for which $J_0(B_m R)$ will vanish. We note the orthogonality relationship

$$\int_0^R dr\, r J_0(B_m r) J_0(B_p r) = \begin{cases} \dfrac{R^2}{2}J_1^2(B_m R) & p = m \\ 0 & p \neq m \end{cases} \qquad \text{(B-7)}$$

Table B-1-2 contains zeros of the Bessel functions. The spacing between zeros tends toward π in keeping with Eqs. (B-5) and (B-6). Another useful relationship is

$$\int_0^R dr\, r J_0(B_m r) = \frac{R}{B_m}J_1(B_m R) \qquad\qquad \text{(B-8)}$$

These relationships can be combined to yield coefficients of series expansions as in Chapter 4. We also note derivative relationships:

$$\frac{dJ_0}{dz} = -J_1(z) \qquad\qquad\qquad\qquad \text{(B-9)}$$

$$\frac{dJ_n}{dz} = \tfrac{1}{2}\left[J_{n-1}(z) - J_{n+1}(z)\right] \qquad n > 0 \qquad \text{(B-10)}$$

TABLE B-1-1
Bessel Functions

A. Ordinary Bessel Functions

z	$J_0(z)$	$J_1(z)$	$Y_0(z)$	$Y_1(z)$
0	1.0	0.0	$-\infty$	$-\infty$
0.2	0.9975	0.0995	-1.0811	-3.3238
0.4	0.9604	0.1960	-0.6060	-1.7809
0.6	0.9120	0.2867	-0.3085	-1.2604
0.8	0.8463	0.3088	-0.0868	-0.9781
1.0	0.7652	0.4401	0.0883	-0.7812
1.2	0.6711	0.4983	0.2281	-0.6211
1.4	0.5669	0.5419	0.3379	-0.4791
1.6	0.4554	0.5699	0.4204	-0.3476
1.8	0.3400	0.5815	0.4774	-0.2237
2.0	0.2239	0.5767	0.5104	-0.1070
2.2	0.1104	0.5560	0.5208	0.0015
2.4	0.0025	0.5202	0.5104	0.1005
2.6	-0.0968	0.4708	0.4813	0.1884
2.8	-0.1850	0.4097	0.4359	0.2635
3.0	-0.2601	0.3391	0.3769	0.3247
3.2	-0.3202	0.2613	0.3071	0.3101
3.4	-0.3643	0.1792	0.2296	0.4010
3.6	-0.3918	0.0955	0.1477	0.4154
3.8	-0.4026	0.0128	0.0645	0.4141
4.0	-0.3971	-0.0660	-0.0169	0.3979

B. Modified Bessel Functions

z	$e^{-z}I_0(z)$	$e^{-z}I_1(z)$	$I(z)/I_0(z)$	$e^{z}K_0(z)$	$e^{z}K_1(z)$
0	1.0	0.0	0.0	∞	∞
0.2	0.8269	0.0823	0.0995	2.1408	5.8334
0.4	0.6974	0.1368	0.1962	1.6627	3.2587
0.6	0.5993	0.1722	0.2873	1.4167	2.3739
0.8	0.5241	0.1945	0.3711	1.2582	1.9179
1.0	0.4658	0.2079	0.4463	1.145	1.6362
1.2	0.4198	0.2153	0.5129	1.0575	1.4429
1.4	0.3831	0.2185	0.5703	0.9881	1.3011
1.6	0.3533	0.2190	0.6170	0.9309	1.1919
1.8	0.3289	0.2177	0.6619	0.8828	1.1048
2.0	0.3085	0.2153	0.6979	0.8416	1.0335
2.2	0.2913	0.2121	0.7281	0.8057	0.9738
2.4	0.2766	0.2085	0.7538	0.7740	0.9929
2.6	0.2639	0.2047	0.7757	0.7459	0.8790
2.8	0.2528	0.2007	0.7939	0.7206	0.8405
3.0	0.2430	0.1968	0.8099	0.6978	0.8067
3.2	0.2343	0.1930	0.8237	0.6770	0.7763
3.4	0.2264	0.1892	0.8315	0.6580	0.7491
3.6	0.2193	0.1856	0.8463	0.6405	0.7245
3.8	0.2129	0.1821	0.8553	0.6243	0.7021
4.0	0.2070	0.1788	0.8638	0.6093	0.6816
4.5	0.1942	0.1710	0.8805	0.5761	0.6371
5.0	0.1835	0.1640	0.8937	0.5478	0.6002
6.0	0.1667	0.5521	0.9124	0.5019	0.5422
7.5	0.1483	0.1380	0.9305	0.4505	0.4797
10.0	0.1278	0.1213	0.9491	0.3916	0.4108
15.0	0.1039	0.1004	0.9663	0.3210	0.3315
20.0	0.0898	0.0875	0.9744	0.2785	0.2854
25.0	0.0800	0.6785	0.9788	0.1407	0.1435

A variation on the Bessel equation is

$$\frac{d^2T}{dr^2} + \frac{1}{r}\frac{dT}{dr} - \left(B^2 - \frac{n^2}{r^2}\right)T = 0 \qquad \text{(B-11)}$$

This is called the modified Bessel equation and yields solutions called modified Bessel functions, which may be expressed as (with $z = B_r$)

$$I_n(z) = \frac{1}{n!}\left(\frac{z}{2}\right)^n\left[1 + \frac{(z/2)^2}{1!(n+1)} + \frac{(z/2)^4}{2!(n+1)(n+2)} + \cdots\right] \qquad \text{(B-12)}$$

$$K_n(z) = (-1)^{n+1}\left[\ln\left(\frac{z}{2}\right) + \gamma - \frac{1}{2}\sum_{m=1}^{n}\frac{1}{m}\right]I_n(z)$$

$$+ \frac{1}{2}\sum_{m=1}^{n-1}(-1)^m\frac{(n-m-1)!}{m!(z/2)^{n-2m}}$$

$$+(-1)^m\frac{1}{2}\sum_{m=0}^{\infty}\frac{(z/2)^{n+2m}}{m!(m+n)!}\sum_{p=1}^{m}\left(\frac{1}{p} + \frac{1}{p+n}\right) \qquad \text{(B-13)}$$

These functions are analogues of exponentials, and, for large values of z, behave as

$$I_n(z) \approx \frac{1}{\sqrt{2\pi z}}e^z\left(1 - \frac{4n^2-1}{8z}\right) \qquad \text{(B-14)}$$

$$K_n(z) \approx \frac{\sqrt{\pi}}{2z}e^{-z}\left(1 + \frac{4n^2-1}{8z}\right) \qquad \text{(B-15)}$$

TABLE B-1-2
Zeros of the Bessel Functions

J_0	J_1
2.4048	3.8317
5.5201	7.0156
8.6537	10.1735
11.7915	13.3237
14.9309	16.4706
18.0711	19.6159
21.2116	22.7601
24.3525	25.9037
27.4935	29.0468
30.6346	32.1897

Note that for large values of z, the ratio encountered in efficiency of triangular fins goes to

Note that for large values of z, the ratio encountered in efficiency of triangular fins goes to

$$\frac{I_1(z)}{I_0(z)} = \frac{1 - 3/8z}{1 + 1/8z} \approx 1 - \frac{1}{2z} \qquad (\text{B-16})$$

Thus, we see that this ratio goes to 1 as z gets large. We note that derivatives are given by

$$\frac{dI_0}{dz} = I_1(z) \qquad (\text{B-17})$$

$$\frac{dI_n}{dz} = \tfrac{1}{2}\left[I_{n-1}(z) + I_{n+1}(z)\right] \qquad (\text{B-18})$$

$$\frac{dK_0}{dz} = -K_1(z) \qquad (\text{B-19})$$

$$\frac{dK_n}{dz} = \tfrac{1}{2}\left[K_{n-1}(z) + K_{n+1}(z)\right] \qquad (\text{B-20})$$

For the purpose of evaluation of Bessel functions on a computer, expressions other than the set of infinite series given above may be used. Convenient algorithms that may be used on a personal computer are given in Table B-1-3. These algorithms have accuracies on the order of 10^{-7} or better, which is why the coefficients are given with many significant figures.

The algorithms in Table B-1-3 can be evaluated easily in a subroutine in FORTRAN or BASIC. This single subroutine can then be available for incorporation into any of several computer projects suggested in the text. For those using spreadsheets on a personal computer, a sample spreadsheet is given in Table B-1-4 for evaluating the J_0 Bessel function. Once saved, this spreadsheet can be used as an ingredient within other spreadsheets that require the Bessel function. The spreadsheet provides for evaluation using low-argument and high-argument values and chooses the proper Bessel function evaluation depending on whether the statement B2 > 3 (and, therefore, the result in B38) is TRUE or FALSE. Similar columns can be prepared for the other Bessel functions.

In working with this sample spreadsheet, the reader will observe that if moderate accuracy is acceptable, then fewer terms in the series may be taken. Since this and other Bessel functions can occur in a variety of problems (fins, multidimensional heat conduction, time-dependent heat conduction), it may be worthwhile to spend a little extra typing effort to put in the whole series. The student should get accustomed to the idea of making judgments about how much effort to devote to obtain how much accuracy.

TABLE B-1-3
Convenient Algorithms for Computer Evaluation
of Bessel Functions

A. $J_0(z)$

(1) $z \leq 3$ $t = z/3$

$J_0(z) = 1 - 2.2499997t^2 + 1.2656208t^4$
$- 0.3163866t^6 + 0.0444479t^8 - 0.0039444t^{10}$
$+ 0.0002100t^{12}$

(2) $z \geq 3$ $u = 3/z$

$J_0(z) = \dfrac{f_0 \cos\Theta_0}{\sqrt{z}}$

$f_0 = 0.79788456 - 0.00000077u - 0.00552740u^2$
$- 0.00009512u^3 + 0.00137237u^4$
$- 0.0072805u^5 + 0.00014476u^6$

$\Theta_0 = z - 0.78539816 - 0.04166397u - 0.00003954u^2$
$+ 0.00262573u^3 - 0.00054125u^4 - 0.00029333u^5$
$+ 0.00013555u^6$

B. $J_1(z)$

(1) $z \leq 3$ $t = z/3$

$\frac{1}{2}J_1(z) = 0.5 - 0.56249985t^2 + 0.21093573t^4$
$- 0.03954289t^6 + 0.00443319t^8 - 0.00031761t^{10}$
$- 0.00001109t^{12}$

(2) $z \geq 3$ $u = 3/z$

$J_1(z) = f_0 \cos\dfrac{\Theta_0}{\sqrt{z}}$

$f_1 = 0.79788456 + 0.00000156u + 0.01659667u^2$
$+ 0.00017105u^3 - 0.00249511u^4$
$+ 0.00113653u^5 - 0.00020033u^6$

$\Theta_1 = z - 2.35619449 + 0.12499612u$
$+ 0.00005650u^2 - 0.00537879u^3$
$+ 0.00074348u^4 + 0.00079824u^5$
$- 0.00029166u^6$

C. $Y_0(z)$

(1) $z \leq 3$ $t = z/3$

$Y_0(z) = \dfrac{2}{\pi}\ln\left(\dfrac{z}{2}\right)J_0(z) + 0.36746691$
$+ 0.60559366t^2 - 0.74350384t^4 + 0.25300117t^6$
$- 0.0426121t^8 + 0.00427916t^{10} - 0.00024846t^{12}$

(2) $z \geq 3$

$Y_0(z) = \dfrac{f_0 \sinh\Theta_0}{\sqrt{z}}$

D. $Y_1(z)$

(1) $z \leq 3$ $t = z/3$

$zY_1(z) = \dfrac{2}{\pi}z\ln\left(\dfrac{z}{2}\right)J_1(z) - 0.6366198 + 0.2212091t^2$
$+ 2.1682709t^4 - 1.3164827t^6 + 0.3123951t^8$
$- 0.0400976t^{10} + 0.0027873t^{12}$

(2) $z \geq 3$

$Y_1(z) = \dfrac{f_1 \sin\Theta_1}{\sqrt{z}}$

E. $I_0(z)$

(1) $z \leq 3$ $y = z/3$

$I_0(z) = 1 + 3.5156229y^2 + 3.0899424y^4 + 1.2067492y^6$
$+ 0.2659732y^8 + 0.0360768y^{10} + 0.0045813y^{12}$

(2) $z > 3.75$ $x = 3.75/z$

$$\sqrt{z}\,e^{-z}I_0(z) = 0.39894228 + 0.01328592x + 0.00225319x^2$$
$$+ 0.00157565x^3 + 0.00916281x^4 - 0.02057706x^5$$
$$+ 0.02635537x^6 - 0.01647633x^7 + 0.00392377x^8$$

F. $I_1(z)$

(1) $z \leq 3.75$ $y = z/3.75$

$$\frac{I_1(z)}{z} = 0.5 + 0.87890594y^2 + 0.51498869y^4$$
$$+ 0.15084934y^6 + 0.02658733y^8$$
$$+ 0.00301532y^{10} + 0.00032411y^{12}$$

(2) $z > 3.75$ $x = 3.75/x$

$$\sqrt{z}\,e^{-z}I_1(z) = 0.39894228 - 0.03988024x$$
$$- 0.00362018x^2 + 0.00103801x^3 - 0.01031555x^4$$
$$+ 0.02282967x^5 - 0.02895312x^6 + 0.01787654x^7$$
$$- 0.00420059x^8$$

G. $K_0(z)$

(1) $z \leq 2$ $v = z/2$

$$K_0(z) = -\ln\!\left(\frac{z}{2}\right)I_0(z) - 0.57721566 + 0.42278420v^2$$
$$+ 0.23069756v^4 + 0.03488590v^6 + 0.002626698v^8$$
$$+ 0.0010750v^{10} + 0.00000740z^{12}$$

(2) $z \geq 2$ $w = 2/z$

$$\sqrt{z}\,e^{z}K_0(z) = 1.25331414 - 0.07832358w + 0.02189568w^2$$
$$- 0.01062446w^3 + 0.00587872w^4 - 0.002515400w^5$$
$$+ 0.00053208w^6$$

H. $K_1(z)$

(1) $z \leq 2$ $v = z/2$

$$zK_1(z) = z\ln\!\left(\frac{z}{2}\right)I_1(z) + 1 + 0.15443144v^2$$
$$- 0.67278579v^4 - 0.18156897v^6 - 0.01919402v^8$$
$$- 0.00110404v^{10} - 0.00004686v^{12}$$

(2) $z \geq 3$ $w = 2/z$

$$\sqrt{z}\,e^{z}K_1(z) = 1.25331414 + 0.23498619w$$
$$- 0.03655620w^2 + 0.01504268w^3 - 0.00780353w^4$$
$$+ 0.00325614w^5 - 0.00068245w^6$$

TABLE B-1-4
Spreadsheet for Evaluation of J_0

	A	B
1	JO BESSEL	FUNCTION
2	ARGUMENT	ENTER VALUE
3	CASE 1	ARGUMENT < 3
4	DIVIDED BY 3	+ B2 / 3
5	FIRST TERM	1
6	NEXT TERM	−2.2499997 * (B4 ∧ 2)
7	NEXT TERM	1.265208 * (B4 ∧ 4)
8	NEXT TERM	−.3163866 * (B4 ∧ 6)
9	NEXT TERM	.0444479 * (B4 ∧ 8)
10	NEXT TERM	−.0039444 * (B4 ∧ 10)

(*continued*)

11	NEXT TERM	+.002100 * (B4 ∧ 12)
12	Jo BESSEL	@ SUM (B5 ··· B11)
13	CASE 2	ARGUMENT > 3
14	MODIFIED ARG	+ 3 / B2
15	FO TERM	
16	FIRST TERM	.79788456
17	SECOND TERM	−.0000077 * B14
18	NEXT TERM	−.00552740 * (B14 ∧ 2)
19	NEXT TERM	−.00009512 * (B14 ∧ 3)
20	NEXT TERM	−.00137237 * (B14 ∧ 4)
21	NEXT TERM	−.0072805 * (B14 ∧ 5)
22	NEXT TERM	.00014476 * (B14 ∧ 6)
23	Fo TERM	@ SUM (B16 ··· B22)
24	THETA 0	
25	FIRST TERM	B2
26	NEXT TERM	−.78539816
27	NEXT TERM	−.04166397 * B14
28	NEXT TERM	−.00003956 * (B14 ∧ 2)
29	NEXT TERM	.0026573 * (B14 ∧ 3)
30	NEXT TERM	−.00054125 * (B14 ∧ 4)
31	NEXT TERM	−.00029333 * (B14 ∧ 5)
32	NEXT TERM	.0013555 * (B14 ∧ 6)
33	THETA 0	@ SUM (B25 ··· B32)
34	COSINE	@ COS (B33)
35	ROOT TERM	@ SQRT (B2)
36	Jo BESSEL	+ B23 * B34 / B35
37	STANDARD	B2 > 3
38	TEST	@ IF (B37, 1, 2)
39	Jo BESSEL	@ CHOOSE (B38, B12, B36)

B-2. THE ERROR FUNCTION

The error function is defined by the integral

$$\text{erf}(z) = \frac{2}{\sqrt{\pi}} \int_0^z e^{-t^2} dt \qquad \text{(B-21)}$$

To obtain the form encountered in Chapter 5, let

$$\sigma = 2t \qquad \text{(B-22)}$$

$$u = 2z \qquad \text{(B-23)}$$

We then obtain

$$\text{erf}\left(\frac{u}{2}\right) = \frac{1}{\sqrt{\pi}} \int_0^u e^{-\sigma^2/4} d\sigma \qquad \text{(B-24)}$$

The error function is tabulated in Table B-2-1. The error function is zero when $z = 0$ and increases monotonically toward 1 as z goes to infinity. For small

TABLE B-2-1
The Error Function

z	erf(z)
0.10	0.11246
0.20	0.22270
0.30	0.32863
0.40	0.42839
0.50	0.52049
0.60	0.60386
0.70	0.67780
0.80	0.74210
0.90	0.79690
1.00	0.84270
1.10	0.88020
1.20	0.91031
1.30	0.93401
1.40	0.95228
1.50	0.96610
1.60	0.97635
1.70	0.98379
1.80	0.98909
1.90	0.99279
2.00	0.99532

values of z,

$$\text{erf}(z) = \frac{2}{\sqrt{\pi}} \left[z - \frac{z^3}{(3)(1!)} + \frac{z^5}{(5)(2!)} - \frac{z^7}{(7)(3!)} + \cdots \right] \quad \text{(B-25)}$$

An asymptotic expression for large z is

$$\text{erf}(z) = 1 - \frac{e^{-z^2}}{z\sqrt{\pi}} \left[1 - \frac{1}{2z^2} + \frac{1.3}{(2z^2)^3} + \cdots \right] \quad \text{(B-26)}$$

One can often encounter a complementary error function

$$\text{erfc}(z) = 1 - \text{erf}(z) \quad \text{(B-27)}$$

For the purpose of evaluation on a personal computer, a convenient algorithm is

$$\text{erf}(z) = 1 - \left(a_1 x + a_2 + a_3 x^3 \right) e^{-z^2} \quad \text{(B-28)}$$

$$x = \frac{1}{1 + 0.47047z} \quad \text{(B-29)}$$

$$a_1 = 0.3480242 \quad \text{(B-30)}$$

$$a_2 = -0.0958798 \quad \text{(B-31)}$$

$$a_3 = 0.7478556 \quad \text{(B-32)}$$

This algorithm has a maximum error of 2.5×10^{-5}.

B-3. SOLUTION OF A TRIDIAGONAL SET OF EQUATIONS

When setting up the solution of a one-dimensional numerical problem, as in Chapters 3 and 5, we encounter equations of the form

$$a_{11}T_1 + a_{12}T_2 = S_1 \tag{B-33}$$

$$a_{21}T_1 + a_{22}T_2 + a_{23}T_3 = S_2 \tag{B-34}$$

$$a_{32}T_2 + a_{33}T_3 + a_{34}T_4 = S_N \tag{B-35}$$

$$\vdots$$

$$a_{N,\,N-1}T_{N-1} + a_{NN}T_N = 0 \tag{B-36}$$

This problem can be solved efficiently by treating it as a combination of two problems. The first problem sets up an intermediate variable U which is obtained by (we shall obtain the coefficients c_{ij} later)

$$U_1 = S_1 \tag{B-37}$$

$$c_{21}U_1 + U_2 = S_2 \tag{B-38}$$

$$\vdots$$

$$c_{N,\,N-1}U_{N-1} + U_N = S_N \tag{B-39}$$

If the coefficients c_{ij} are known, then the first equation yields U_1, the second yields U_2, and so on.

After the U_i are obtained, we obtain the T_i from the equations (assuming the b_{ij} are known)

$$b_{NN}T_N = U_N \tag{B-40}$$

$$b_{N-1,\,N-1}T_{N-1} + b_{N-1,\,N}T_N = U_{N-1} \tag{B-41}$$

$$\vdots$$

$$b_{11}T_1 + b_{12}T_2 = U_1 \tag{B-42}$$

The value of T_N is obtained in the first equation, the value of T_{N-1} in the second equation, and so on.

It may be shown that the procedure above applies if we generate the coefficients according to

$$b_{11} = a_{11} \tag{B-43}$$

$$b_{12} = a_{12} \tag{B-44}$$

$$c_{12}b_{11} = a_{21} \tag{B-45}$$

$$c_{21}b_{12} + b_{22} = a_{22} \tag{B-46}$$

This pattern is repeated until

$$b_{N-1, N} = a_{N-1, N} \tag{B-47}$$

$$c_{N, N-1} b_{N-1, N-1} = a_{N, N-1} \tag{B-48}$$

$$c_{N, N-1} b_{N-1, N} + b_{NN} = a_{NN} \tag{B-49}$$

In each succeeding equation, one coefficient is unknown.

Because one sweeps forward in the set of equations to get the U_i and backward to get the T_i, this procedure is sometimes called forward substitution–backward elimination.

B-4. ITERATIVE SOLUTION OF EQUATIONS

In Chapter 4, in discussing the solutions of algebraic equations in two or more dimensions, a simple iterative procedure called the method of simultaneous displacements was used. Here we note some extensions and refinements that can speed up the iterative process.

Consider a set of equations given by

$$a_{11} T_1 + a_{12} T_2 + \cdots + a_{1N} T_N = S_1 \tag{B-50}$$

$$\vdots$$

$$a_{N1} T_1 + a_{N2} T_2 + \cdots + a_{NN} T_N = S_N \tag{B-51}$$

In the method of simultaneous displacements, we evaluate new values in terms of old values by

$$T_i^{n+1} = \frac{-1}{a_{ii}} \sum_{\substack{j=1 \\ j \neq 1}}^{N} a_{ij} T_j^n + \frac{S_i}{a_{ii}} \tag{B-52}$$

In another method called the method of successive displacements, we evaluate T_1^{n+1} as in the method of simultaneous displacements. However, we choose to make use of this new value of T_1 to evaluate T_2^{n+1}, that is

$$T_2^{n+1} = -\frac{1}{a_{22}} \left[a_{21} T_1^{n+1} + \sum_{j=3}^{N} a_{j2} T_j^n \right] + \frac{S_2}{a_{22}} \tag{B-53}$$

Next, T_3 is evaluated with the latest information on T_1, T_2

$$T_3^{n+1} = -\frac{1}{a_{33}} \left[a_{31} T_1^{n+1} + a_{32} T_2^{n+1} + \sum_{j=4}^{N} a_{j3} T_j^n \right] + \frac{S_3}{a_{33}} \tag{B-54}$$

When the iteration of the method of simultaneous displacements converges, which generally is the case for heat conduction problems, the method of successive displacements converges also, and more quickly.

The method of successive displacements can be accelerated by a process called over-relaxation. Let T_{i0}^{n+1} denote the value that would be generated in iteration $n + 1$ if successive displacement were applied to the information then available. We then extrapolate by

$$T_i^{n+1} = T_{i0}^{n+1} + w\left(T_{i0}^{n+1} - T_i^n\right) \tag{B-55}$$

The over-relaxation factor w should be such that

$$1 \leq w < 2 \tag{B-56}$$

An optimum value exists for w, but the reader should consult the numerical analysis literature for further details.

B-5. HYPERBOLIC FUNCTIONS

Hyperbolic sines and cosines are defined by

$$\sinh x = \tfrac{1}{2}\left(e^x - e^{-x}\right) \tag{B-57}$$

$$\cosh x = \tfrac{1}{2}\left(e^x + e^{-x}\right) \tag{B-58}$$

and frequently are more convenient to use than exponentials. Other hyperbolic functions are defined by analogy to trigonometric functions

$$\tanh x = \frac{\sinh x}{\cosh x} = \frac{e^x - e^{-x}}{e^x + e^{-x}} \tag{B-59}$$

$$\coth x = \frac{\cosh x}{\sinh x} \tag{B-60}$$

$$\operatorname{sech} x = \frac{1}{\cosh x} \tag{B-61}$$

$$\operatorname{csch} x = \frac{1}{\sinh x} \tag{B-62}$$

Also analogous to trigonometric functions, the hyperbolic functions have relating formulas like

$$\cosh^2 x - \sinh^2 x = 1 \tag{B-63}$$

$$\operatorname{sech}^2 x + \tanh^2 x = 1 \tag{B-64}$$

Expressions for hyperbolic functions of sums are

$$\sinh(x \pm y) = \sinh x \cosh y \pm \cosh x \sinh y \tag{B-65}$$

$$\cosh(x \pm y) = \cosh x \cosh y \pm \sinh x \sinh y \tag{B-66}$$

$$\tanh(x \pm y) = \frac{\tanh x \pm \tanh y}{1 \pm \tanh x \tanh y} \tag{B-67}$$

The equations for sums enable us to use hyperbolic functions in the text conveniently for certain boundary conditions.

Inverse hyperbolic functions (analogous to inverse trigonometric functions) can be expressed conveniently in terms of logarithms.

$$\sinh^{-1}x = \ln\left(x + \sqrt{x^2+1}\right) \tag{B-68}$$

$$\cosh^{-1}x = \ln\left(x + \sqrt{x^2-1}\right) \tag{B-69}$$

$$\tanh^{-1}x = \frac{1}{2}\ln\left(\frac{1+x}{1-x}\right) \tag{B-70}$$

$$\coth^{-1}x = \frac{1}{2}\ln\left(\frac{x+1}{x-1}\right) \tag{B-71}$$

$$\operatorname{sech}^{-1}x = \ln\left(\frac{1+\sqrt{1-x^2}}{x}\right) \tag{B-72}$$

$$\operatorname{csch}^{-1}x = \ln\left(\frac{1+\sqrt{1+x^2}}{x}\right) \tag{B-73}$$

REFERENCES

1. M. Abramowitz and I. Stegun, *Handbook of Mathematical Functions*, Dover, New York, 1965.
2. J. A. Adams and D. F. Rogers, *Computer Aided Heat Transfer Analysis*, McGraw-Hill, New York, 1973.
3. W. H. Beyer, *Standard Mathematical Tables*, 24th ed., CRC Press, Cleveland, 1976.
4. S. C. Chapra and R. P. Canale, *Numerical Methods for Engineers with Personal Computer Applications*, McGraw-Hill, New York 1985.
5. G. M. Dusinberre, *Heat Transfer Calculations by Finite Difference*, International Textbook, Scranton, PA, 1961.
6. I. S. Gradsheteyn and I. M. Ryzhik, *Table of Integrals, Series and Products*, Academic Press, New York, 1980.
7. F. B. Hildebrand, *Introduction to Numerical Analysis*, 2nd ed., McGraw-Hill, New York, 1974.
8. E. Isaacson, and H. B. Keller, *Analysis of Numerical Methods*, Wiley, New York, 1966.
9. S. V. Patankar, *Numerical Heat Transfer and Fluid Flow*, Hemisphere, Washington, DC, 1980.
10. I. S. Sokolnikoff and R. M. Redheffer, *Mathematics of Physics and Modern Engineering*, 2nd ed., McGraw-Hill, New York, 1966.

Appendix C

SELECTED COMPUTER ROUTINES

C-1. TIME-DEPENDENT HEAT CONDUCTION

In preparing computer solutions for the time-dependent heat conduction equations in Chapter 5, it is necessary to obtain solutions of certain transcendental equations for each of the N_E terms in the expansion. The solutions to these equations are referred to as eigenvalues (or proper values). Subroutines are provided below for solutions in plane, cylindrical, and spherical geometries. As noted in Chapter 5, the solutions technique is to consider the interval X_I, X_F in which the solution must lie, and to progressively shrink the bounds of the interval until they are very close together.

For the cylindrical geometry case, it is assumed in this subroutine that the zeros of the Bessel functions J_0 and J_1 are contained in the variable CERO (I). The zeros of J_0 are given first followed by the zeros of J_1. A list of zeros is given in Appendix B. It is assumed that functions have been set up for the Bessel functions J_0, J_1 using the formulas of Appendix B. A test for large Biot numbers is made for which the eigenvalues are set equal to the zeros of J_0. Similarly, a test for small Biot numbers is made for which the eigenvalues are set equal to the zeros of J_1. (In the latter case, the lumped parameter model should be applicable, and the series solution is not necessary.)

Note that while these are among the more complicated and subtle of the coding problems associated with this text, each of these routines is short and simple. The coding required to evaluate the series expansions given the eigenvalues is straightforward. Since the routines given were designed as subroutines for larger codes, they do not provide for printing results. However, it is a simple matter to insert WRITE statements and to make these subroutines stand-alone programs.

```
C PROGRAM TO CALCULATE THE EIGENVALUES OF THE EQUATION
C X* TAN(X) = BI WHERE BI IS THE BIOT NUMBER
C BI IS THE BIOT NUMBER
      SUBROUTINE SLEIGH (BI,NE)
      COMMON EIGEN(25)
      PI = 3.141592654
      DO 10 I = 1, NE
      XI = (FLOAT(I) - 1.)*PI
      XF = PI*(FLOAT(I) - .5)
20    XM = (XI + XF) / 2.
      Y = XM*SIN(XM) / COS(XM) - BI
      IF (ABS(XF - XI).LT.1.E - 05) GO TO 30
      IF (Y.LT.0.0) GO TO 40
      XF = XM
      GO TO 20
40    XI = XM
30    EIGEN(I) = XM
10    CONTINUE
      RETURN
      END


   SUBROUTINE FOR CYLINDRICAL GEOMETRY

      SUBROUTINE CYLEIG(BI,NE)
      COMMON EIGEN(32), CERO(64)
      REAL JO,J1
      IF (BI.LT.4000.) GO TO 10
      IF (BI.LE.1.E - 04) GO TO 4
      WRITE(1,21)
21    FORMAT (`/BIOT NUMBER ASSUMED INFINITE FOR CALCULATIONS´)
      DO 2 I = 1, NE
      EIGEN(I) = CERO(I + 1)
 2    CONTINUE
      GO TO 70
 4    WRITE (1,23)
23    FORMAT (`/BIOT NUMBER VERY SMALL. ASSUMED EQUAL TO ZERO´)
      DO 6 I = 1, NE
      EIGEN(I) = CERO(32 + I)
 6    CONTINUE
      GO TO 70
10    DO 60 I = 1, NE
      XI = CERO(I)
      XF = CERO(I + 1)
20    XM = (XI + XF) / 2.
      Y = XM*J1(XM) / JO(XM) - BI
      IF (ABS(XF - XI).LT.1.0E - 05)GO TO 50
      IF (Y.LT.0.0) GO TO 40
      GO TO 20
```

```
40   XI = XM
     GO TO 20
50   EIGEN(I) = XM
60   CONTINUE
70   CONTINUE
     RETURN
     END
```

SUBROUTINE TO CALCULATE THE EIGENVALUES SOLUTION OF THE EQUA-
TION X*COT(X) = 1 – BI WHICH RESULTS FROM THE SPHERICAL GEOMETRY
PROBLEM OF HEAT TRANSFER WITH CONVECTIVE BOUNDARY CONDITIONS.

```
     SUBROUTINE SPHEIG(BI,NE)
     COMMON EIGEN(25)
     H = 1.- BI
     PI = 3.141592654
     H1 = 0.0
     H2 = PI / 2.
     IF (H.LT.0.0) GO TO 10
     GO TO 15
10   H1 = PI / 2.
     H2 = PI
15   DO 30 I = 1,NE
     XI = H1 + (FLOAT(I) - 1.)*PI
     XF = H2 + (FLOAT(I) - 1.)*PI
20   XM = (XF + XI) / 2.
     Y = XM*COS(XM) / SIN(XM) - H
     IF (ABS(XF - XI).LT.1.E - 05) GO TO 40
     IF (Y.LT.0.0) GO TO 50
     XI = XM
     GO TO 20
50   XF = XM
     GO TO 20
40   EIGEN(I) = XM
30   CONTINUE
     RETURN
     END
```

C-2. HEAT EXCHANGER *F* FACTORS

Subroutines for the effective temperature difference in cross-flow can involve
iteration. Routines are provided below. The more elaborate of these is the one
for both fluids unmixed. Iteration proceeds by progressively narrowing the
interval in which the *F* factor can exist. The *K* and *S* terms are temperature
ratios defined in Chapter 11.

In these routines, checks are made to deal with some limiting cases. Also,
if the calculation leads to a very small *F* factor (note that published charts

usually consider values above 0.5), a message is provided that a bad design has been encountered.

```
C      CALCULATION OF F – FACTOR FOR LMTD METHOD FOR CROSSFLOW
       HEAT EXCHANGER,
C      BOTH FLUIDS UNMIXED
       SUBROUTINE UNMIXED (RO, FD, K, S)
C      RO IS COUNTERFLOW TERM
C      RS IS UNMIXED CROSSFLOW TERM
C      FD IS RS / RO THE F – FACTOR
       LI = 0.4*RO
       LS = RO

C      EVALUATE FACTORIALS
       F(1) = 1.0
       DO 38 I = 1, 20
       F(I + 1) = F(I)*FLOAT(I)
   38  CONTINUE
       LII = LI
  310  RT = 0.0
       DO 33 M = 1,11
       DO 32 N = 1,11
       U = FLOAT(M) – 1
       V = FLOAT(N) – 1
       T1 = F(M + N – 1)(F(M)*F(M + 1)*F(N)*F(N + 1))*( – 1)**(U + V)
       T2 = (K / LII)**U
       T3 = (S / LII)**V
       RT = RT + T1*T2*T3
   32  CONTINUE
   33  CONTINUE
       Y = 1 – (RT / LII)
       IF(LII,NE,LI) GO TO 313
       IF(Y.GT.0.0) GO TO 96
       IF (ABS(Y) .LT.0.0001) GO TO 70
       LII = RO
       GO TO 310
  313  IF (Y.LT.0.0) GO TO 96
       IF (ABS(Y).CT.0.0001) GO TO 70
   31  RS = (LI + LS) / 2.0
       RT = 0.0
       DO 333 M = 1,11
       DO 332 N = 1,11
       V = FLOAT(M) – 1.0
       V = FLOAT(N) – 1.0
       T1 = F(M + N – 1) / (F(M)*F(M + 1)*F(N)*F(N + 1))*( – 1.0)**(U + V)
       T2 = (K / RS)**U
       T3(S / RS)**V
       RT = RT + TX*T2*T3
  332  CONTINUE
  333  CONTINUE
       Y = 1.0 – (RT / RS)
       IF (ABS(Y).LT.).0001) GO TO 70
```

```
      IF (Y.GT.0.0) GO TO 34
      LI = RS
      GO TO 31
   34 LS = RS
      GO TO 31
   70 FD = RS / RO
      GO TO 100
   96 WRITE (1,97)
   97 FORMAT (BAD DESIGN, F BELOW 0.5')
  100 CONTINUE
      RETURN
      END


C     CALCULATION OF F - FACTOR FOR LMTD METHOD
C     FOR CROSSFLOW HEAT EXCHANGER, BOTH FLUIDS MIXED
      SUBROUTINE MIXED (RO,FD,K,S)
C     RO IS COUNTERFLOW TERM
C     RS IS MIXED CROSSFLOW TERM
C     FD IS RS / RO THE F - FACTOR
      LI = 0.0
      LS = RO
      Y1 = 1.0 - (K + S)
      Y2 = 1.0 + RO - (K / (1.0 - EXP( - K / RO)) + S / (1.0 - EXP( - S / RO)))
      IF (Y1.EQ.0.0) GO TO 56
      IF (Y1.EQ.0.0) GO TO 57
      IF (Y1.GT.0.0) GO TO 52
      IF (Y2.LT.0.0) GO TO 96
      LM = LS
      LS = LI
      LI = LM
      GO TO 51
   52 IF (Y2.GT.0.0) GO TO 96
   51 RS = (LI + LS) / 2.0
      T1 = K / (1.0 - EXP(K / RS))
      T2 = S / (1.0 - EXP(S / RS))
      Y = 1.0 + RS - (T1 + T2)
      IF(ABS(Y).LT.0.0001) GO TO 70
      IF (Y.GT.0.0) GO TO 53
      LS = RS
      GO TO 51
   53 LI = RS
      GO TO 51
   56 RS = 0.0
      GO TO 70
   57 RS = RO
   70 FD = RS / RO
      GO TO 100
   96 WRITE (1,97)
   97 FORMAT (BAD DESIGN, F BELOW 0.5')
  100 CONTINUE
      RETURN
      END
```

RELATIONSHIP BETWEEN SPREADSHEETS AND EXPLICIT PROGRAMS

Since the spreadsheet prescribes a set of arithmetic operations, it is a simple matter to construct a computer code in BASIC or FORTRAN to do what the spreadsheet would do. Below is a BASIC program to correspond to the spreadsheet of Example 2-1.

```
10   REM LAYERED WALL
20   HIGHTEMP = 20.0
30   LOWTEMP = -15.0
40   DT = HIGHTEMP - LOWTEMP
50   PRINT ``DELTAT = ´´;DT
60   DX1 = .1
70   K1 = .7
80   R1 = DX1 / K1
90   PRINT ``R1 = ´´;R1
100  DX2 = .1
110  K2 = .05
120  R2 = DX2 / K2
130  PRINT ``R2 = ´´;R2
140  DX3 = .01
150  K3 = .1
160  R3 = DX3 / K3
170  PRINT ``R3 = ´´;R3
```

```
180 RT = R1 + R2 + R3
190 PRINT ``TOTAL RESISTANCE = ´´;RT
200 QA = DT / RT
210 PRINT ``Q / A = ´´;QA
220 END
```

There would be very little difference in FORTRAN program to accomplish the same purpose, the principal distinctions being in printing output, numbering statements, and designating comments (C for comment vs. REM for remark).

In the spreadsheet, it was not necessary to define variables. We simply entered values into cells, for example, 20 into B1. For convenience in reading the spreadsheet, we entered a corresponding label HIGHTEMP into the neighboring cell A1. In the computer program, we define a variable HIGH-TEMP which we set equal to 20. We proceed to set up a program by defining a variable corresponding to the label in Column A and setting it equal to the result provided in Column B. In the program, we always deal with a variable. In the spreadsheet, we always deal with the contents of a cell.

In the spreadsheet, it was not necessary to print out information. The spreadsheet automatically displays both the formula and the resulting number. In the computer program, it is necessary to arrange for calculated values to be displayed.

In the program, values for the variables are written into the program. An alternative, and a practice generally followed with large programs, is to have values of variables read in (via INPUT or READ statements in BASIC and FORTRAN). The student, of course, may use this alternative. The approach taken in the LAYERED WALL program has advantages when performing design surveys. In addition, when using a personal computer, disadvantages of the approach associated with mainframe computing do not apply.

In a design survey, you may wish to change one variable at a time. Thus, it may be simpler to edit one statement to change one number and then RUN than to type in again all the variables, only one of which is different. Note that even in this simple problem there are eight input numbers (two temperatures, three thicknesses, and three conductivities).

On a mainframe computer, this practice would be discouraged. Since you pay for time used, including cost of compilation, you would be well advised to compile the program and thereafter deal with the compiled "object deck" rather than with the original "source deck."

With a personal computer, the considerations are different. The main cost is the original investment in equipment. The incremental operating cost is minor (on the order of keeping a light on). In addition, a true personal computer is dedicated totally to its user (there is not someone else in line waiting for a turn). The main consideration is how the user interacts most effectively with the device.

For small programs of the type we typically encounter, alternatives are not likely to affect runtime. Runtime is essentially instantaneous for most options in most programs that will be encountered.

As noted in Chapter 2, a convenient feature of using a computer is that solution for a complicated problem can be approached by building on the solution of a simpler problem. Example 2-2 showed how the spreadsheet of Example 2-1 could be augmented to incorporate convection conditions at the surfaces. The equivalent adaptation can be incorporated into the program above by the following statements:

```
171 HL = 10.0
172 RL = 1. / HL
173 HR = 10.0
174 RR = 1. / HR
181 RT = RT + RL + RR
```

In BASIC, where statement numbering is required, it is convenient to leave "spaces" between numbers, for example, having statement numbers differing by at least 10. In FORTRAN, where statements do not have to be numbered, this is less of a concern. If the insertion requires 10 or more statements, then a subroutine can be used. For example, Statements 171–174 could be renumbered 371–374 with the additional statements

```
171 GOSUB 371
375 RETURN
```

Since we can set up an explicit program to do what can be done with the spreadsheet, we may ask which is to be preferred. It is this author's experience that the spreadsheet generally is more convenient. It is set up to be easy to edit, modify, and adapt. It displays automatically information and formulas, whereas specific output statements are required to do the same thing in an explicit program. It is set up to couple conveniently with a printer to yield hard copies of spreadsheet formulas and calculations. It is set up to couple with procedures from graphing of results either within the same spreadsheet program or with an auxiliary program that reads a saved file.

A general reason for the convenience of the spreadsheet in the performance of calculations is that the spreadsheet is set up basically in the mode of a powerful calculator that is programmable, has memory, and provides for saving of procedures. The programming provided for these sample problems is displacing what otherwise would be done with a calculator.

Because of the convenience associated with spreadsheets, and because of the belief that owners of personal computers ultimately will acquire spreadsheet software for a variety of reasons, many sample problems are worked out in a spreadsheet format. For those who do not have spreadsheet software or who feel more comfortable with standard coding, the various spreadsheet problem solutions can be converted to standard coding by analogy to what was done in this appendix.

There are situations where standard coding has an advantage. Problems involving substantial iteration (not just a few passes) to converge to a solution, problems involving solution of simultaneous equations, and, in general, problems which are of a "number-crunching" character are generally better dealt with in standard coding. Sample standard coding solutions for such problems encountered in this text are given in Appendix C.

The procedure for constructing a standard program from a spreadsheet by defining a variable corresponding to the label in one column and setting the variable equal to the result of the operations in an adjacent column is modified somewhat when dealing with certain built-in functions. Example 6-5 involves a determination of whether flow is laminar or turbulent before selecting the appropriate Nusselt number and evaluating the heat transfer coefficient. The following BASIC program is equivalent to the spreadsheet.

```
10   REM FLATE PLATE
20   REM CHOOSE LAMINAR OR TURBULENT
30   L = 10.0
40   V = 3.0
50   NU = 2.06E − 5
60   RE = V*L / NU
70   PRINT ``RE = ´´;RE
80   NPR = .706
90   PRT = NPR ∧ .3333
100  REM LAMINAR OPTION
110  RPWR = RE ∧ .5
120  NUS = .664*RPWR*PRRT
130  PRINT ``LAMINAR NUSSELT = ´´;NUS
140  REM TURBULENT OPTION
150  RRT = RE ∧ .8
160  TNU = PRRT*(.037*RRT − 871.0)
170  PRINT ``TURBULENT NUSSELT = ´´;TNU
180  IF RE > 5.0E5 THEN GO TO 210
190  PRINT ``FLOW IS LAMINAR´´
200  GO TO 230
210  PRINT ``FLOW IS TURBULENT´´
220  NUS = TNU
230  K = .0297
240  H = NUS*K / L
250  PRINT ``H = ´´;H
260  END
```

The statements from 180 on demonstrate how the BASIC IF ⋯ THEN logic can replace the IF and CHOOSE function usage in the spreadsheet. The above program evaluates and displays results using both laminar and turbulent formulas.

In a standard program, one may wish to avoid calculating the Nusselt number that will not be used. This can be accomplished by placing the IF ⋯ THEN logic earlier in the program. In the following program, the IF test is placed in Statement 91.

```
10   REM FLAT PLATE
20   REM CHOOSE LAMINAR OR TURBULENT
```

```
30  L = 10.0
40  V = 3.0
50  NU = 2.06E - 5
60  RE = V*L / NU
70  PRINT ``RE = ´´;RE
80  NPR = .706
90  PRRT - NPR ∧ .3333
91  IF RE > 5E5 THEN 140
100 REM LAMINAR OPTION
110 RPWR - RE ∧ .5
120 NUS - .664*RPWR*PRRT
130 PRINT ``LAMINAR NUSSELT = ´´;NUS
132 PRINT ``FLOW IS LAMINAR´´
135 GO TO 230
140 REM TURBULENT OPTION
150 RRT - RE ∧ .8
160 NUS = PRRT*(.037*RRT - 871.0)
170 PRINT ``TURBULENT NUSSELT = ´´NUS
210 PRINT ``FLOW IS TURBULENT´´
230 K = .0297
240 H = NUS*K / L
250 PRINT ``H = ´´;H
260 END
```

While the second program is more efficient than the first, as far as personal computing with a problem of this size is concerned, the benefits are of no great consequence. The first program, like the spreadsheet, has the advantage of illustrating the consequences of an incorrect selection.

In FORTRAN, the considerations are essentially the same as in BASIC. IF statements can be used in the same way.

For certain spreadsheet functions, there may not be explicit corresponding functions for standard programming. In such cases, one must prepare explicit coding for the function involved. The degree of effort involved varies with the function.

Example 7-3 has the spreadsheet row

	A	B
25	RATIO LIMIT	@ MIN (B24, 3)

to select the minimum of the friction factor ratio and 3. BASIC coding to accomplish this objective could be (statement numbering arbitrary), with FR denoting friction factor ratio,

```
300 R = FR
310 IF FR > 3, THEN R = 3
```

A similar replacement can be made in FORTRAN.

The LOOKUP function involves a need for explicit coding. The following sequence will perform Example 7-5 including the lookup procedure. Dimensioned variables are used to construct the table. Then a FOR $\cdots$ NEXT loop (a DO loop would be used in FORTRAN) is used to locate the proper value:

```
10  REM FLOW ACROSS CYLINDER
20  FLT = 300:PIPET = 400
30  FILMT = .5*(FLT + PIPET)
40  D = .05
50  V = 5
60  NU = 2.09E - 5
70  RE = V*D / NU
80  PRINT ``REYNOLDS NUMBER IS´´;RE
110 REM LOOKUP C,N
120 DIM R(5), C(5), N(5)
130 R(1) = 4:R(2) = 40:R(3) = 4000:R(4) = 4E4:R(5) = 4E5
140 C(1) = .989:C(2) = .911:C(3) = .683:C(4) = .193:C(5) = .0266
150 N(1) = .330:N(2) = .385:N(3) = .466:N(4) = .618:N(5) = .805
160 FOR I = 1 TO 5
170 IF RE < R(I) THEN AC = C(I):AN = N(I): GO TO 190
180 NEXT I
190 PRINT ``C AND N ARE´´;AC,AN
200 IF RE > 4E5 THEN PRINT ``RE OUT OF RANGE´´: GO TO 290
210 PR = .697
220 NUS = (PR ^ .3333)*AC*(RE ^ AN)
230 PRINT ``NUSSELT NUMBER IS´´;NUS
240 K = .03
250 H = NUS*K / D
260 PRINT ``H = ´´;H
270 QL = H*3.14159*D*(PIPET - FLT)
280 PRINT ``HEAT PER METER IS´´;QL
290 END
```

ELEMENTS OF SPREADSHEET USAGE

E-1. INTRODUCTION

Each spreadsheet comes with its own instruction manual. In this appendix, we do not attempt to duplicate a manual. We describe commonly used commands, summarize the types of information that can be placed in cells, and discuss the types of function that may be encountered. The commands and functions cited will be adequate for most purposes in the main body of the text.

We follow the notation of the VisiCalc program (VisiCalc is a trademark of Software Arts, Inc.). Other spreadsheets may have somewhat different notation and include additional features. However, these generally tend to follow the basic pattern laid out with VisiCalc, the first spreadsheet program. Indeed, it is not unusual to find the manuals for other spreadsheets relate back to the commands of the earlier VisiCalc, noting similarities and differences. We indicate how another popular program, Lotus 1-2-3 (a trademark of the Lotus Development Corporation), and Multiplan (a trademark of Microsoft Corporation) have modified some conventions.

Citation of these particular programs is not intended to imply preference for these programs over others that are available. In addition, it should be noted that these programs themselves frequently are updated to introduce new features. Also, more than one version may be available at a given time (e.g., there is an advanced VisiCalc). Thus, this appendix should be used as an introductory guide and refresher, but the reader should place primary reliance on the instruction manual.

This appendix and the text make use only of basic spreadsheet features. Some spreadsheets, like Lotus 1-2-3 have integrated software features. In Lotus 1-2-3 there is capability to create graphic displays that adjust automatically as "what-if" variations are made in parameters. There is also capability to use database management commands which can be useful in looking up data. Other elaborate programs may include word processing to facilitate integrating results into report write-ups. You should consult your instruction and manual to see what features you have.

E-2. SPREADSHEET COMMANDS

Commands are used to instruct the program to perform particular functions. You might wish to save a spreadsheet for your diskette, you may wish to copy a column in your spreadsheet, you may wish to insert a row to introduce additional information and so on.

When you load your spreadsheet program, you are likely to be faced with a blank spreadsheet. You may then wish to construct a calculation by entering information into individual cells as discussed in Section E-3. At some point you may determine that you wish to issue a command. You then have to "tell" the program that a command will be forthcoming. In VisiCalc, this is done with the symbol / (the divide sign). Upon pressing that key, you will be faced with a menu or list of available commands or command categories. You then select the menu option corresponding to the command you wish to issue.

Unlike VisiCalc and Lotus 1-2-3, Multiplan does not require the use of the symbol / before issuing a command. The / symbol is used to distinguish a command from text entries in cells. VisiCalc and Lotus 1-2-3 place a symbol before a command. As we shall see in Section E-3, Multiplan places a symbol with text.

If you wish to issue a command related to file storage, you would, in VisiCalc, type S for storage. Other spreadsheets may use a different convention. Lotus 1-2-3 uses F for file and Multiplan uses T for transfer. Suppose you wish to save a spreadsheet that you have just prepared. You may just have typed in the spreadsheet for Example 2-1 and now wish to save it. Upon typing S for storage, you will be given another menu of choices. One of these will be S for save. Upon typing S for save you will be asked to provide a name for your file. You may choose to call it LAYERS, since it deals with layers of wall. After you type in the name and press ENTER or RETURN (depending on what the analogous key is called on your computer), the spreadsheet file will be saved on a diskette.

Suppose on the next day you wish to work with the spreadsheet LAYERS. You would then type / SL. The / S tells the computer that you wish to issue a file storage command. The L tells the computer that you wish to load a file. You will then be asked for the name of the file to be loaded. You should keep a directory of the files you have created, although spreadsheet programs typically allow you to scan the list of files on the diskette. Again, specific notations can differ in other programs, for example, Lotus 1-2-3 uses R for retrieve instead of L for load.

In addition to being able to save and load files, you are likely to want to copy sections of spreadsheets. This will be true when you wish to prepare a table of numbers based on different values of input information. In Example 2-1, you may wish to tabulate heat loss as a function of insulation thickness. After copying the section, you would then change the insulation thickness in the new section.

To copy, you would place the cursor at the beginning of the range to be copied. You would then type / R for replicate (in Lotus 1-2-3, you would type / C for copy). You would be asked to complete the range. You would type a period and move the cursor to the end of the range and press RETURN. You would then be asked where you want the copied section to be placed. You would move the cursor to the beginning of that "target range" and press RETURN.

Different spreadsheets have different capabilities in regard to copying. For example, some are restricted to copying from within a single column or row at a time. Others, like Lotus 1-2-3, can copy a range consisting of several columns and rows. You should consult your own instruction manual.

When copying, distinction must be made between relative and absolute copying if formulas are involved. For example, in Example 2-1, cell B3 contains the formula $+B1 - B2$. If the contents of Column B are copied and place in Column C, then the computer has to know whether you intend cell C3 to contain $+B1 - B2$ (absolute copy) or $+C1 - C2$ (relative copy). VisiCalc will ask you. Lotus 1-2-3 will assume that you mean relative unless you have written the formula a certain way ($+\$B\$1 - \$B\$2$) to denote a desire to have absolute copying. You should check the convention for your own spreadsheet in your instruction manual.

Another command you are likely to use is INSERT. You may have created a spreadsheet that uses certain specified (input) information. In another situation, you may find it necessary to calculate that information and you thus would like to have a column available within your spreadsheet to perform the required calculations. You would type / I for insert. You would then be asked to type R or C for row or column. When you type C, a blank column will appear where your cursor is located. The column previously there and all columns to the right of it will have been moved one column to the right, with all formulas adjusted automatically. In Lotus 1-2-3, an intermediate command category W for worksheet is needed, so the sequence would be / W I C. Lotus 1-2-3 also gives you the opportunity to insert several blank columns at once.

You may find occasion to combine individual spreadsheets. For example, you may encounter a problem involving both natural convection and radiation for which you have individual spreadsheets. You now wish to put these two spreadsheets into one and then introduce a column to add their effects. Let us assume that each spreadsheet contains four columns.

TABLE E-2-1
Selected Commands in VisiCalc[a]

Symbol	Meaning
/	Call for command
/ S	Call for menu of file storage commands
/ SS	Save the spreadsheet on diskette
/ SL	Load a spreadsheet from diskette
/ R	Replicate a range of the spreadsheet
/ I	Call for menu of insert commands
/ IC	Insert a column
/ IR	Insert a Row

[a]Consult your instruction manual for analogous command symbols for your spreadsheet on your computer.

You may proceed by loading the first spreadsheet, placing the cursor in the first column, and inserting four column, thereby moving the first set of spreadsheet instructions to Columns E–H. You then load the second spreadsheet which will appear in Columns A–D. You now have a combined spreadsheet. You can now add another column to add the effects of the two modes of heat transfer.

Note that when VisiCalc loads a file, it does not erase what is already on the sheet in cells not used by the new file. To clear out an old sheet, it is necessary to type / C for clear first. In Lotus 1-2-3, the / FR for file retrieve does erase what is already on the sheet, so one would use / FC for file combine. You should check the instruction manual for your spreadsheet for the conventions that apply.

The above commands will satisfy most of your needs for the types of application in the text. There may be alternate means of accomplishing some goals (e.g., using a MOVE command to make room for another spreadsheet instead of INSERT commands). There may be items that may make your time at the screen more efficient (e.g., by using manual instead of automatic recalculation). You should consult your instruction manual to become familiar with additional commands. The commands discussed here are summarized in Table E-2-1.

E-3. CELL CONTENTS

A spreadsheet is an array of cells. A cell can contain text (in which case, it usually is called a label), a number, a formula, or a logical statement. The spreadsheet will interpret information to be text if it starts with a letter or quotation mark. It will interpret a number as a number. If the cell content contains one of the formula symbols in Table E-3-1 and begins with either a number or one of these symbols, the spreadsheet will interpret the contents to be a formula. If the cell begins with a number or one of the symbols in Table

E-3-1 and contains, in addition, an equality (=) or inequality (> or <) symbol, the statement is a logical statement (which is either TRUE or FALSE). Combinations of these symbols can also be used (< = for less than or equal, > = for greater than or equal, < > for not equal). Combinations of logical statements can be linked in a single statement with logical @ AND or @ OR functions. Two logical assertions linked by @ AND will yield TRUE if both assertions are true. If linked by @ OR, TRUE will result if either assertion is true. The reason for including a quotation to signal a label is to permit having labels that start with numbers or formula symbols. Some sample cell contents are given in Table E-3-2.

Cell references in VisiCalc are made to column by letter and to row by number. Thus, the formula 3 + B2 instructs the spreadsheet to add 3 to the contents of cell B2 and to place the result in the current cell. Spreadsheets have conventions (see your instruction manual) as to order of performance of operations when a formula appears. This author has found it desirable to use parentheses liberally so as to avoid reliance on recall of the order of operations.

The logical statements are helpful in setting up criteria and making choices, for example, whether the Reynolds number is high enough for turbulent flow to prevail. The statement

+B1 > 2

TABLE E-3-1
Formula Symbols

+	Addition
−	Subtraction
*	Multiplication
/	Division
∧	Exponentiation
( )	Parentheses
@	Function sign

TABLE E-3-2
Sample Cell Contents

Content	Type
TEMPERATURE	Label
3	Number
3 + 4	Formula
3 + B2	Formula
'3 + B2	Label
(B1 − B2) / (B3 + B4)	Formula
B1 + B2	Label
+B1 + B2	Formula
+B1>2	Logical statement

is either TRUE or FALSE. VisiCalc assigns the number 1 to correspond to TRUE and 2 to correspond to FALSE. Lotus 1-2-3 uses 0 for false. On the basis of the value indicating TRUE or FALSE, choices using logical functions can be made as discussed in Section E-4.

References to cells in Multiplan are made by row and column number identification. Thus, the formula 3 + B2 in Table E-3-1 would be 3 + R2C2, since B is Column 2 and the 2 in B2 indicates Row 2. Two other important differences are present in Multiplan. One is that the @ sign is not used with functions. The other is that text requires use of quotation marks to begin and end the text field (or the use of the ALPHA command). As noted in Section E-2, the convenience of omitting the symbol / for commands in Multiplan implies that we cannot assume automatically that a letter implies text.

The simple categories of cell content provide the basis for spreadsheet usage. The numbers are used for input information. The formulas are used to perform a sequence of calculations. The logical statements are used to make choices. The labels are used to explain what you are doing.

E-4. BUILT-IN FUNCTIONS

The symbol @ in Table E-3-1 indicated the use of a built-in function. Multiplan, although not using the symbol @, has similar built-in functions. Spreadsheets, like computer programming languages in general and like electronic calculators, provide for easy evaluation of certain functions. A list of built-in functions in VisiCalc is provided in Table E-4-1. More elaborate spreadsheet programs may contain more functions or some variations on these functions. Lotus 1-2-3, for example, has more elaborate LOOKUP functions.

Many of the functions are of a type that might be found on a calculator. The formula

$$B1 + @ \ SIN \ (B2)$$

would take the sine of the contents of cell B2 and add it to the contents of cell B1. The @ SIN is a function that applies to a single number.

A second type of function in the spreadsheet is one that applies to a range of numbers. The @ SUM (B1 ... B4) will add the contents of cells B1, B2, B3, and B4. Other functions will select minimum and maximum values from a range or look up a value within a range. You should consult your instruction manual to see how to use these individual functions.

A third type of function is a logical function. The statement

$$@ \ IF \ (B1, \ B2, \ B3)$$

says that if B1 is 1 (corresponding to TRUE), assign the value that is in cell B2. Otherwise, assign the value that is in cell B3. Logical choices can also be made with the @ CHOOSE function.

The arguments of function can be numbers, cell contents, formulas, or logical statements, for example,

$$@ \text{ SIN (B1*B2)}$$

Whether you evaluate the argument first is a matter of individual preference. This author frequently finds it convenient to have the argument evaluated separately and displayed.

This author particulary recommends separate evaluation in connection with logical statements. It is possible, for example, to write a logical statement

$$@ \text{ IF (A5>6, B2, B3)}$$

With this statement, it requires cross-referencing to other cells to see if the condition A5 > 6 has been satisfied.

In Chapter 6, where a spreadsheet was used to evaluate convection from a plate, three steps were used to make a choice after the Reynolds number was

TABLE E-4-1
VisiCalc Build-in Functions

Symbol	Description
@ ABS (v)	Absolute value of v
@ ACOS (v)	Arc cosine of v
@ ASIN (v)	Arc sine of v
@ ATAN (v)	Arc tangent of v
@ COS (v)	Cosine of v
@ EXP (v)	e^v
@ INT (v)	Integer portion of v
@ LN (v)	Natural logarithm (base e) of v
@ LOG (v)	Logarithm base 10 of v
@ SIN (v)	Sine of v
@ SQRT (v)	Square root of v
@ TAN (v)	Tangent of v
@ PI	PI (3.1415926536)
@ AVERAGE (range)	Average of the nonblank entries in the range of cells
@ COUNT (range)	Number of nonblank entries in range
@ MAX (range)	Maximum value contained in the range
@ MIN (range)	Minimum value contained in the range
@ NPV (dr, range)	Net present value of entries in range, discount rate dr
@ SUM (range)	Sum of entries in the range
@ LOOKUP (v, range)	Find first value in range larger than v, select value from neighboring range
@ IF (arg 1, arg 2, arg 3)	Assigns argument arg 2 if arg 1 is TRUE; otherwise, assigns arg 3
@ CHOOSE (arg 1, N1, N2,...)	Assigns N1 if arg 1 is 1, N2 if arg 1 is 2, etc.
@ NOT (arg)	Assigns FALSE if arg is TRUE and vice versa
@ AND	Used for simultaneous conditions in logical statements
@ OR	Used for alternate conditions in logical statements

evaluated. The first step was a logical statement asserting that the Reynolds number exceeded 5×10^5. The TRUE or FALSE response provides a clear indication as to whether the flow is turbulent or laminar. The @ IF and @ CHOOSE functions then were used to select the appropriate Nusselt number. Combining steps would have been possible.

The built-in functions provide for a large variety of situations that you are likely to encounter. In some programs like Lotus 1-2-3, provision is made for the user to create additional functions through what are called macros. You should consult your instruction manual to see if your spreadsheet provides such an option.

REFERENCES

1. Anonymous, *Multiplan Software Library Manual*, Texas Instruments Inc., 1982.
2. E. M. Baras, *The Osborne/McGraw-Hill Guide to Using Lotus 1-2-3*, Osborne/McGraw-Hill, New York, 1984.
3. D. Bricklin, and B. Frankston, *VisiCalc Computer Software Program*, Personal Software, Inc., Sunnyvale, CA, 1979.
4. J. Posner et. al., *Lotus 1-2-3 User's Manual*, Lotus Development Corporation, Cambridge, MA, 1983.

SUMMARY OF PARAMETERS, FORMULAS, AND EQUATIONS

This appendix summarizes definitions of dimensionless parameters cited in the text and lists formulas, equations, and correlations used in the text. These are provided for convenience. For explanations associated with terms in the formulas, and so on, the reader should consult the body of the text.

F-1. DIMENSIONLESS PARAMETERS

Name	Symbol	Formula
Biot number	Bi	$\dfrac{hl}{k}$ (k for solid)
Bond number	Bo	$\dfrac{g\,\Delta\rho\,L^2}{\sigma}$
Eckert number	Ec	$\dfrac{u^2}{c_p\,\Delta T}$
Euler number	Eu	$\dfrac{\Delta p}{\frac{1}{2}\rho u^2}$
Fourier modulus	Fo	$\dfrac{\alpha t}{L^2}$
Galileo number	Ga	$\dfrac{g\rho_l(\rho_l-\rho_v)L^3}{\mu_l^2}$

(continued)

Name	Symbol	Formula
Grashof number	Gr	$\dfrac{g\beta\,\Delta T\,L^3}{\nu^2}$
Modified Grashof number	Gr*	$\mathrm{Gr}\,\mathrm{Nu}$
Graetz number	Gz	$\mathrm{Re}\,\mathrm{Pr}\dfrac{D}{L}$
j Factor (Colburn)	*j*	$\mathrm{St}\,\mathrm{Pr}^{2/3}$
Jakob number	Ja	$\dfrac{c_p\,\Delta T}{h_{fg}}$
Mach number	M	$\dfrac{u}{u_s} = \dfrac{u}{\sqrt{\gamma RT}}$
Nusslet number	Nu	$\dfrac{hL}{k}$ (*k* for fluid)
Peclet number	Pe	$\dfrac{uL}{\alpha} = \mathrm{Re}\,\mathrm{Pr}$
Prandtl number	Pr	$\dfrac{c_p\mu}{k}$
Rayleigh number	Ra	$\mathrm{Gr}\,\mathrm{Pr}$
Reynolds number	Re	$\dfrac{\rho uL}{\mu}$
Stanton number	St	$\dfrac{h}{\rho u c_p} = \dfrac{\mathrm{Nu}}{\mathrm{Re}\,\mathrm{Pr}}$
Weber number	We	$\dfrac{\rho u^2 L}{\sigma}$
Mass Transfer Parameters		
Lewis number	Le	$\dfrac{\alpha}{D}$
Sherwood number	Sh	$\dfrac{hmL}{D}$
Schmidt number	Sc	$\dfrac{\nu}{D}$

F-2. FORMULAS—CHAPTER 2

Fourier's Law

$$q = -kA\,\nabla T$$

Thermal Resistance

$$R_{th} = \dfrac{\Delta T}{q}$$

Plane Geometry—One Layer

$$R_{th} = \dfrac{\Delta x}{kA}$$

$$R_{th} = \sum_{i=1}^{I} \left(\frac{\Delta x}{kA} \right)_i$$

Convection at Surface

$$R_{th} = \frac{1}{hA}$$

Cylindrical Geometry—One Layer

$$R_{th} = \frac{\ln(r_o/r_i)}{2\pi kL}$$

Cylindrical Geometry—Multiple Layers

$$R_{th} = \sum_{n=1}^{N} \frac{\ln(r_o/r_i)_n}{2\pi k_n L}$$

Overall Heat Transfer Coefficient

$$U = \frac{q}{A\,\Delta T}$$

Critical Radius of Insulation—Cylindrical Geometry

$$r_o = \frac{k}{h}$$

F-3. FORMULAS—CHAPTER 3

Heat Conduction Equation

$$\rho c \frac{\partial T}{\partial t} = \frac{1}{r^n} \frac{\partial}{\partial r} \left(kr^n \frac{\partial T}{\partial r} \right) + q''' \qquad n = \begin{cases} 0 & \text{plane} \\ 1 & \text{cylinder} \\ 2 & \text{sphere} \end{cases}$$

Temperature in Slab with Uniform Source

$$T(x) = T_w + \frac{q'''}{2k}(H^2 - x^2)$$

Temperature in Cylinder with Uniform Source

$$T(r) = T_w + \frac{q'''}{4k}(R^2 - r^2)$$

Temperature in a Rectangular Fin

$$T(x) - T_\infty = (T_0 - T_\infty)\frac{\cosh m(L_c - x)}{\cosh mL_c}$$

$$m^2 = \frac{hP}{kA}$$

$$L_c = L + \tfrac{1}{2}t$$

Efficiency of a Rectangular Fin

$$\eta = \frac{\tanh mL_c}{mL_c}$$

Heat Transfer from a Rectangular Fin

$$q = hPL_c\eta(T_0 - T_\infty)$$

Efficiency of a Triangular Fin

$$\eta = \frac{1}{mL}\frac{I_1(2mL)}{I_0(2mL)}$$

Approximate Efficiency of a Triangular Fin

$$\eta = \frac{\tanh fmL}{mL}$$

$$f = e^{-0.1471}(mL)^2$$

Efficiency of a Circumferential Fin

$$\eta = \frac{2\sqrt{kt/2h}\,K_1(mr_1) - I_1(mr_1)K_1(mr_{2c})/I_1(mr_{2c})}{L_c(1 + r_{2c}/r_1)K_0(mr_1) + I_0(mr_1)K_1(mr_{2c})/I_1(mr_{2c})}$$

Approximate Efficiency of a Circumferential Fin

$$\eta = \frac{\tanh m'L}{m'L}$$

$$m' = m\sqrt{\frac{1 + r_{2c}/r_{1c}}{2}}$$

Conduction Shape Factor

$$S = \frac{q}{k\,\Delta T}$$

Relationship Between Shape Factor and Thermal Resistance

$$R_{th} = \frac{1}{kS}$$

Individual Shape Factor Formulas—Specific Geometrics (see Fig. 4-2-1)
Temperature in a Two-Dimensional Block
Uniform Heat Flux Specified at Left Face
Uniform Temperature Specified at Other Faces

$$T(x, y) - T_w = \sum_{n=0}^{\infty} \frac{(-1)^n}{2n+1} \frac{4q''}{\pi} \frac{\sinh B_n(a-x)}{kB_n\cosh B_n a} \cos B_n y$$

$$B_n = \frac{(2n+1)\pi}{2b} \qquad n = 0, 1, \ldots, \infty$$

Temperature in a Two-Dimensional Cylinder
Uniform Heat Source Specified at One End
Uniform Temperature Specified at Other Faces

$$T(r, z) - T_w = \sum_n \frac{2q''}{B_n R J_1(B_n R)} \frac{\sinh B_n(H-z)}{kB_n\cosh B_n H} J_0(B_n r)$$

B_n are solutions of $J_0(B_n R) = 0$

Two-Dimensional Block, Internal Heat Source, Uniform Surface Temperature

$$T(x, y) - T_w = \sum_{n=1}^{\infty} \sum_{m=1}^{\infty} A_{nm}\cos B_n x \cos C_m y$$

$$B_n = (2n+1)\frac{\pi}{2a}$$

$$C_m = (2m+1)\frac{\pi}{2b}$$

$$A_{nm} = \frac{Q_{nm}}{B_n^2 + C_m^2}$$

$$Q_{nm} = \frac{(1/k)\int_0^a dx \int_0^b dy\, q'''(x, y)\cos B_n x \cos C_m y}{\int_0^a dx \cos^2 B_n x \int_0^b dy \cos^2 C_m y}$$

F-5. FORMULAS—CHAPTER 5

Temperature Variation—Lumped Capacity Model

$$\Theta_{av}(t) = \Theta_{av}(0)e^{-(hA/\rho cV)t}$$

Effective Length Dimension for Biot and Fourier Numbers

$$L_{eff} = \frac{V}{A}$$

Approximate Criterion for Validity of Lumped Capacity Model

$$Bi \le 0.1$$

Temperature Variation in Plane Geometry

$$\frac{\Theta(x,t)}{\Theta_0} = \sum_{n=0}^{\infty} \frac{2\sin B_n L}{B_n L + \sin B_n L \cos B_n L} \cos B_n x \left(e^{-Fo(B_n L)^2}\right)$$

B_n is the solution of $B_n L \tan B_n L = Bi$

Approximate Criterion for First Term in Series to Be Sufficient

$$Fo > 0.2$$

Temperature Variation in Cylindrical Geometry

$$\frac{\Theta(r,t)}{\Theta_0} = \sum_{n=0}^{\infty} \frac{2J_1(B_n R)}{B_n R \left[J_0^2(B_n R) + J_1^2(B_n R)\right]} J_n(B_n r) e^{-Fo(B_n R/2)^2}$$

B_n is the solution of $B_n R J_1(B_n R)/J_0(B_n R) = 2Bi$

Temperature Variation in Spherical Geometry

$$\frac{\Theta(r,t)}{\Theta_0} = \sum_{n=D}^{\infty} \frac{2(\sin B_n R - B_n R \cos B_n R)}{B_n R - \sin B_n R \cos B_n R} \frac{\sin B_n r}{B_n r} e^{-Fo(B_n R/3)^2}$$

B_n is the solution of $-BR \cot BR = -1 + 3Bi$

Temperature in Semi-infinite Wall, Sudden Surface Temperature Change

$$T - T_0 = (T_1 - T_0)\left[1 - erf\left(\frac{x}{2\sqrt{\alpha t}}\right)\right]$$

Heat Flux, Same Case

$$q = \frac{T_1 - T_0}{\sqrt{\pi \alpha t}} e^{-x^2/4\alpha t}$$

$$T - T_0 = \frac{2q_0}{kA}\sqrt{\frac{\alpha t}{\pi}}\, e^{-x^2/4\alpha t} - \frac{q_0 x}{kA}\left[1 - \mathrm{erf}\left(\frac{x}{2\sqrt{\alpha t}}\right)\right]$$

Temperature in Semi-infinite Wall, Sudden Application of Convection

$$\frac{T - T_0}{T_\infty - T_0} = 1 - \mathrm{erf}\left(\frac{x}{2\sqrt{\alpha t}}\right) - e^{(h/k)[x+(h/k)\alpha t]}\left[1 - \mathrm{erf}\left(\frac{x}{2\sqrt{\alpha t}} + \frac{h\sqrt{\alpha t}}{k}\right)\right]$$

Response to Burst of Energy, Infinite Medium

$$T - T_0 = \frac{Q^{(n)}}{\rho c (4\pi\alpha t)^{n/2}}\, e^{-r^2/4\alpha t} \qquad n = \begin{cases} 1 & \text{plane} \\ 2 & \text{line} \\ 3 & \text{point source} \end{cases}$$

F-6. FORMULAS—CHAPTER 6

General Conservation Equation

$$\nabla \cdot \mathbf{J}_p + \frac{\partial N_p}{\partial t} = S_p$$

Conservation of Mass, Steady State, Constant Properties, Two Dimensions

$$\frac{\partial u}{\partial x} + \frac{\partial v}{\partial y} = 0$$

Conservation of x Momentum, Constant Properties

$$\frac{\partial}{\partial x}(u^2) + \frac{\partial}{\partial y}(uv) = \frac{\mu}{\rho}\frac{\partial^2 u}{\partial y^2} - \frac{1}{\rho}\frac{\partial p}{\partial x}$$

Conservation of Energy for Flat Plate Analysis

$$\frac{\partial}{\partial x}(uT) + \frac{\partial}{\partial y}(uT) = \frac{k}{\rho c}\frac{\partial^2 T}{\partial y^2}$$

Speed Profile in Boundary Layer—Approximate

$$\frac{u}{u_\infty} = \frac{3}{2}\frac{y}{\delta} - \frac{1}{2}\left(\frac{y}{\delta}\right)^3$$

Boundary Layer Thickness—Approximate

$$\frac{\delta}{x} = \frac{4.64}{\sqrt{Re_x}}$$

Boundary Layer Thickness—Location at which Exact $u/u_\infty = 0.99$

$$\frac{\delta}{x} = \frac{4.92}{\sqrt{Re_x}}$$

Ratio of Thermal and Velocity Boundary Layer Thicknesses

$$\frac{\delta_t}{\delta} = \frac{Pr^{-1/3}}{1.025}$$

Heat Transfer at a Flat Plate, Laminar Flow

$$Nu_x = 0.332\, Re_x^{1/2} Pr^{1/3}$$

Average Heat Transfer Coefficient, Flat Plate, Laminar Flow

$$\bar{h} = 2h(L)$$

Coefficient of Friction, Flat Plate, Laminar Flow

$$C_f = \frac{0.664}{\sqrt{Re_x}}$$

Heat Transfer Friction Relationship

$$St\, Pr^{2/3} = \tfrac{1}{2}C_f$$

Heat Transfer, Flat Plate, Laminar Flow, Liquid Metal

$$Nu_x = 0.564\sqrt{Pe_x}$$

Heat Transfer, Laminar Flow, Flat Plate, General Fluid

$$Nu_x = \frac{0.3387\, Re_x^{1/2} Pr^{1/3}}{\left[1 + (0.0468/Pr)^{2/3}\right]^{1/4}}$$

Friction Coefficient, Turbulent Flow, Flat Plate

$$C_f = 0.0592 Re_x^{-1/5} \qquad 5 \times 10^5 < Re_x < 10^7$$

$$C_f = 0.370\left[\log(Re_x)\right]^{-2.584} \qquad 10^7 < Re_x < 10^9$$

Heat Transfer, Turbulent Flow, Flat Plate

$$\mathrm{St\,Pr}^{2/3} = 0.0296\,\mathrm{Re}_x^{-1/5} \qquad 5 \times 10^5 < \mathrm{Re}_x < 10^7$$

$$\mathrm{St\,Pr}^{2/3} = 0.185\big[\log(\mathrm{Re}_x)\big]^{-2.584} \qquad 10^7 < \mathrm{Re}_x < 10^9$$

Average Heat Transfer, Flat Plate, Turbulent Flow

$$\overline{\mathrm{Nu}} = \mathrm{Pr}^{1/3}\big(0.037\,\mathrm{Re}_L^{0.8} - 871\big) \qquad 5 \times 10^5 < \mathrm{Re}_L < 10^7$$

Stagnation Temperature from High-Speed Flow, Reversible

$$T_0 = T_\infty + \frac{u_\infty^2}{2c_p g_c}$$

Adiabatic Wall Temperature

$$T_{aw} = T_\infty + r(T_0 - T_\infty)$$

Recovery Factor

$$r = \begin{cases} \mathrm{Pr}^{1/2} & \text{laminar flow} \\ \mathrm{Pr}^{1/3} & \text{turbulent flow} \end{cases}$$

Reference Temperature for Property Evaluation

$$T^* = \tfrac{1}{2}(T_w + T_\infty) + 0.22(T_{aw} - T_\infty)$$

Mass Transfer, Laminar Flow

$$\mathrm{Sh} = 0.332\,\mathrm{Re}^{1/2}\mathrm{Sc}^{1/3}$$

Mass Transfer, Friction Relation

$$\mathrm{St}_M \mathrm{Sc}^{2/3} = \tfrac{1}{2}C_f$$

F-7. FORMULAS—CHAPTER 7

Velocity Profile, Laminar Flow, Long Tube

$$u = u_{max}\frac{R^2 - r^2}{R^2}$$

$$u_{max} = -\frac{R^2}{4\mu}\frac{dp}{dx}$$

Temperature Profile, Laminar Flow, Long Tube

$$T_w - T(r) = \frac{u_{\max}}{\alpha R^2}\frac{\partial T}{\partial x}\left(\frac{R^4}{4} - \frac{R^2 r^2}{4} - \frac{R^4}{16} + \frac{r^4}{16}\right)$$

Heat Transfer, Laminar Flow, Long Tube, Uniform Heat Flux

$$\text{Nu} = 4.364$$

Hydraulic Diameter

$$D_H = \frac{4A}{P}$$

Entrance Effects on Heat Transfer, Laminar Flow, Developed Velocity Profile

$$\text{Nu} = 3.66 + \frac{0.0668\,\text{Gz}}{1 + 0.04\,\text{Gz}}$$

Entrance Effects on Both Heat Transfer and Flow, Laminar Flow

$$\text{Nu} = 1.86\,\text{Gz}^{1/3}\left(\frac{\mu}{\mu_s}\right)^{0.14}$$

Turbulent Flow, Heat Transfer–Friction Relationship

$$\text{St}\,\text{Pr}^{2/3} = \frac{f}{8}$$

Friction Factor—Simple Correlation

$$f = 0.184\,\text{Re}^{-1/5}$$

General Friction Factor Equation with Roughness

$$f = \frac{1}{3.24\log_{10}^2\left[(\varepsilon/3.7D)^{1.11} + 6.9/\text{Re}\right]}$$

Heat Transfer in Turbulent Flow: Dittus–Boelter Equation

$$\text{Nu} = 0.023\,\text{Re}^{0.8}\text{Pr}^n$$

$$n = \begin{cases} 0.4 & \text{heating} \\ 0.3 & \text{cooling} \end{cases}$$

Heat Transfer in Turbulent Flow with Entrance Effect

$$\text{Nu} = 0.036\,\text{Re}^{0.8}\text{Pr}^{1/3}\left(\frac{D}{L}\right)^{0.055}$$

$$Nu = 0.027\,Re^{0.8}Pr^{1/3}\left(\frac{\mu}{\mu_w}\right)^{0.14}$$

Heat Transfer in Turbulent Flow—Petukhov Equation

$$Nu = \frac{(f/8)Re\,Pr}{1.07 + 12.7\sqrt{f/8}\,(Pr^{2/3} - 1)}\left(\frac{\mu}{\mu_w}\right)^{n}$$

$$n = \begin{cases} 0.11 & T_w > T_b \\ 0.25 & T_w < T_b \\ 0 & \text{constant heat flux} \end{cases}$$

Effect of Roughness on Heat Transfer

$$\frac{Nu}{Nu_{smooth}} = \left[\min\left(\frac{f}{f_{smooth}}, 3\right)\right]^{0.68\,Pr^{0.215}}$$

Heat Transfer to Liquid Metals in Tubes, Constant Wall Temperature

$$Nu = 5 + 0.025\,Pe^{0.8}$$

Heat Transfer to Liquid Metals in Tubes, Constant Heat Flux

$$Nu = 4.82 + 0.185\,Pe^{0.827}$$

Drag Force over a Bluff Body

$$F = C_D A\,\frac{\rho u_\infty^2}{2g_c}$$

Drag Coefficient for a Cylinder in Cross-flow

$$C_D = \begin{cases} 11.3\,Re^{-0.623} & 0.1 < Re < 8 \\ 5.2\,Re^{-0.239} & 8 < Re < 10^3 \\ 0.7 + 0.1\log Re & 10^3 < Re < 10^5 \end{cases}$$

Heat Transfer in Cross-flow

$$Nu = C\,Re^n Pr^{1/3}$$

See Tables 7-5-1 and 7-5-2 for values of C and n

Heat Transfer in Cross-flow, Cylinders

$$Nu = 0.3 + \frac{0.62\,Re^{1/2}Pr^{1/3}}{\left[1 + (0.4/Pr)^{2/3}\right]^{1/2}}\left[1 + \left(\frac{Re}{282,000}\right)^{5/8}\right]^{4/5}$$

Modified Formula for 20,000 < Re < 400,000

$$Nu = 0.3 + \frac{0.62\,Re^{1/2}Pr^{1/3}}{\left[1 + (0.4/Pr)^{2/3}\right]^{1/4}}\left[1 + \left(\frac{Re}{282,000}\right)^{1/2}\right]$$

Heat Transfer in Cross-flow, Liquid Metals (Pe < 0.2)

$$Nu = \frac{1}{0.8237 - 0.5\ln Pe}$$

Heat Transfer for Flow over a Sphere

$$Nu = 2 + (0.4\,Re^{1/2} + 0.06\,Re^{2/3})Pr^{0.4}\left(\frac{\mu_\infty}{\mu_w}\right)^{1/4} \qquad \begin{array}{l} 3.5 < Re < 8\times10^4 \\ 0.7 < Pr < 380 \end{array}$$

Heat Transfer for Flow Across a Bank of Tubes

$$Nu = C\,Re^n Pr^{1/3}$$

See Tables 7-6-1 and 7-6-2 for values of C and n

Correction for Fewer than 10 Rows

$$Nu(M) = Nu(10)\left(\frac{M}{10}\right)^{0.12}$$

Heat Transfer for Flow Across a Bank of Tubes

$$Nu = C\,Re^n Pr_\infty^{0.36}\left(\frac{Pr_\infty}{Pr_w}\right)^{1/4}$$

For $10^2 < Re < 2\times10^5$
  (1) Rectangular Array

$$C = 0.27 \qquad n = 0.63$$

  (2) Triangular Array

$$n = 0.6$$

$$C = \begin{cases} 0.35\left(\dfrac{p_t}{p_p}\right)^{0.2} & \dfrac{p_t}{p_p} < 2 \\[3ex] 0.4 & \dfrac{p_t}{p_p} > 2 \end{cases}$$

$$n = 0.84$$

$$C = \begin{cases} 0.021 & \text{rectangular array} \\ 0.022 & \text{triangular array} \end{cases}$$

Pressure Drop, Bank of Tubes with N rows

$$\Delta p = \frac{2f'G_{\max}^2 N}{\rho}\left(\frac{\mu_w}{\mu_b}\right)^{0.14}$$

Friction Factor, Triangular Array of Tubes

$$f' = \left\{0.25 + \left[\frac{0.118}{(p_t - D)/D}\right]1.08\right\}\mathrm{Re}^{-0.16}$$

Friction Factor, Rectangular Array of Tubes

$$f' = \left\{0.044 + \frac{0.08p_p/D}{[(p_t - D)/D](0.43 + 1.13D/p_p)}\right\}\mathrm{Re}^{-0.15}$$

F-8. FORMULAS—CHAPTER 8

Momentum Equation, Natural Convection at Vertical Plate

$$\frac{\partial}{\partial z}(\rho u^2) + \frac{\partial}{\partial y}(\rho uv) = g\rho\beta(T - T_\infty) + \mu\frac{\partial^2 u}{\partial y^2}$$

Velocity Profile at Vertical Plate

$$u = u_0\frac{y}{\delta}\left(1 - \frac{y}{\delta}\right)^2$$

$$u = 5.17\frac{\nu}{z}(0.952 + \mathrm{Pr})^{-1/2}\mathrm{Gr}^{1/2}$$

Temperature Profile at Vertical Plate

$$\frac{T - T_\infty}{T_w - T_\infty} = \left(1 - \frac{y}{\delta}\right)^2$$

Heat Transfer, Natural Convection, Vertical Plate, Simple Model, Laminar

$$\mathrm{Nu} = 0.508\,\mathrm{Pr}^{1/2}(0.952 + \mathrm{Pr})^{-1/4}\mathrm{Gr}^{1/4}$$

Heat Transfer, Vertical Plate, Laminar, General Fluid

$$Nu = \frac{0.75\,Pr^{1/4}}{\left(2.435 + 4.884\sqrt{Pr} + 4.953\,Pr\right)^{1/4}}(Gr\,Pr)^{1/4}$$

Simple Natural Convection Correlation

$$Nu = C\,Ra^n$$

See Table 8-4-1 for Values of C and n

General Form for Natural Convection Correlations

$$Nu = \left\{ A_1 + \frac{A_2\,Ra^{n_1}}{\left[1 + (A_3/Pr)^{n_2}\right]^{n_3}} \right\}^{n_4}$$

See Table 8-4-2 for values of coefficients

Natural Convection, Vertical Plate, Uniform Heat Flux, Laminar

$$Nu = 0.60(Gr{*}Pr)^{1/5}$$

Natural Convection, Vertical Plate, Uniform Heat Flux, Turbulent

$$Nu = 0.17(Gr{*}Pr)^{1/4}$$

Heat Transfer from Vertical Cylinder

$$\overline{Nu}_{cyl} = \overline{Nu}_{plate}\left[1 + 1.43\left(\frac{L}{D\,Gr^{1/4}}\right)^{0.9}\right]$$

Heat Transfer, Inclined Surface, Heated Surface Facing Up

$$Nu = 0.56(Gr\,Pr\cos\theta)^{1/4} \qquad 15° < \theta < 75° \qquad Gr < Gr_c$$

$$Nu = 0.14\left[(Gr\,Pr)^{1/3} - (Gr_c Pr)^{1/3}\right] + 0.56(Gr\,Pr\cos\theta)^{1/4}$$

$$Gr > Gr_c \qquad 10^5 < Gr\,Pr\cos\theta < 10^{11}$$

See Tables 8-5-1 for values of Gr_c

Heat Transfer, Inclined Surface, Heated Surface Facing Down

$$Nu = (0.58\,Gr\,Pr\cos\theta)^{1/5} \qquad \theta < 88°$$

Heat Transfer in Enclosure, Hot Floor, Insulated Walls

$$Nu = 0.069\,Ra^{1/3}Pr^{0.074} \qquad 3 \times 10^5 < Ra < 7 \times 10^9$$

$$\text{Nu}_L = 1 + 1.44\left(1 - \frac{1708}{\text{Ra}_L\cos\theta}\right)\left\{1 - \frac{1708[\sin(1.8\theta)]^{1.6}}{\text{Ra}_L\cos\theta}\right\}$$

$$+ \left[\left(\frac{\text{Ra}\cos\theta}{5830}\right)^{1/3} - 1\right]$$

Nonpositive factors should be set to zero.

Effective Conductivity, Natural Convection Between Concentric Cylinders and Spheres

$$\frac{k_{\text{eff}}}{k} = A\left(\frac{\text{Pr}}{0.861 + \text{Pr}}\right)^{1/4}\text{Ra}^{1/4}\zeta^{1/4}$$

$$A = \begin{cases} 0.386 & \text{cylinders} \\ 0.74 & \text{spheres} \end{cases}$$

$$\zeta = \begin{cases} \dfrac{[\ln(D_o/D_i)]^4}{(L/D_o)^3[1 + (D_o/D_i)^{3/5}]^5} & \text{cylinders} \\[3ex] \dfrac{(L/D_i)(D_o/D_i)^3}{[1 + (D_o/D_i)^{7/5}]^5} & \text{spheres} \end{cases}$$

Combined Free and Forced Convections, Aiding Flow

$$\text{Nu} = \left(\text{Nu}_{fc}^3 + \text{Nu}_{nc}^3\right)^{1/3}$$

F-9. FORMULAS—CHAPTER 9

Condensation Heat Transfer, Vertical Plate, Laminar

$$\bar{h} = 1.13\left[\frac{g\rho_l(\rho_l - \rho_v)k_l^3 h_{fg}}{\mu_l(T_{\text{sat}} - T_s)L}\right]^{1/4}$$

Condensation Heat Transfer, Vertical Plate, Turbulent

$$\bar{h} = \left\{0.0077\left[\frac{g\rho_l(\rho_l - \rho_v)k_l^3}{\mu_l^2}\right]^{1/3}\left[\frac{4L(T_{\text{sat}} - T_s)}{\mu_l h_{fg}}\right]^{0.4}\right\}^{5/3}$$

Condensation Heat Transfer, Horizontal Tubes, N Vertical Rows

$$\bar{h} = 0.729 \left[\frac{g\rho_l(\rho_l - \rho_v)k_l^3 h_{fg}}{\mu_l(T_{sat} - T_s)ND} \right]^{1/4}$$

Condensation Inside Tubes, Re < 35,000

$$\bar{h} = 0.555 \left[\frac{g\rho_l(\rho_l - \rho_v)k_l^3 h'_{fg}}{\mu_l(T_{sat} - T_s)D} \right]^{1/4}$$

$$h'_{fg} = h_{fg} + 0.375 C_{pl}(T_{sat} - T_s)$$

Condensation Inside Tubes, High Re

$$\mathrm{Nu} = 0.026\, \mathrm{Re}_m^{0.8} \mathrm{Pr}_l^{1/3}$$

$$\mathrm{Re}_m = \frac{D}{\mu_l} \left[\rho_l u_l + \rho_v u_v \left(\frac{\rho_l}{\rho_v} \right)^{1/2} \right]$$

Pool Boiling

$$\mathrm{Nu} = \frac{1}{C_{sf}^3} \frac{\mathrm{Bo}^{1/2}\mathrm{Ja}^2}{\mathrm{Pr}^{3n-1}}$$

Maximum Heat Flux, Pool Boiling

$$q''_{max} = 0.15 \rho_v^{1/2} h_{fg} \left[g(\rho_l - \rho_v)\sigma \right]^{1/4}$$

Forced Convection Boiling over Hot Surface

$$q'' = \left[q''^2_{Fc} + q''^2_{Pb} - q''^2_i \right]^{1/2}$$

Maximum Heat Flux, Forced Convection over Hot Surface

$$\frac{q_{max}}{\rho_v h_{fg} u_\infty} = \frac{1}{\pi} \left[1 + \left(\frac{4}{\mathrm{We}} \right)^{1/3} \right] \qquad \frac{q_{max}}{\rho_v h_{fg} u_\infty} > \frac{0.275}{\pi} \left(\frac{\rho_l}{\rho_v} \right)^{1/2} + 1$$

$$\frac{q_{max}}{\rho_v h_{fg} u_\infty} = \frac{1}{\pi} \left(\frac{\rho_l}{\rho_v} \right)^{1/2} \left[\frac{(\rho_l/\rho_v)^{1/4}}{169} + \frac{1}{19.2\,\mathrm{We}^{1/3}} \right]$$

$$\frac{q_{max}}{\rho_v h_{fg} u_\infty} < \frac{0.275}{\pi} \left(\frac{\rho_l}{\rho_v} \right)^{1/2} + 1$$

Forced Convection Boiling in Tubes

$$h = h_1 + h_2$$

$$h_1 = 0.023 \frac{k_l}{D} \mathrm{Re}^{0.8} \mathrm{Pr}^{0.4} F$$

$$\mathrm{Re} = \frac{G(1-x)D}{\mu_l}$$

$$F = \begin{cases} 2.608 \left(\dfrac{1}{X_{tt}} \right)^{0.7434} & 0.7 < \dfrac{1}{X_{tt}} < 100 \\[2em] 2.2709 \left(\dfrac{1}{X_{tt}} \right)^{0.3562} & 0.1 < \dfrac{1}{X_{tt}} < 0.7 \end{cases}$$

$$X_{tt} = \left(\frac{1-x}{x} \right)^{0.9} \left(\frac{\rho_l}{\rho_v} \right)^{1/2} \left(\frac{\mu_v}{\mu_l} \right)^{0.1}$$

$$h_2 = 0.00122 \frac{k_l^{0.79} c_{pl}^{0.45} \rho_l^{0.44}}{\sigma^{0.5} \mu_l^{0.29} h_{fg}^{0.24} \rho_v^{0.24}} T_e^{0.24} \Delta p^{0.75} S$$

$$\Delta p = p_{\mathrm{sat}}(T_s) - p_{\mathrm{sat}}(T_{\mathrm{sat}})$$

$$S = \begin{cases} 3.6278 - 0.64557 \log_{10} \mathrm{Re}_{TP} & 2 \times 10^4 < \mathrm{Re}_{TP} < 3 \times 10^5 \\ 0.1 & \mathrm{Re}_{TP} > 10^5 \end{cases}$$

$$\mathrm{Re}_{TP} = \mathrm{Re}\, F^{1.25}$$

F-10. FORMULAS—CHAPTER 10

Black Body Radiation per Unit Wavelength

$$\frac{dE_b}{d\lambda} = \frac{2\pi h c^2 \lambda^{-5}}{e h c / \lambda k T - 1}$$

Total Black Body Radiation

$$E_b = \sigma T^4$$

Radiation Shape Factor

$$A_1 F_{12} = \frac{1}{\pi} \int dA_1 \int dA_2 \frac{\cos \phi_1 \cos \phi_2}{r^2}$$

For Specific Configurations, see Table 10-3-1

Reciprocity

$$A_1 F_{12} = A_2 F_{21}$$

Conservation and Shape Factors

$$\sum_{n=1}^{N} F_{in} = 1$$

Energy Balance Among Black Bodies

$$E_{bi} A_i = Q_i + \sum_{j=1}^{N} E_{bj} A_j F_{ji}$$

Internal Resistance—Grey Body

$$R_{int} = \frac{1 - \varepsilon}{\varepsilon A}$$

Geometric Resistance Between Grey Bodies

$$R_{geom} = \frac{1}{A_1 F_{12}} = \frac{1}{A_2 F_{21}}$$

Internal Resistance—Transmitting Body (Diffuse Reflection)

$$R_{int} = \frac{\rho}{\varepsilon A (1 - \tau)}$$

Geometric Resistance Between Grey Body 1 and Transmitting Body 2

$$R_{geom} = \frac{1}{A_1 F_{12} (1 - \tau_2)}$$

Geometric Resistance Between Grey Bodies Separated by Transmitting Body 2
(a) Transparent (Specular) Transmission

$$R_{geom} = \frac{1}{A_1 F_{13} \tau_{2t}}$$

(b) Diffuse (Translucent) Transmission

$$R_{geom} = \frac{1}{A_1 F_{12} F_{23} \tau_{2d}}$$

$$J_{\text{eff}} = \frac{J'}{1 - \tau} \qquad J' \text{ excludes transmitted radiation}$$

Internal Resistance—Specular Reflection

$$R_{\text{int}} = \frac{\rho_D}{\varepsilon A (1 - \rho_s)}$$

Geometric Resistance Between Specular Surface 1 and Grey Surface 2

$$R_{\text{geom}} = \frac{1}{A_1 F_{12}(1 - \rho_{1s})}$$

Geometric Resistance Between Two Grey Surfaces Interacting with Specular Surface 1 (Primes Denote Mirror Image Surface)

$$R_{\text{geom}} = \frac{1}{A_2 F_{23} + \rho_{1s} A_{2'} F_{2'3}}$$

Modified Radiosity—Specularly Reflecting Surface

$$J_{\text{eff}} = \frac{J_D}{1 - \rho_s}$$

$$T_{\text{eff}} = (\text{sunlight}) = 5800^\circ K$$

$$T_{\text{eff}} = (\text{atmosphere}) = 0.0552 T_{\text{air}}^{1.5}$$

F-11. FORMULAS–CHAPTER 11

Heat Exchange, Parallel or Counterflow

$$q = UA \, \Delta T_m$$

$$\Delta T_m = \frac{\Delta T_2 - \Delta T_1}{\ln(\Delta T_2/\Delta T_1)}$$

Heat Exchange, Other Heat Exchangers

$$q = UAF\Delta T_m \qquad \Delta T_m \text{ based on counterflow}$$

F Factor, Shell and Tube

$$F = \frac{\sqrt{1 + R^2}\ln[(1 - P)/1 - RP]}{(R - 1)\ln\left\{\left[2 - P(R + 1 - \sqrt{1 + R^2})\right]/\left[2 - P(R + 1 + \sqrt{1 + R^2})\right]\right\}}$$

$$R = \frac{C_t}{C_s} = \frac{T_{\text{out}, s} - T_{\text{in}, s}}{T_{\text{in}, t} - T_{\text{out}, t}}$$

$$P = \frac{T_{\text{in}, t} - T_{\text{out}, t}}{T_{\text{in}, t} - T_{\text{in}, s}}$$

F Factor, Cross-Flow, One Fluid Mixed (Denoted by T)

$$F = \frac{V^{\cdot}}{V_{cF}}$$

$$V_{cF} = \frac{K - S}{\ln[(1 - S)/(1 - K)]}$$

$$V = \frac{S}{\ln \dfrac{1}{1 - (S/K)\ln[1/(1 - K)]}}$$

$$K = \frac{T_1 - T_2}{T_1 - t_1}$$

$$S = \frac{t_2 - t_1}{T_1 - t_1}$$

Heat Exchange, Effectiveness

$$q = \varepsilon C_{\min}(T_{h, \text{in}} - T_{c, \text{in}})$$

Effectiveness Formulas—See Tables 11-5-1, and 11-5-2

Temperature Difference versus Position, Counterflow

$$T_h - T_c = (T_h - T_c)_1 e^{-(1/C_h - 1/C_c)UPx}$$

Temperature Difference Versus Position, Parallel Flow

$$T_h - T_c = (T_h - T_c)_1 e^{-(1/C_h + 1/C_c P)UPx}$$

INDEX